T0138315

For and Against Method

FOR and AGAINST METHOD

Including Lakatos's Lectures on Scientific Method and the Lakatos-Feyerabend Correspondence

Imre Lakatos and Paul Feyerabend

Edited and with an Introduction by

Matteo Motterlini

The University of Chicago Press / Chicago & London

IMRE LAKATOS was professor of logic at the London School of Economics. He was author of *Proofs and Refutations* and the two-volume *Philosophical Papers.* PAUL FEYERABEND taught at many universities, principally the University of California, Berkeley, and the Federal Institute of Technology, Zurich. Among his books are *Against Method, Science in a Free Society, Farewell to Reason,* and *Killing Time: The Autobiography of Paul Feyerabend,* the last published by the University of Chicago Press. MATTEO MOTTERLINI is lecturer in the Department of Economics at the University of Trento and visiting research fellow at the Center for Philosophy of Natural and Social Sciences at the London School of Economics. He is author of *Imre Lakatos: Science, Mathematics, and History,* in press.

The University of Chicago Press, Chicago 60637
The University of Chicago Press, Ltd., London

08 07 06 05 04 03 02 01 00 99 1 2 3 4 5

ISBN: 0-226-46774-0 (cloth)

Library of Congress Cataloging-in-Publication Data

Lakatos, Imre.
For and against method : including Lakatos's lectures on scientific method and the Lakatos-Feyerabend correspondence / Imre Lakatos and Paul Feyerabend ; edited and with an introduction by Matteo Motterlini.
p. cm.
Includes bibliographical references and index.
ISBN 0-226-46774-0 (alk. paper).
1. Science—Methodology. 2. Science—Philosophy. 3. Lakatos, Imre—Correspondence. 4. Feyerabend, Paul K., 1924- —Correspondence. I. Feyerabend, Paul K., 1924- . II. Motterlini, Matteo. III. Title.
Q175.3.L35 1999
501—dc21 99-13581
CIP

∞ The paper used in this publication meets the minimum requirements of the American National Standard for Information Sciences—Permanence of Paper for Printed Library Materials, ANSI Z39.48-1992.

To my parents

*Were I to choose an auspicious image for the new millennium,
I would choose that one: the sudden agile leap of the philosopher who
raises himself above the weight of the world, showing that with all his
gravity he has the secret of lightness.*

ITALO CALVINO

Contents

Preface

This is how Paul Feyerabend remembers Imre Lakatos in his autobiography, *Killing Time:*

> I lectured at . . . the London School of Economics. . . . "Science has many holes," I said in passing. "A Popperian triviality," shouted Imre Lakatos. That shut me up; but I soon smiled at the incident. . . . The lecture hall at the London School of Economics was directly opposite Imre's office window. In spring and summer, when the windows were open, Imre could hear every word I said. Feeling outraged, or simulating outrage at the drift of my story—with Imre you were never sure—he left the Great Thinkers who happened to be with him, came over, and tried to set things right. Imre and I . . . differed in outlook, character, and ambition; yet we became really good friends.

The stronger their friendship became, the more challenging their intellectual conflict. The work which was to determine Feyerabend's fame and notoriety, *Against Method,* stemmed from Imre's challenge to Paul:

> In 1970 Imre cornered me at a party. "Paul," he said, "you have such strange ideas. Why don't you write them down? I shall write a reply, we publish the whole thing and I promise you—we shall have a lot of fun." I liked the suggestion and started working.

Lakatos died suddenly of a heart attack on 2 February 1974 without having written his reply; yet the reader has here the chance to reconstruct Lakatos's original counterarguments and, at the same time, enjoy the "fun" the two authors had in matching wits with each other.

This volume shows Lakatos and Feyerabend's intellectual relationship through a selection of hitherto unpublished writings from the Archive of Professor Imre Lakatos at the British Library of Political and Economic Science. The first of these is the transcript of a series of lectures on scientific method held by Lakatos at the London School of Economics from January to March 1973, thus giving in effect his final view on the subject. Next come Feyerabend's "Theses on Anarchism," written in 1973, containing a sketch of the attack on the rational position that Feyerabend would later expound in his *Against Method.* The latter was—in Feyerabend's words—"a long and rather personal *letter* to Imre and every wicked phrase it contains was written in anticipation of an even more wicked reply from the recipient." The correspondence Feyerabend and Lakatos exchanged between December 1967 and February 1974 clearly shows that Feyerabend's remark is far from being merely rhetorical, for each of his views is here constantly contested and debated by Lakatos up to his very last days.

The whole material is thus of considerable importance to anyone concerned with the development of the philosophical views of Lakatos and Feyerabend, or indeed, with the philosophy of science in general.

For those unfamiliar with the subject, the introduction to this volume—in the form of an imaginary dialogue between Lakatos and Feyerabend—aims to give the main themes of the debate. Lakatos's lectures, in turn, provide a fascinating insight into the problems of philosophy of science and their relevance to pedagogical and socio-political matters—one that remains accessible to the general reader. Feyerabend's "Theses" challenge Lakatos's "rationalistic cliffhanger" from an anarchist standpoint. Finally, the letters included here reveal how much the two friends relished the flavour of philosophical controversy, regarding it as the antidote to that conformism so widely present in the world of academics, educationalists, specialists, and professional politicians.

Information regarding the history of the material published in this volume and the editing process it has undergone appears in the introductory note to each item. All works cited in this volume are listed in the Bibliography. Editorial footnotes are indicated by Arabic numerals.

Acknowledgments

This book would have not appeared were it not for Grazia Borrini Feyerabend, Spiro Latsis, Gillian Page, John Watkins, and John Worrall: to them and to the Imre Lakatos Memorial Fund I should like to express my special gratitude. I am grateful to Alex Bellamy, Nancy Cartwright, Marco Del Seta, Donald Gillies, Giulio Giorello, Colin Howson, Ladislav Kvasz, Brendan Larvor, Axel Leijonhufvud, Jancis Long, Marco Mondadori, John Preston, Elio Sindoni, Mauricio Suárez, Peter Urbach, Thomas Uebel, and Elie Zahar for many useful suggestions. I should also like to express my thanks to Susan Abrams for unfailing help and encouragement. I am indebted to Francesco Guala, with whom I discussed most of the problems raised in the editing process, and to Benedetta Tiana for stylistic improvements. The Centre for the Philosophy of the Natural and Social Sciences and the University of Ferrara are also gratefully acknowledged. Any mistakes, of course, are my own responsibility.

Introduction: A Dialogue

Matteo Motterlini

A dialogue is a discourse consisting of question and answer on some philosophical or political subject, with due regard to the characters of the persons introduced and the choice of diction. The dialectic is the art of discourse by which we either refute or establish some proposition by means of question and answer on the part of the interlocutors.

Diogenes Laertius

The following dialogue between Lakatos and Feyerabend is obviously nothing more than fiction, but over the years a real dialogue did take place between the two friends. It consisted in a genuine, lengthy, continuous, and outspoken exchange of letters and papers which shows the two men taking stands in the discussion *for* and *against* method. My fictitious reconstruction mirrors their own contributions, but paraphrases them for stylistic reasons. I refer to the original texts in the footnotes.

The rhetorical form of the dialogue is well described in the above fragment by Diogenes Laertius. The reason for adopting it here is given by the two imaginary interlocutors explicitly at the beginning of their discourse.

Paul Feyerabend: Rumour has it, dear *Imre,* that while one can freely discuss ideas in a loose way, in letters, phone calls, and at dinner, academics will always prefer an essay or a book. And any paper of this kind has a beginning, a middle, and an end. There is an exposition, a development, and a result. After that the idea is as clear and well-defined as a dead butterfly in a collector's box.[1]

Imre Lakatos: Plato thought that the *gulf* between *ideas* and *life* could be bridged by *dialogue*—not by a written dialogue, which he considered but a superficial account of past events, but by a real, spoken exchange between people of different backgrounds. I agree that a *dialogue* reveals more than an essay. It can show the effect of arguments on outsiders. It makes explicit the loose ends which an essay tries to conceal by showing the *inconclusiveness of "conclusions"*[2] . . .

1. See Feyerabend 1991, 163–64.

2. Lakatos wrote his masterpiece in the philosophy of mathematics, *Proofs and Refutations,* in dialogue form; it started from a nonproblematical situation and gradually evolved into BETA's final remark: "I had no problems at the beginning, and now I have nothing *but* problems!"

Paul: . . . and, above all, it can demonstrate the chimaerical nature of what we believe to be the most solid parts of our lives. And with this, we have already reached our topic. In particular, I would like to discuss the tremendous *gulf* that exists between the various *images* of science and the 'real thing'. I cannot but think that the fine dividing line between scientific truth and epistemological 'castles in the air' is in fact very similar to the line we draw between 'normal' and 'insane' people: a trait which recurs among the latter is the tendency to detach themselves further and further from reality.[3]

Imre: I agree that any attempt to reform science by bringing it closer to the abstract image philosophers have of it is bound to damage and may even destroy it.[4]

Paul: The point on which we disagree, though, is your attempt to save both 'Progress' and 'Reason'. You claim there are standards which are flexible enough that they leave science leeway in which to *progress* and at the same time substantial enough to let *reason* survive.[5]

Imre: As a matter of fact, my standards apply to series of theories (research programmes) and not to individual theories; they judge the evolution of a programme over a period of time, and not its shape in a given instant; they compare its growth with that of rival programs. These criteria are therefore open both to the history and to the practice of science. Moreover, I term 'progressive' any programme which predicts events confirmed by subsequent research, thereby leading to the discovery of 'new' facts. I term 'degenerating' any programme which makes no such predictions, but simply 'saves' data discovered by its rival. Since I do not believe there exists any 'natural saturation point' in a programme, I can also distinguish between *falsification* and *rejection,* something Popper could not do. In this way I am entitled to 'shift' his initial problem—the demarcation between science and pseudoscience—to the new one of demarcating between *good* science and *bad* science (i.e., between progressive and degenerating programmes).[6]

3. See Feyerabend 1973; 1975a, chap. 16.

4. Lakatos expresses his worry that philosophers, or theologians, or party civil servants, or indeed, any kind of religious or state authorities, in attempting to frame scientific research in preordained guidelines, are in fact responsible for its sluggishness. See lecture 1, this volume.

5. See Feyerabend, "Theses on Anarchism," this volume.

6. Lakatos (1970) claims that "there are no such things as *crucial experiments,* at least if these are meant to be experiments which can instantly overthrow a research programme." The idea of *"instant rationality"* is thus utopian. We may claim an experiment crucial only *"with long insight,"* when one programme suffers defeat and it is superseded by another one: "Kepler's ellipses were generally admitted as crucial evidence for Newton

Paul: I'm with you, but one question keeps bothering me, and that is whether there is any *pragmatic* implication in evaluating theories with your kind of standards.

Imre: Methodological standards act like teachers: they give marks to theories. Moral criteria used in judging individuals have grave practical implications in education; similarly, scientific criteria used in judging theories have deep consequences for scientific method.[7]

Paul: Are you saying that if a research programme is judged better than a rival one, scientists *ought* to work on the allegedly superior one?

Imre: I am actually injecting some Popperian elements into the judgement of whether a programme progresses or degenerates, or whether it is overtaking another one. I am giving you criteria for progress and stagnation within a programme, and rules for the 'elimination' of entire programmes. Should a programme explain in a progressive way more than a rival programme accounts for, then it 'supersedes' the latter, and the rival one may be 'rejected' or simply 'shelved'. You cannot at this point fail to understand what the *pragmatic* meaning of 'rejecting' a programme is: very simply, it means the decision to *cease working on it.*[8]

Paul: OK, but it is easy to see that standards of your kind have practical force only if combined with some *time limit* after which to keep working on a degenerating programme would be 'irrational'. If you accept the idea of the time limit, then unfortunately, arguments very similar to the ones you used against naive falsificationism backfire against your own standards. Consider that if it is unwise to reject faulty theories the moment they are born because they might grow and improve, then it is also unwise to reject research programmes on a downward trend because they might recover and attain unforseen splendour: a butterfly emerges when the caterpillar has reached its lowest state of degeneration.[9]

---

and against Descartes only about one hundred years after Newton's claim. The anomalous behaviour of Mercury's perihelion was known for decades as one of many yet unsolved difficulties of Newton's programme; but only the fact that Einstein's theory explained it better transformed a dull anomaly into a brilliant 'refutation' of Newton's research programme." The rationality of science is therefore slower than most people tend to think, like "Minerva's owl, it flies at dusk" (72, 86–87).

7. See Lakatos 1968a, 343. Lakatos's view on the pragmatic relevance of theological and philosophical speculation comes out explicitly in lecture 1, this volume, in which he claims that the demarcation between science and pseudoscience is not a problem of armchair philosophy, but is of vital social and political importance.

8. See Lakatos 1971a, 112; 1970, 70.

9. See Feyerabend 1970c, 215; 1976a, 214.

Imre: Don't get me wrong here. My methodology deals exclusively with fully fledged research programmes, but has no intention of handing out *advice* to the scientist on how to arrive at good theories or on which of two rival theories he should work on. The standards of appraisal I put forward explain why it is *rational* to accept Einstein's theory rather than Newton's, but they do not force the scientist to work on the Einsteinian programme rather than the Newtonian one. I can only judge what scientists have done: I can say whether they have progressed or not. But *I cannot give them any advice—nor do I wish to.*[10]

Paul: And yet, at the beginning, the bold project of "the *logic* of scientific discovery" was aimed at describing those rules which govern the *acceptance* and *rejection* of scientific theories. Rules that should have functioned as a *code of intellectual honesty* whose violation was intolerable.[11] What, then, is the point of laying down rules which may be either followed or ignored? You're like the author of a cookbook who describes the recipe for making *good* pizza and then remarks: "Of course, I am not telling you what to do, but whatever you do, keep a record of it."[12] Your *standards* are only *verbal ornaments:* a remembrance of past happier times when it was still thought possible to run a complex and often catastrophic business like science by following a few simple and 'rational' rules. As a matter of fact, your flexible scientific 'method' is nothing but a disguised version of my *anything goes.*[13]

Imre: There is freedom ('anarchy', if you like) in choosing which programme to work on, but the products *must be judged.* You are conflating *methodological appraisal* of a research programme with *heuristic advice* on what to do. One may rationally stick to a degenerating programme until it is overtaken by another, and even after. What one must not do is ignore its poor record. Playing a risky game is perfectly rational (and honest): what is irrational (and dishonest) is to pretend the risk isn't there, or to belittle it. Everyone is free to follow his own peculiar inclinations, but only as long as he publicly admits the state of open competitiveness.[14]

Paul: I still think you are not clear enough in your distinction of *rationality* and *honesty:* a person can easily be rational and dishonest (or irrational and honest). Dillinger was surely dishonest, but it would be

10. See Lakatos 1971b, 174, 178. On the different positions held by Lakatos concerning the link between appraisals and advice, see Motterlini 1995.

11. See, for example, Popper [1934] 1959, chaps. I and II; see also Lakatos 1971a, 103.

12. See Musgrave 1978, 475.

13. See Feyerabend 1970c, 215–16.

14. See Lakatos 1971a, 117.

hard to show that he was irrational with regard to his research programme, which just happened to be organised crime.[15] If your only piece of *advice* is to be *honest* in judging the evidential pros and cons of the various research programmes, then consider how *futile* is the point of view which allows a thief to steal as much as he wants, and yet be praised by the police and by everybody else as an honest man provided he admits to stealing. If your methodology differs from anarchism in this sense *only,* then I'm ready to become one of its fans. Who would prefer criticism to praise, if all he has to do is describe his actions in the language of a particular school?[16]

Imre: Wait a second. I'm not saying that people who support a degenerating research programme should enjoy as much freedom as you seem to imply. In fact, they should *not* be allowed to publish their papers, which contain, in general, solemn iterations of their positions or attempts to reabsorb counterevidence by ad hoc adjustments. Editors *should* refuse publication, research foundations *should* refuse them funds.[17]

Paul: And here we come across yet another "strange case of Dr Jekyll and Mr Hyde." *First* you give full rein to your *anarchic* vein, maintaining that the only irrational kind of behaviour consists in denying the state of the programme one is working on. *Then* you entreat publishers and societies to refuse printing and funds! Let me also add that I was not at all confusing 'methodological appraisal' with 'heuristic advice', as you seem to suggest. I was rather insisting that there is a legitimate link between them. You are the first to admit this when, betraying your *authoritarian* nature, you strengthen your standards *not* on an argumentative level, but by shaping a historical and social situation which renders it difficult, *in practice,* to cultivate a degenerating programme. Taken by themselves your standards are incapable of ruling out the most outrageous behaviour; taken in conjunction with a certain kind of conservatism, on the other hand, they have a subtle but firm influence on the scientific community. You want it both ways: you're making the omelette (you have more liberal standards), keeping the eggs (you have them used in a conservative way), and even passing as a rationalist![18]

Imre: I'm not 'passing' as a rationalist, I'm a full-blooded rationalist!

15. See Hall 1971, 152; Musgrave 1978, 487.
16. See Feyerabend 1976a, 216.
17. See Lakatos 1971a, 117.
18. See Feyerabend, 1976a, 213.

Paul: You abhor irrationality, that's true. And yet you can exclude it only by adopting measures which turn out to be irrational when set against your very own standards! This of course doesn't make you a willing anarchist, but it makes you a rationalist who by misadventure ends up in irrationality.[19]

Imre: But still there is a considerable difference between us.

Paul: There is a considerable difference in *rhetorics.* Combining commonsense standards of scientists with the methodology of scientific research programmes, you utilize the intuitive plausibility of the former to support the latter: a splendid Trojan horse that can be used to smuggle real, honest (a word you hold so dear) anarchism into the minds of our most dedicated rationalists. You are much better at this than I am, since rationalists are constitutionally incapable of accepting anarchism when it is offered to them undisguised. One day, of course, they will realise that this is what happened. That will be the day they will finally be ready for anarchism, pure and simple.[20] I have to admit that your plan is diabolical. But remember, my name is *Lucifer,* so it is *I* who bring the *light, not* you![21]

Imre: Yes, but Lucifer denotes the chap who brings *false* light, while I am shrouding them *in the darkness of truth.*[22] My methodology is a theory for characterising real cases of growth of knowledge and distinguishing them from impostures. Its appraisals are retrospective: they tell us only that a programme has been better than its rival *up till now,* without in any way deciding anything for the future.

Paul: This, however, means that any piece of advice based on past performance will be totally *arbitrary,* and we are back at the start.[23]

Imre: No, we aren't. If the methodology of research programmes aims to be something more than a descriptive account of the past performance of theories, then it must provide its methodological rules with an extramethodological support of a conjectural kind. I once asked that Sir Karl Popper admit a 'whiff' of inductivism in order to relate the scientific gambit of pragmatic acceptances and rejections to verisimilitude.[24]

Paul: A 'whiff'? I would rather say a full-blown storm.[25]

19. See Feyerabend 1978a, 185.
20. See Feyerabend, "Theses on Anarchism," this volume; 1975a, 187–200.
21. See letter, PF to IL, 20 January 1972, this volume.
22. See letter, IL to PF, 25 January 1973, this volume.
23. Cf. Hacking 1979, 169.
24. See Lakatos 1974a, 154–59.
25. The remark here attributed to Feyerabend is actually Newton-Smith's (1981, 68).

Imre: Call it what you wish. The point is that only a similar 'inductive principle' can turn science from a mere game into an epistemologically rational activity; from a set of lighthearted sceptical gambits pursued for intellectual fun into a serious fallibilistic venture of approximating the 'Truth of the Universe'.[26]

Paul: But what have we gained?

Imre: I can now give a *positive* answer to your previous question concerning the value of any *practical indications* based on judgements which refer exclusively to scientists' *past* performances. Thanks to our '*conjectural* principle of induction', the fact that our appraisals may in the future be contradicted does not constitute a good reason for not relying on them now. Even though the future is unpredictable, programmes chosen at random are not all equally promising. Thus, from an appraisal such as "programme A has been degenerating up till now whereas programme B has been progressing," one may possibly derive a *piece of practical advice,* such as that the scientific community should devote most of its intellectual and economic resources to programme B (and note that *most* is not equivalent to *all!*). This solution certainly offers 'all the advantages of honest theft over dishonest toil'; but it might be that in this area 'honest theft' is our only option.[27]

Paul: So what's left of the anti-inductivist bequest of Popper, who is commonly known as the slayer of Logical Positivism and as the one who solved (in a negative way) 'Hume's problem'?[28]

Imre: It seems to me that Popper has to admit that methodological appraisals are interesting primarily because of a *hidden* inductive assumption—that is, that if we act in conformity with these appraisals, we have a better chance to get nearer to the Truth than otherwise.[29] This reminds me of Columbus when the "sea current carries exotic plants, animal carcases, finely carved wooden objects, and he visualizes the far-off and yet unknown land from which these objects come."[30]

26. See Lakatos 1971a, 101.

27. See Lakatos 1968a, 181–91; Worrall 1978b, 326.

28. Reference is obviously to the problem of the justification of inductive inferences, which, along with 'Kant's problem' (also known as the problem of demarcation), is, according to Popper ([1932–1933] 1979), the fundamental problem of the theory of knowledge.

29. See Lakatos 1974a, 158. Popper would seem finally to admit this in his *Replies to My Critics:* "there may be a 'whiff' of inductivism [which] enters with the vague realist assumption that reality, though unknown, is in some respects similar to what science tells us, or, in other words, with the assumption that science can progress towards greater verisimilitude" (1974b, 1193).

30. Mach 1905, 171. Lakatos does not quote him directly, but puts forward a similar idea in his "Popper on Demarcation and Induction" (1974a, 158).

Paul: One of the examples Ernst Mach loved to use when he wanted to show the vital importance of conjectures, even the most speculative ones.[31]

Imre: Neither can we do without bold hypotheses in the theory of knowledge. The fact that one particular assumption is put forward as pure speculation shows that we are conscious both of its lack of proof and of its necessity. There is nothing wrong with *fallible* and *speculative* metaphysics, but only with interpreting such metaphysical statements as infallible inductive principles.[32]

Paul: I like your candid fallibilism, which is surely a step in the right direction. I mean towards releasing our most deeply rooted beliefs from their putrid foundations. Yet the task of scientists no longer lies in "searching for the truth" or "improving predictions," but rather, in the words of the Sophists, "in making the weaker case the stronger one, thereby to sustain the motion of the whole."[33]

Imre: So, from your point of view, as I understood it, it is not "the *truth* [that] will make you free"?[34]

Paul: The truth, whatever it is, *be damned.* Play, fun, and fiction will make you free. Someone who laughs looks intelligent (much more so than someone who explains her 'profound convictions'). She seems magically lifted out of the sea of fear, poverty, and egoism into which fate threw her and in which she is kept by the 'truth'. What we need is to take things *lightly.*[35] Were I to choose an auspicious image for the new millennium, I would choose this one: the sudden agile leap of the philosopher who raises himself above the weight of the world, showing that with all his gravity he has the secret of lightness.[36] (Of course, I would be talking about the lightness of thoughtfulness rather than the lightness of frivolity. In fact, the thoughtful kind of lightness can make frivolity seem dull and heavy.) You have a gift for laughter, even where your own position is concerned; thus for me you are a good guy.[37]

Imre: But we live in a world moulded by science: isn't this reason enough to study science?

31. See the whole of chapter 14 ("The Hypothesis") in Mach 1905. For an evaluation of Mach far from the positivistic cliché, see Feyerabend 1987, chap. 7.

32. See Lakatos 1968a, 186–87.

33. See Feyerabend 1993, 21.

34. Evidently, this is not Lakatos, but the New Testament (John 8:32).

35. See Feyerabend 1973.

36. This is not Feyerabend, but Italo Calvino, *Six Memos for the Next Millennium* (London: Vintage, 1993), 12.

37. See letter, PF to IL, 16 April 1971, this volume.

Paul: Of course it is. But when there is an invasion of locusts, people study locusts *in order to be free of them,* not so as to turn them into new-found gods![38]

Imre: So here's my idea of pesticide. I have claimed that the unit of appraisal for the growth of knowledge is a series of theories, in which each one is obtained by adding some auxiliary clauses in order to accommodate certain anomalies and produce new predictions. But we must also require that, at least once in a while, the increase in content should be corroborated: the programme as a whole should also display an *intermittently* progressive empirical shift.[39]

Paul: And this is where your recipe fails to work: how should we conclude that the research programme in question has 'run out of steam' and should therefore be abandoned? In fact, what appears to be a sequence of degenerating adjustments may just happen to be the initial phase of a long progressive development. After Aristotle and Ptolemy, the idea of the Earth moving—that weird, ancient, and 'entirely ridiculous' Pythagorean view—was dumped into history's rubbish heap, until Copernicus breathed new life into it and forged it into a weapon to defeat its defeaters.[40]

Imre: All programmes are at first only "excrescences of imagination fighting for existence by trying to outgrow each other." However, such "flowers of phantasy" must be destroyed by merciless criticism, before a single one develops further and attains some permanence. Lacking the *role of criticism,* science would be reduced to "a witches' sabbath of adventurous ideas."[41]

Paul: But my objection returns: if you don't specify a *time limit* for a (degenerating) programme, 'criticism' won't guide the growth of knowledge. How could you then distinguish scientists doing science from witches in 'sabbath'?

Imre: It would not be very wise to assert, in the abstract, a time limit valid for all situations. In fact, any appraisal of an *individual* case should not be applied mechanically, but should rather follow from general principles allowing for some *Spielraum.* And I chose the term 'intermittently' in my proposal to give sufficient *rational* scope for dogmatic adherence to a programme in the face of prima facie 'refutations'.[42]

Paul: I can't tell how your idea of the growth of knowledge differs

38. See Feyerabend 1980a, chap. 11.
39. See Lakatos 1970, 49.
40. See Feyerabend 1975a, 41.
41. Mach 1905, 77.
42. See Lakatos 1970, 49.

from mine. It resembles an ever-increasing *ocean of alternatives:* every single theory, every fairy tale, every myth forces the others into greater articulation and via a competitive process they all contribute to the development of our consciousness.[43]

Imre: But I'm lucky enough to own a compass that enables me to *navigate that ocean in many ways.* I still think it's useless to indicate a *time limit* in the abstract relating to a research programme, deciding, say, on its thirtieth or fiftieth degenerating version that it must be rejected. Notwithstanding, I still think it possible, sensible, and practical to give that time limit *indirectly,* by comparatively evaluating two or more programmes and their respective states of progress.

Paul: Where, then, does the 'objective' (as opposed to socio-psychological) reason to reject a programme lie?

Imre: The objective reason to reject a programme is supplied by *another* programme which explains the previous success of its rival and supersedes it by a further display of *heuristic power.*[44]

Paul: But my objection can spring back against your time limit in 'comparative terms' . . .

Imre: . . . alright, but I have *no* intention of claiming any direct inference from: "programme A is currently most favored by evidence" to: "the *only rational* course of action is that of working on programme A" (or: "it would be *irrational* to try to develop any alternative programme"). If we were to accept this criterion, we would be claiming that all the great scientists in history have acted irrationally! The wave theory of light, for example, was not unambiguously the best theory available when Fresnel decided to work on it in the early nineteenth century; it was Fresnel's work that turned it into overwhelmingly the best available theory.[45]

Paul: This seems to me simply another way of saying that there is no *rational* way of showing that the choices taken by a scientist who works inside a degenerating research programme are necessarily irrational.[46]

Imre: You're right, and you're wrong. If we are satisfied with 'deductive rationality', then *you're right:* if we are referring to mere *logical possibilities,* then of course there is nothing illogical in believing and hoping that however badly a programme has behaved in the past, it may

43. See Feyerabend 1993, 21.

44. That is, the ability of the programme to anticipate 'novel' facts during its growth; see Lakatos 1970, 69. On the notion of the 'heuristic power' of a programme, see also Urbach 1978. On the concept of 'novelty', see Worrall 1978a, 1985; Zahar 1973; and also "Editor's Note: The Value of Novelty," this volume.

45. See Worrall 1990, 332.

46. See Feyerabend 1976a, 389.

still recover and reach unsuspected peaks of splendour. But Duhem had already shown that deductive logic alone when coupled with crude observational results can supply only a very weak theory of rationality.[47] On the other hand, if we will not be satisfied with the weak requirements that logic alone demands from scientific practice, then *you're wrong*. No doubt there are cases of very general metaphysical ideas that have had a chequered history; once absorbed into a steadily degenerating programme, they have then much later been brought back to life as elements of a progressive programme. Atomism is often cited as a good example of this. *But* if we take a look at specific cases, if we consider individual research programmes, then it becomes apparent that in the history of physics no one who has stuck to a highly degenerating research programme when a progressive alternative was available has ever managed to reverse the situation. Thus, although I agree that there is nothing *illogical* in choosing to work on a degenerating programme, that choice is indeed *irrational* (unscientific) simply in the sense that it does not follow a *procedure* that seems to have invariably *paid off* in science.[48]

Paul: And what on earth would these 'procedures' be? The new astronomy of Copernicus, Kepler, and Galileo took root; witch hunts came to an end: these facts were brought about because independent thinkers resolved to introduce and defend obsolete theories *in spite of all the traditional methodological procedures*. The theory of witchcraft, far from being a mere outburst of folly, had a well-defined structure between the seventeenth and eighteenth centuries; it was 'rationally' formulated and 'empirically' confirmed. The Copernican theory, on the

47. Cf. Duhem 1906:

> *Pure logic is not the only rule for our judgements;* certain opinions which do not fall under the hammer of the principle of contradiction are in any case perfectly unreasonable. These motives which do not proceed from logic and yet direct our choices, these 'reasons which reason does not know' and which speak to the ample 'mind of finesse' but not the 'geometric mind', constitute what is appropriately called good sense. . . . After Foucault's experiment had shown that light traveled faster in air than in water, Biot gave up supporting the emission hypothesis; strictly, *pure logic would not have compelled him to give it up,* for Foucault's experiment was not the crucial experiment that Arago thought he saw in it, *but* by resisting wave optics for a longer time Biot would have been *lacking in good sense.* (218–19, emphasis added)

48. See Worrall 1990, 346, 349, 350. Worrall claims further that such a "stronger" theory of scientific rationality "would have nothing resembling an a priori justification, but would simply rest on an inductive extrapolation from past to future—this time an inductive extrapolation of a methodological kind" (46). Cf. above the Popperian 'solution' of the problem of induction advocated by Lakatos.

other hand, contradicted some of the most convincing observations of the time, as well as reasonable principles of physics which had produced surprising results in physiology, psychology, and even theology.[49]

Imre: I disagree, and I'll explain why. It seems to me that while there has been no general agreement concerning a theory of scientific rationality, there has been considerable agreement concerning the *basic value judgements* about specific achievements of science, that is, whether a *particular move* in the game was scientific or crankish, or whether a particular gambit was played correctly or not.[50]

Paul: And yet the 'common scientific wisdom' you give so much weight to is *not very common* and certainly *not very wise.* On the one hand, basic value judgements are not as uniform as you assume. 'Science' is split into numerous disciplines, each of which may adopt a different attitude towards a given theory, and single disciplines into schools, heresies, and so forth. The basic value judgements of an experimentalist will differ from those of a theoretician. A faithful Bohrian will regard modifications of quantum theory with a different eye than will the faithful Einsteinian. On the other hand, basic value judgements are only rarely made for good reasons. Everyone agrees now that Copernicus's hypothesis was a big step forward but hardly anyone can give a halfway decent account of it, let alone enumerate the reasons for its excellence. Newton's theory of gravitation was "highly regarded by the greatest scientists," most of whom were unaware of its difficulties and some of whom believed that it could be derived from Kepler's law. Whatever unity remains is dissolved during revolutions. *Revolutions* leave no theory unturned and, above all, no principle unchallenged. Now: if revolutions challenge *all* the ideas born in connection with those procedures, *including 'basic' value judgements,* how can you decide to reject, say, the standards of Aristotelian philosophy *along with* its 'basic' value judgements in order to replace them with the standards and the basic value judgements of Galileo's or Newton's science?[51]

Imre: On the basis of a 'rational reconstruction'.

Paul: On the basis of a rational reconstruction *of what?*

Imre: On the basis of the rational reconstruction of science from the point of view of *modern science.*

Paul: But in this way you are assuming what has still to be proved: the methodological supremacy of modern science. And you are also

49. See Feyerabend 1975a, chaps. 6–12.

50. See Lakatos 1971a, 124.

51. See Feyerabend 1976a, 207–10.

condemning the Aristotelians from 'our' point of view, without showing that *ours* is better than *theirs*.

Imre: Your position is just a colourful version of Pyrrhonian scepticism. You should look at the excellent book by Dick Popkin: *The History of Scepticism.*[52] From a sceptic's point of view, scientific theories are a set of beliefs which have equal epistemological ranking to so many other sets of beliefs. There may be *change* in belief systems but no *progress.* It follows that *any* system is free to grow and influence any other; but none can claim epistemological superiority. You deny any possibility of producing any theory of appraisal whatsoever. Your only piece of advice is: *do your own thing.* This is your only code of intellectual honesty.[53]

Paul: Be careful, I'm not simply a Pyrrhonian, but rather a 'cultural relativist'. I think *the validity of ideas depends on the tradition against which we compare them.* Einstein is better than Newton from a modern scientist's point of view, worse from a Dinglerian's, and the problem is of no interest at all from a Hopi Indian's point of view.[54]

Imre: I'm *very* careful, and to my mind there is a basic weakness in your position. I can show that you are two-faced: one face is the face of a *sceptic,* the other is the face of an *authoritarian.*[55] Let me explain: the tolerant sceptics believed that utopian dogmatism was responsible for the worst suffering of mankind. They pointed out that those who claimed to possess moral, political, and religious 'truth', those who boasted of knowing which way progress lies, used the Inquisition and torture, bloody wars, and genocide in order to realize their predictions in practice. Tolerant sceptics dreamt of a social contract that would decide how to restrain the human animal and minimise suffering. From their point of view—and yours—happiness and welfare replaced Truth. They argued that *betrayal of reason (or rather 'reason') by man was better than the betrayal of man by reason.* However, in the face of any controversy, a sceptic has no choice: he either turns into a dogmatist or he resorts to force without argument. Thus, ultimately, there is only one type of political philosophy consistent with scepticism: the philosophy which equates *right* with *might.* This is why many sceptics became well-paid courtiers of the bloodiest tyrants in history.[56]

52. The reference is here to Popkin 1979. See also letter, IL to PF, 21 February 1969, note 55, this volume.

53. See Lakatos, lectures 1 and 7, this volume; 1973, 107–8.

54. See Feyerabend 1978a, part II, chap. 3; 1987. On Feyerabend's relativism, see also Couvalis 1989, appendix 1.

55. See letter, IL to PF, 2 March 1973, this volume.

56. See Lakatos, "The Intellectuals' Betrayal of Reason," appendix C, this volume.

Paul: I still think that scepticism *is not going far enough.* If the sceptic does not know anything, then he may well do whatever he wants to do; that is, he may engage in propaganda, he may defend the status quo, he may oppose it: 'anything goes'. Anything goes, on the other hand, clearly does not mean scepticism. It means: *anything goes,* therefore also law and order, argument, irrationalism, et cetera. But one point has to be made clear, I would not *hurt a fly*—let alone a human being.[57]

Imre: I do not mind your anything goes, but when it comes to *moral* theory I even make mincemeat pieces of Pyrrho. Of course you would not hurt a fly, as you put it. The question is what you do when you are in a position in which you can hurt either one fly or another and you are bound to hurt *one.* Would you commit suicide? You may remember that before I started on research programmes, I discovered that I had to replace the question of the acceptance and *rejection* of theories with the *preference* of one theory over another. And this of course also applies to ethics and politics. Thus I am faced with a problem which *I* am willing to face but *you* cannot.[58]

Paul: I admit I'm wrong, and you are right. But I don't mind being wrong here and there.[59]

Imre: Well, there are two different kinds of betrayal of reason, and yours is certainly the worst. The first consists in mitigated scepticism which originates as a sort of blind reaction to the outrages of dogmatism. It is the ancient betrayal of reason and I regard it as a venial sin. The second is radical scepticism. Undeterred by the long series of successes of Newtonian science, radical scepticism has tried to show that they were sham successes and even the best theories of the exact sciences were nothing other than irrational beliefs. The hallmark of the modern betrayal of reason is the intellectual attack on the objective epistemological value of the exact sciences. I regard this betrayal of reason as criminal.[60]

Paul: I hold that any inquiry into a theory of 'rationality' should try to answer *two* main questions: (1) *What is science?* How does it proceed, what are its results, how do its procedures, standards, and results differ from procedures, standards, and results of other enterprises? (2) *What is so great about science?* What, for instance, makes it preferable to the form of life of the Azande? What makes modern science preferable to

57. See letter, PF to IL, 7 August 1972, this volume.
58. See letter, IL to PF, 14 August 1972, this volume.
59. See letter, PF to IL, August 1972, this volume.
60. See Lakatos, appendix C, this volume.

the science of the Aristotelians? You, on the contrary, along with all the other Friends of Reason, do not show, but simply assume, that modern science is 'objectively' better than the basic wisdom of witches and warlocks. In this way you take (mis-)possession of the term 'rationality' for ideological purposes, and you equate it with the standards characteristic of a certain intellectual community: that of scientists of 'the past three hundred years'. To define as 'rational' whatever is consistent with those standards implies you have already answered the second question. *But you have not:* you don't argue, you simply use the alleged superiority of science to justify those same standards you have already encoded in your methodology of scientific research programmes.[61] You are forgetting that the strength of scepticism lies in the fact that, together with the particular results, the *criteria to assess them are also changing*—you're the one who should leaf through the pages of Popkin's book! What would you do if faced with a 'new style of reasoning' capable of producing yet another particular knowledge? What future Lakatos will inveigh against the hypothetical-deductive method and the theory of research programmes to which it has given birth?[62]

Imre: I'm amused by your suggestion that scientific revolutions are revolutions in standards. This is of course the story I encapsuled in my announced book "The Changing Logic of Scientific Discovery."[63] I might agree that methods in science (and mathematics) change and can be expected to change. The important thing is to try to ensure that such methodological changes are for the better. However, we can take charge of this only if we succeed in rationally reconstructing change in standards as we reconstruct change in scientific theories. From this point of view my "Changing Logic" aims at grasping the "unfolding of reason" and presenting it "cut and dry," after its process of formation has been completed.[64]

Paul: But it might be the opposite of what you claim! I mean that *the better a methodology seems to capture the* rationality *of science, the*

61. See Feyerabend 1975a, 181; 1976a, 203; 1978a, 99–107.

62. Cf. Hacking 1979, 185–87.

63. "The Changing Logic of Scientific Discovery" is the title of the book Lakatos had in mind to write, but which was never completed; see also letter, IL to PF, 10 January 1974, and appendix C, this volume. On a few possible ways in which Lakatos might have carried out this project, see Worrall 1988, 1989b; Larvor 1998, 205–10; Motterlini 1999.

64. Larvor 1998, 9–17. Note the Hegelian language. Together with Popper's critical philosophy, Hegel's dialectic was indeed one of the major "ideological sources" of Lakatos's philosophy. See also Kadvany 1995.

greater its actual mystification of science. After all, if the most arbitrary moves often coincide with the main radical turns in the growth of knowledge, then putting forward a 'theory of rationality' and using it to rule our ('internal') reconstruction of history is a tyrannical act of the intellect which damages both science and society. Nor is it of any use to claim stubbornly that science is superior to other forms of life. Science today reigns supreme not because of its comparative merits, but because the whole show has been rigged in its favour.[65]

Imre: Alright, alright. Attaching a label with 'rational' or 'irrational' written on it to researchers' strategies is not, after all, so significant for my methodological appraisals. But I'm not waving a white flag: the fundamentally valid idea that a programme is adopted by researchers not only for its explanatory power but especially for its heuristic power remains. It is adopted and retained for its ability both to put on the table new and interesting problems and to point to possible solutions. And my exhortation towards a *rational reconstruction* of individual historical cases should be taken as a *historiographical* programme, an encouragement towards defining the reasons and strategies which have produced new ideas. There is, therefore, nothing wrong in appraising past beliefs according to a given norm or theory of rationality. On the contrary, such judgements lead to historical data which is not easily obtainable in other ways, and also allow one to outline and explain the whole process. This does not imply that there is any need to pry into the brains of scientists in order to assess the 'reasons' or peculiar aversions which have governed their choices, but only that we should try to analyse and evaluate the case we are faced with *in the light of our methodological standards.* Any appraisal of 'rationality' of this kind is doubly desirable: the historian learns 'new' facts, the philosopher checks his own standards.

Paul: I'm ready to admit that as an instrument for carrying out research in the history of ideas your theory is vastly more sophisticated than Kuhn's, and so it will definitely lead to more detailed research, and to more discoveries. The discoveries may in the end turn *against you,* but that does not discredit you today, when no other theory provides an equally detailed inventory of suggestions.[66]

Imre: And I concede that one has to be sceptical with regard to an

65. See Feyerabend 1978a, 103.

66. See Feyerabend 1975c, 17. Feyerabend was not alone in thinking this: many case studies have been inspired by the methodology of scientific research programmes since then, both in the social and the natural sciences; see lecture 8, note 11, and "Editor's Note: The Value of Novelty," note 9, this volume.

immutable statute law. This is why I advocate a *pluralistic* system of authority, one that would allow the particular authority of 'basic' statements to criticise the general authority of the theory of rationality, and vice versa.[67] Only in this way can the proliferation of differing points of view, the comparison between different 'rational' reconstructions, the awareness of local strategies and of the reasons behind researchers' moves specify how we can *learn from history* and, especially, how we can escape from the influence of the 'worst' philosophies or, to put it in the words of John Maynard Keynes, "to emancipate ourselves from old ideas which ramify, for those brought up as most of us have been, into every corner of our minds."[68]

Paul: You've struck a blow for your side, so let me answer back in the same way, and quote from Lenin: "History in general, and the history of revolutions in particular, is always richer in content, more varied, more many-sided, more lively and ingenious than even the best parties, the most conscious vanguards of the most advanced classes can imagine. This leads to two important practical consequences. First, that in order to fulfill its task, the revolutionary class[69] must be able to master all forms or aspects of social activity without exception.[70] Second, it must be ready to pass from one to another in the quickest and most unexpected manner."[71] Lenin, of course, is addressing parties and the revolutionary vanguards rather than scientists and methodologists, but the lesson is the same: methodological rules should be adapted to the circumstances and reinvented anew each time. This increases freedom, the sense of humanity, and the hope of succeeding. After all, you are the only philosopher of science who secretly imbibes the forbidden brew of Hegelian dialectic, and the results are evident in your magnificent work "Proofs and Refutations." All that is required now is that you confess your 'vices' openly.[72]

Imre: I'm afraid that some ardent Popperite may already be rejecting all that I am about to say, but I confess that even the *poverty of*

67. See Lakatos 1971a, 136–38.

68. Keynes 1936, xxiii.

69. i.e., the class of those who want to change either a part of society, such as science, or society as a whole.

70. i.e., it must be able to understand, and to apply, not only a particular methodology, but any methodology, and any variation it can imagine.

71. Lenin, "Left-Wing Communism—An Infantile Disorder." Quoted in Feyerabend 1975a, 17–18.

72. See Feyerabend 1970a, 101.

historicism is better than the complete absence of it. Always providing, of course, that it is handled with the care necessary in dealing with any explosives . . . [73]

Paul: . . . and is placed under the right targets.

73. See Lakatos 1959–1961, 61. For an overall assessment of Lakatos's philosophy "between the Hegelian Devil (i.e. historicism) and the Popperian deep blue sea (i.e. fallibilism)," see Motterlini 1999.

ONE

Lectures on Scientific Method

Imre Lakatos

Lakatos gave a course of eight lectures at the London School of Economics in the Lent term of 1973. These lectures were recorded and then transcribed by Sandra Mitchell with the help of Gregory Currie. The transcript was deposited in the Archive of Professor Imre Lakatos at the British Library of Political and Economic Science, catalogued by Michael Hallet in 1979 and updated by Sue Donnelly in 1995 (henceforth cited as Archive). Lecture 5 was accidentally not recorded.

I am well aware of certain difficulties and ambiguities present in this material—this is particularly true of lecture 6. I felt it advisable, however, to leave unchanged the content of what Lakatos told his students, choosing to clarify certain points and provide additional information in the footnotes only.

My general approach to matters of presentation has been to leave the transcribed material unchanged. Exceptions include a few linguistic mistakes and passages in which the structure of arguments was particularly cumbersome. The characteristic colloquial form of the lectures has been retained as far as this was possible, but the style has been refined in minor ways and the structure of the presentation polished to meet the requirements of published work.

I have taken the liberty of supplying a title for each lecture and of making some alterations where the original typescript was incomplete, or where minor changes seemed to produce a major increase in readability. Brief omissions are signalled by ellipses within square brackets. Biographical and bibliographical references have also been provided. Some illustrations have been chosen in accordance with Lakatos's own blackboard drawings, and others added where the text seemed to call for them.

Lakatos always took a great deal of care over the presentation of any work of his which he intended to publish. It is very likely that he would not have regarded the material published here for the first time as satisfactory. I am, nonetheless, very pleased to have had the opportunity of presenting it. These lectures are sharp in style, original in thought, and forceful in tone, and yet are often ambivalent and open to discussion. They reflect Lakatos's lively interest in ethical, religious, social, and political issues connected both to the general problems of the philosophy of science and to the specific question of the standards of scientific appraisal.

Right from his early contributions to Hungarian literary and academic journals, Lakatos's interest in mathematics, science, and history was always closely linked to educational and socio-political issues. In 1956, on the eve of his departure from

Hungary, Lakatos delivered an impassioned speech at the Petöfi Circle pedagogy meeting, in which he dealt with the problem of the education of a new generation of scholars. He argued in favour of talent, curiosity, original thinking, autonomy, the right to doubt and the right to dissent, demand for proofs, and respect for facts (see this volume, appendix A). One of the main features of Lakatos's attitude towards teaching was his effort to force students to criticise his own ideas. This is why his lectures are so openly provocative and also why many of his remarks on academics, philosophers, scholars, intellectuals, and politicians are outrageous, and intentionally so.

Ernst Gellner remembers that Lakatos's lectures were "intelligible, fascinating, dramatic, and above all conspicuously amusing even for non-specialists" (*The Times,* 8 February 1974). John Watkins recalls that when Lakatos lectured, "the room [was] crowded, the atmosphere electric, and from time to time there would be a gale of laughter" ("Lakatos Obituary," *The Times,* 6 February 1974).

Lecture One

The Demarcation Problem

The demarcation problem may be formulated in the following terms: what distinguishes science from pseudoscience? This is an extreme way of putting it, since the more general problem, called the *Generalized Demarcation Problem,* is really the problem of the appraisal of scientific theories, and attempts to answer the question: when is one theory better than another? We are, naturally, assuming a continuous scale whereby the value zero corresponds to a pseudoscientific theory and positive values to theories considered scientific in a higher or lesser degree.

This is not an esoteric problem just for armchair philosophers, and I will give you a few historical examples—which you all know—where such demarcation criteria could have helped.

First, let us take the 1616 banning by the Catholic Church of the works of Copernicus, and particularly of his *De Revolutionibus Orbium Caelestium.* A few of you may know the grounds on which the Inquisition tribunal based its judgement: the book was considered pseudoscientific and rejected because it contained too much speculation. The text of the ban reads that the book will be banned until it is proven. In the 1820s the Church removed the book from the Index, maintaining that the theories contained in the book had been proven. As you know, Einstein came much later!

Let me mention another example which concerns an authority similar to the Inquisition: in 1949 the Central Committee of the Soviet Communist Party discussed the merits of Mendelian genetics as opposed to Lysenko's environmentalist theory. They decided that Lysenko's theory was better, and the leading authority of the rival school, N. I. Vavilov, was killed in a concentration camp.[1]

Let me point out that Galileo, Copernicus's main supporter, while leading an extremely comfortable life in comparison to many of his contemporaries, continues to this day to enjoy the reputation of a suffering hero, when he actually was not, whereas Vavilov, who most certainly was, does not. Another difference between our two examples is that although the Church banned the book, it did not ban its discussion, whereas in the Soviet Union even the discussion of Mendelian ideas was banned.

Our third example: last year [1972] the American Philosophical Association discussed the following proposal put forward by Hilary Putnam, professor of philosophy at Harvard University. It said:

> The American Philosophical Association should endorse the American Anthropological Association's condemnation of the racist, sexist and anti-working class theories of Richard Hernstein, William Shockley and Arthur Jensen as dangerous and unscientific.[2] This is not a question of non-scientists interfering with a scientific controversy, and it is our duty to point out that an ideology with no specific scientific merit is trying to disguise itself as science. We condemn the irresponsible support given to such unfounded theories by the *Atlantic Monthly* and *The New York Times* whose publication invites serious reflection of the issues concerned, especially in view of the destructive political uses to which such theories could be put . . .

and so on. The motion was passed.

The main point is that the conclusions of these theories were 'unfounded'. This is exactly the same reason given by the Church in Copernicus's case: his theories were not sufficiently proven from facts. Let us move on to a less dramatic case which you probably all know about: I am referring to Velikovsky, who in the 1940s proposed a certain

1. The works of Trofim Denisoviç Lysenko (1898–1964) provided the guidelines of Soviet scientific research right up to Stalin's death in 1953. Nikolaj Ivanoviç Vavilov (1887–1943), a botanical scholar opposed to Lysenko's theories and authoritarian attitude, was deported to Siberia following Lysenko's accusations.

2. See Jensen 1969 and Shockley 1971a,b. Cf. also below, lecture 6.

astronomical theory. The publisher who published it had reason to regret his decision soon enough, when he found that the leading scientists who usually published with him refused to submit further texts, refusing to work with someone who published science together with pseudoscience.[3]

These examples are enough to show that the demarcation problem is far from being an esoteric problem, and that its solution could affect our lives, especially at a time when research grants are state-controlled, science is subject to a monopolistic patronage, the mass of publications is such that it is impossible for us to judge everything ourselves, and intellectual parasitism is on the rise, making us feel we would enjoy having a few people thrown out of their jobs—I mean chairs.

The demarcation problem as I have illustrated it is, of course, very similar to the problem I inherited from Popper, who founded his philosophical reputation on his so-called "falsifiability solution." I would like to read out a few passages from Karl Popper so that you can see what bothered him and his generation. Let us go back to the Vienna of the 1920s. The chairs in philosophy in Austrian and German universities were held primarily by Hegelian philosophers at first and later by Heideggerians. Vienna was quite a lively town: it was host to the Vienna Circle of Logical Positivism, and Sigmund Freud lived there, as well as Alfred Adler, the socialists of the Second and Third International, and Kurt Gödel, the leading mathematical logician of the time. It was not the provincial outpost on the European border it is today, but one of Europe's intellectual capitals. It is here, too, that we come across Karl

3. In *Worlds in Collision* Immanuel Velikovsky set forth the theory of "cosmic catastrophism," according to which the planets in the solar system were not fixed in their orbits. Earth, especially, was subject to impressive cataclysms caused by gigantic comets. One of these was thought to have passed very close to Earth during the captivity of the people of Israel in Egypt, thereby causing the division of the Red Sea. Velikovsky's theory would seem to imply that peoples who lived at the time of any similar cataclysm would have recorded it, which is certainly not the case, since many peoples who lived at the time of the events referred to in the Bible do not seem to have even noticed them. Velikovsky attempted an explanation of this fact by referring to a sort of "collective amnesia." Cataclysms, he argued, would have had such traumatic consequences that the peoples involved would have tried to forget them, and consequently would have avoided any mention of them in their historical accounts. Velikovsky's book, first published in 1950, climbed to the top of the best-seller list within a few days after its appearance, provoking furious reactions among scientists. Notwithstanding its huge success, the book was withdrawn from the market—for the first and only time in the history of American publishing—and reappeared soon after, thanks to another publisher. This whole story is the subject of *The Velikovsky Affair,* edited by Alfred de Grazia (1966).

Popper and Ludwig Wittgenstein, both school teachers in Vienna in the 1920s. In fact for some years Popper could not even get a job as a school teacher and became a cabinet maker, which is no doubt why he got interested in the demarcation problem as a kind of special machine in which one feeds, for example, a few inaugural lectures by these fashionable philosophers and the machine flashes up "meaningful" or "meaningless," and if the response is the latter, then both the chair and its holder go up in flames. This was, as a matter of fact, Leibniz's original idea. In the seventeenth century Leibniz wanted to construct a machine which would flash up the word "false" if you fed in a mistake.

I will read out two passages from Hegel, just to give you an idea of what Popper and his colleagues in the Vienna Circle were up against: Hegel was as much at the centre of study as, let's say, Newton's theory of colours. Hegel's philosophy of nature was considered proper science. In his *Open Society* (vol. II, page 290), Popper quotes Hegel's definition of heat: "Heat is the self-restoration of matter in its formlessness. Its liquidity [is] the triumph of its abstract homogeneity over specific definiteness; its abstract, purely self-existing continuity as negation of negation is here set as activity." Let us follow Popper to page 332 of his *Conjectures and Refutations* where he quotes Hegel's definition of electricity: "Electricity [. . .] is the purpose of the form from which it emancipates itself, it is the form that is just about to overcome its own indifference; for electricity is the immediate emergence, or the actuality just emerging, from the proximity of the form [. . .]." I will not go on. From the point of view of proper science today, this is nonsense, and we would most probably not appoint a man like Hegel to a chair!

But when we come across cases like Velikovsky's, it turns out that Velikovsky is a pseudoscientist simply because quite a few astronomers tell us so. If we go on to ask according to what *criteria* we think that Velikovsky's theory is, in actual fact, pseudoscientific, we realise that according to Popper's criteria this theory is falsifiable. So, what is wrong with this? This is not an easy question. But as I shall show later on, the problem in reference to natural science is easier than with the social sciences; as luck would have it, most chairs here at the London School of Economics are in the social sciences! To illustrate the deplorable state of the social sciences in Germany in the 1930s, I shall quote Popper once again. He is in his turn quoting Heidegger, *Selbstbehauptung der deutschen Universität—Das Rektorat* (1933–1934): "To will the essence of German universities is to will the essence of science itself; it is to will the historical, spiritual mission of the German nation, as a nation which has reached its full consciousness in the experience of its State. Science and the German destiny must together attain power in the essential will."

This is the background against which one has to place the Vienna Circle's attempt to divide meaningful from meaningless. If I compare these remarks with today's sociology textbooks, for example, I have a difficult time spotting the difference. So, as you can see, the demarcation problem is far from being unimportant because it is a question of determining what criteria should be satisfied in order to have a moral justification to burn down the LSE—and I assure you that I am sorry the revolutionary '60s are over!

Before moving on to the social sciences, I would like to describe the three main schools of thought on the demarcation problem.

Let us call the first school *militant positivism;* you will understand why later on. The problem of this school was to find certain demarcation criteria similar to those I have outlined, but these also had to satisfy certain boundary conditions, as a mathematician would say. I am referring to a definite set of people to which most scientists as well as Popper and Carnap would belong. These people think that there are goodies and baddies among scientific theories, and once you have defined a demarcation criterion, you should divide all your theories between the two groups. You would end up, for example, with a *goodies* list including Copernicus's (Theory$_1$), Galileo's (T_2), Kepler's (T_3), Newton's (T_4) . . . and Einstein's (T_5), along with—but this is just my supposition—Darwin's (T_6). Let me just anticipate that nobody to date has yet found a demarcation criterion according to which Darwin can be described as scientific, but this is exactly what we are looking for. And I suppose that if I do not want to be on bad terms with the mathematical economists at LSE, I should perhaps include in this list Walras and Marshall. But Marx certainly does not belong here.

Let us now group the *baddies:* Velikovsky would be a pseudoscientist, and Marx too. Popper would probably put Marx with the goodies, as you may know from his *Open Society,* but Soviet Marxism, as practised in the Soviet Union, would go with the baddies. Then of course we would have Freud, Adler, Marcuse and Lévi-Strauss. But this is not exactly correct. Why? Simply because demarcation criteria judge theories and not people, nor their collected works. We know, for example, that Newton wrote certain theological works that contained funny ideas about fallen angels; Kepler, too, had some ideas about musical harmonies in the heavens—so we are really talking about very specific and well-articulated theories, and not about people in general, as human beings or as they are represented by their collected works.

Briefly, according to militant positivism we have a programme to find a definition which puts every theory in its proper place. However, before dealing with this research programme, I would like to mention two other schools of thought.

One of them is known under different names: *scepticism, epistemological anarchism* and *cultural relativism* are all synonyms for the line of thought which goes back to the ancient sceptic, Pyrrho, and has its main supporter today in Paul Feyerabend, whose work "Against Method" appeared in volume 4 of the Minnesota Studies.[4] If we follow the sceptics, the demarcation problem is unsolvable. The reason is very simple: there is no demarcation line; epistemologically speaking, all theories are on a par. Scepticism regards scientific theories as just one family of beliefs which ranks equal to thousands of other families of different beliefs. Any one of these families—or systems—is not more *right* than any other, even though some have more *might* than others. While there may be changes in systems of belief, there cannot be any progress. According to the Positivists, there is a line of progress leading up at least to Einstein; the epistemological anarchists, on the other hand, do not admit any such progress: what they see are only changing fashions, 'bandwagon effects'. This school of thought, temporarily silenced by the stunning success of Newtonian science, is today regaining momentum. According to this view, philosophy of science is a perfectly legitimate activity: one can have a demarcation criterion (what a relief!), but it may or may not influence people. Note that epistemological anarchism has nothing to do with Mao's "let a hundred flowers bloom," since 'flowers' is, of course, a normative term. Rather, flowers and weeds may bloom together—there is no demarcation line between them. This presents a very important problem. Feyerabend has absolutely no intention of imposing a subjective distinction between flowers and weeds on anybody. Any system of beliefs—including Popper's philosophy of science—is free to grow and influence any other, but none can claim epistemological superiority.

And what about *intellectual honesty?* This too is a very interesting problem. According to militant positivism, it is simply dishonest to pub-

4. Lakatos is referring to "Against Method: Outline of an Anarchist Theory of Knowledge," written by Feyerabend in 1968 and published in 1970; all following mentions of Feyerabend in these lectures refer to this work.

lish anything pseudoscientific. Of course one can disagree about what science is, depending on whether you follow the Church or Professor Putnam. The Church says that you cannot publish any unproven theory if it is in conflict with the Bible, but are free to do so once it has been proved. Professor Putnam, on the other hand, maintains you cannot publish any unproven scientific theory, unless it is in the interest of the American proletariat, as seen through his own visionary eyes.

According to Popper, intellectual dishonesty means putting forward a theory without specifying the experimental conditions under which it could be given up. I remember when back in my Popperian days I used to put this question to Marxists and Freudians: "Tell me, what specific historical or social events would have to occur in order for you to give up your Marxism?" I remember that this was usually accompanied by either stunned silence or confusion. But I was very pleased with the effect.

Much later I put the same question to a prominent scientist, who could not give any answer because, he said, "of course anomalies always spring up, but somehow sooner or later we always solve them." This is why, according to Feyerabend, who follows in Popper's footsteps, all these criteria for intellectual honesty have one and the same function: they are empty rhetoric to frighten school children. Feyerabend's only piece of advice is to remain faithful to yourself—to do your own thing and not let yourself be judged. He adds that along with his logic courses he also gives a course in Black Magic and other similar alternatives. Feyerabend's stand is not exactly a laughing matter. We may, for instance, choose to consider the mortality rate among Christian Scientists who do not accept artificial medical help and only wish to be healed by God. The interesting thing in this case is that their mortality rate is not perceptibly higher than among academics. I came across an interesting case in *The Observer* yesterday—perhaps you saw it, too—about how a couple of dozen psychiatrists went under assumed names to other psychiatrists and were subsequently put into mental hospitals.

So the sceptics can always point to the fallibility of scientific theories, and occasionally they are right. It does not necessarily follow that the more scientific the beliefs, the happier people are: consider the Swedes, for example, who have the most perfect socialist society in western Europe, and appear to be very satisfied with themselves on moral grounds if you take into account the intensity of their anti-Vietnam activities, and yet have a higher suicide rate than, say, the Azande people.

The Freudian theories and Marxism also appear to have something which Einstein's theory of quantum electrodynamics lacks. If you happen to meet someone who works in these latter fields, you usually find you are facing a person subject to frequent headaches brought on by the number of problems he has to solve, someone who has many doubts about the whole theory, who doesn't really know if he is coming or going, who sees puzzles everywhere. Now consider a committed Freudian or Marxist: he lives in a state of happiness, he can explain everything, and enjoys that happy, relaxed state of mind called 'understanding'. If you go to a theoretical physicist, he usually says: "I do not understand what is going on in the universe, but I have some theories and occasionally my experiments work; but I still do not understand what God meant with this chaos." Approach a Freudian or a Marxist, on the other hand, and everything falls into place. This reminds me how impressed I was by Popper's phrase: "all-explanatory theories have an irresistible effect on the weak mind." The phrase is, of course, sarcastic; however, some beliefs make people happier than others. [. . .]

According to epistemological anarchism, intellectual influence is directly proportional to the vocal energy, the faith and the propaganda skills of competing groups. I think I had better come back to this school later on. But I would like you to think that I have dealt fairly with this solution to the demarcation problem: it has certain attractions. Sceptics—not all of them, of course—usually say: "Truth? There is no such word." If one mentions the word "truth," they say: "well, what *is* truth? Truth is what the victor believes." If a struggle between beliefs occurs, and one emerges victorious, that belief is what we call "truth." I have thought a lot about that: for instance, we know now that Copernicus—Galileo actually—really earned a sort of propaganda victory over Ptolemy. However, most people think that Galileo's theory is true, but that judgement is not held in the light of Einstein. Another example: most people believe the years 1917 or 1789 to be moments of progress in history. However, if we take another point of view, one can regard—as I do—1917 as the year the new Dark Ages began. In my view, 1945 is the year the only great colonial empire of the twentieth century, namely, the Soviet one in Eastern Europe, was established; but it is also the year of the glorious victory over the Nazis. As you can see, there are many beliefs, and many points of view: who can tell which is better? Of course, if this theory is true, we shall never come up with a definition of demarcation. Let us leave the New Left for now and come to the Right.

Let us now come to another school of thought: *élitist authoritarianism.* It believes that there *is* a demarcation between the goodies and the baddies. For the benefit of the sociologists who are in the audience, I shall immediately add that this view is currently represented in the philosophy of science by Polanyi, Kuhn and Merton.[5] This is actually the standard view institutionalised today in scientific organizations such as the Royal Society. According to this view, there is a *demarcation,* but there are no demarcation *criteria.* Why? Because the difference between science and pseudoscience is inarticulable. Only a wise judge—a great scientist, for instance—can see the difference. So you can have a jury but you cannot have a law. A law, of course, gives the jury guidance, but the jury has to interpret it. And yet, according to this theory, there can be no laws, there is only the jury of wise men.

The great scientist sits in judgement and decides what is good science and what is bad science. Fellows of the Royal Society appoint further fellows, and professors appoint further professors. In conclusion: yes to demarcation, no to demarcation criteria.

London University has a beautiful way of preserving confidentiality in filling chairs, which is very much in this tradition. Let us suppose the University is about to appoint a very important professor: say six candidates are short-listed. The referees report to the committee and an official from Senate House reads out the reference report to them. Now, supposing there are three reference reports for each candidate, the committee is faced with eighteen reports. The man reads them all out and when the original reports are passed around the table each referee can look at them very quickly, but no copies are allowed. A decision is reached in less than one hour. As you can see, the appointed professor emerges like a conservative leader, and there are no criteria with which to judge his theories, and the whole thing is clouded in esoteric mystery.

Those of you who are with the Department of Government here at LSE will know all about Oakeshott's[6] theory of politics, which is very similar to the case we have examined. According to this group of people, demarcation criteria are hybris. How can a mere philosopher devise criteria for distinguishing between good and bad science, knowing it is an unutterable mystic secret of the Royal Society? In Professor Oakeshott's view, politicians can make politics, but political philosophy is impossible—this is what his philosophy aims to point out.

History is, of course, very interesting for these people, whatever their criterion. Historians of science belonging to this tradition are bound to show that Galileo was better than Copernicus, that Kepler was better

5. See Polanyi 1958; Kuhn [1962] 1970; Merton 1973.

6. See Oakeshott 1993. For a critical analysis of Oakeshott's political philosophy, see Watkins 1952.

than Galileo, that Einstein was better than Newton, and so on. It lies with the historian to examine all this in detail. According to scepticism, certain influences in history may bring sudden changes in intellectual fashions. However, authoritarianism maintains that the Royal Society may finance a history of science, but not a philosophy of science. Similar to a royal court, it keeps its court historians. It could, for example, give you £2,000, or whatever amount, to set sail from England and follow in Darwin's path and find out he vomited somewhere in the Indian Ocean. This is more or less all a non-scientist can really say about great scientists; in this tradition, as we have seen, we are landed with court historians.

The same people who argue in favour of this tacit dimension (*Tacit Dimension* is the title of one of Polanyi's books) also argue in the following way: suppose you wanted to learn to ski, would you go to the library and read a textbook, or look up the rules of skiing in a book? Of course not. You go straight to the best skiing instructor and try to imitate him. We are talking about two different things here: the question in this case is how one learns to *do* science. The answer is: sit at the feet of the great scientists—sit at the Mandarin's or the great Rabbi's feet and you too will become a great Mandarin or a great Rabbi, or a great scientist; or a great politician, if you first become private secretary to a great minister.

The interesting thing, however, is not only to learn to ski, but to learn to *judge* who is the better skier. In order to do this, you have to sit at the feet of the great scientist for decades. There is a lot of evidence for this view; if you check the statistics it turns out that Nobel Prize winners are usually disciples of other Nobel Prize winners. This is a curious characteristic of the genealogy of science. For instance—and we are now talking about the field of science but of course further on we will want to generalize this demarcation criterion to all fields of intellectual exercise—we might find that a brilliant young man goes to Oxford and comes out five years later a complete idiot; whereas if a mediocre student goes to a good department, he will not only embark on a good career, but he may also reach some quite creditable achievements. So you can see there is quite a lot to say for Polanyi, and the principle of academic autonomy—non-interference from government or students—actually stems from his philosophy. Outsiders cannot judge scientific achievement and should not have a say in matters scientific.

Polanyi put forward his ideas in the 1930s primarily to counter Soviet-like intervention in academic life. I think that this was a very worthy exercise, and one has to have some sympathy for this theory. Do not forget that this academic autonomy is also saying, in a negative way, that the Catholic Church, the Central Committee for the Communist Party, the Nazi Party and so on, should leave scientists alone. This is the "truth content" of this school of thought. Let me briefly add that the sociology of science as it exists today actually latches on in some way to this school of thought. Merton, who is a leading sociologist of science of our time, raised the problem in the thirties: it is terribly difficult for a young man to choose his master, since at the age of eighteen, according to Polanyiite philosophy, he cannot tell the master from the fraud, and yet he must decide at whose feet he will sit. It is a sort of existential philosophy: a leap into darkness. Merton tried, also against Soviet and Nazi meddling with science, to find the norms of a scientific community. We should at least try to recognise the differences between a scientific community and, say, the Communist Party. So he had his famous criterion: namely, Communism, in the sense that there is no private property of knowledge in science, and only priority disputes exist, but they play a weak role in Merton's sociology of science. [. . .]

To sum up, if élitist authoritarianism is right, instead of a demarcation between science and pseudoscience, we would have a demarcation between a *scientific* community and a *non-scientific* community.

The interesting thing is of course that this school of thought cannot explain two matters: (1) how it is that rival schools can come off the same production line, the same scientific community, and yet remain rival, with one perhaps calling the other pseudoscientific. Is it that the same school produces both good and bad science, or rival scientific theories? The other problem is: (2) how scientific revolutions come about. Is a sudden change in the scientific community progress or degeneration?

Thomas Kuhn solved these two problems with two postulates. Remember that Russell calls postulates "intellectual theft." Kuhn postulated that in each scientific community there is always a dominant "paradigm" (which is of course sheer factual nonsense) and brings up another problem: how is it that people believed in the paradigm in the first place? Secondly, he postulated that if scientific revolutions occur in a community which satisfies Mertonian socio-psychological standards, they are always progressive. That means that within a scientific community *might is right.* So if a revolution occurs, this implies increased knowledge. This is really Kuhn's contribution to the debate, and you can see that he is really a rather ad hoc footnote to Polanyi and Merton.

In this case, therefore, if we want to judge the results we have to look at the people who produced them and not at the scientific theories themselves. So we have this socio-psychological demarcation between scientific and non-scientific communities, but we still have not come up with the answers to the two questions.

One question is: What if a scientific community degenerates, namely, it changes its views always for the worse? Because logically, of course, there is nothing impossible in a scientific community's first holding Einstein's views, and then undergoing a Kuhnian conversion to Newton's—this is actually what happened with elementary particle physics. The second point is: What shall we do in the social sciences? In the social sciences, in order for fellows of the Royal Society to elect other fellows, we actually need a Royal Society. But there is no Royal Society in the social sciences. How do we go about electing the first fellows? The question, therefore, is: how do we start the social sciences? Do we first judge their programmes or do we first establish a community of social scientists along Mertonian or other standards? What is the first task? First institutionalise doctrines which are not there, or start by producing doctrines and forget about the institutionalisation? If we decide to forget the institutionalisation, how can we judge the product, since our only criterion is to judge the institutionalised community? This is why the social sciences are considered such a plague and also, as we shall see, why they play such a central role in this whole problem.

I think I shall stop here; next time—I shall outline the course in a single sentence—I shall show you how the different demarcation criteria were proposed, how they failed and were replaced by better ones; I shall then arrive at my own scientific research programmes, and then come back to Feyerabend and Polanyi.

Lecture Two

The Theological Nature of Scientific Standards

In Popper's writings you will find that he is an antipositivist, that positivism for him is actually evil: something which will strike you as puzzling after what I have said. How can this be? Popper talks about quite a different sort of positivism from what we have been talking about: we could call the two kinds $positivism_1$ and $positivism_2$.[1]

Suppose I am dealing with empiricists who want to found empirical knowledge on certainty, and I tell them that there is a chalk in my hand; they would immediately ask the question: "How do you know that you have a chalk in your hand? Can you prove it?" At best all I can say is that I have the idea or the feeling, the combined visual and tactile perception, that there is a chalk in my hand; but it may all be in my brain because, after all, with suitable nervous excitation you can have the same idea introduced into your brain that makes you think you have a chalk in your hand. Thus, you would have to draw back from these materialist statements to an empirio-criticist[2] statement (in Marxist terminology) such as "I *see,* or *feel,* a chalk in my hand."

Of course you can ask yourself: does this mean that the state of affairs in my brain now is similar to the one which would occur were my brain excited by my actually seeing and feeling the chalk? Suppose you were just an artificial brain in a laboratory, and not a human brain: how would you know you had a body? All this gets you into trouble and leads to Descartes's "universal doubt."[3] Whoever has followed Professor Watkins's lectures will of course know what Descartes's solution is.[4] Descartes thinks that once you have gone through this process step by step, once you have gone through the psychotherapy of doubt, God will reward your inflicted self-punishment, extend his guiding hand and lead you up the ladder from this doubt to the brain states, then to the chalk, and even to inductive generalizations and universal laws.

1. Lakatos often uses numerical subscripts to indicate differing points of view. See, for instance, the distinctions between $Popper_0$, $Popper_1$, and $Popper_2$ in his "Falsification and the Methodology of Scientific Research Programmes" (1970, 93), describing, respectively, "the dogmatic falsificationist who never published a word invented—and criticised—first by Ayer and then by many others," "the naive falsificationist," and "the sophisticated falsificationist." Feyerabend makes fun of this habit when he threatens to distinguish between $Lakatos_1$ and $Lakatos_2$, and when he refers to $Popper_{-\infty}$: see letter, PF to IL, 28 February 1970, this volume.

2. Richard Avenarius used the term "empiriocriticism" to refer to his *Critique of Pure Experience* (1888–1890). The same term was used to indicate Mach's point of view from his *Analysis of Sensations* (1886) onwards. Avenarius and Mach, both considered "subjectivist idealists," are the targets of Lenin's attack right from the opening pages of *Materialism and Empirio-Criticism* (1908).

3. See the "First Meditation" in Descartes's *Dialogue on Method* (1637). For a contemporary version of the subject, see Putnam 1981, chap. 1.

4. John Watkins, professor of philosophy at LSE since 1966 (now professor emeritus), gave the course on the History of Modern Philosophy: Bacon to Kant in 1973. On the subject see his *Science and Scepticism* (1984).

Popper is actually following Lenin, whose *Materialism and Empirio-Criticism* he translated from Russian to German in 1919 (Popper translated the book with a Hungarian Politburo member whose Russian was very good, but not so his German. Popper, on the other hand, did not know Russian but his German was good—so they worked on the translation jointly), when he claims that Cartesian scepticism is all nonsense and one can start directly with the chalk. Only positivists have these doubts, not materialists.

As is apparent, the word *positivism* is used in completely different contexts. We might just mention Auguste Comte, for instance: he is known by some people as the founder of positivism. Frankly, I cannot possibly list all the people who call themselves either proponents or enemies of positivism, but I can assure you that they all use the word in a somewhat different sense from the two uses I have mentioned. Perhaps we can cross out the word 'positivism', which was devised to irritate German audiences, and use 'demarcationism', which is an ugly word. Anyway, I do not mind what we call it: just remember the constant confusion surrounding the word 'positivism' in the last few hundred years. Whenever you hear the word, ask for a definition, substitute the definition and forget about the word.

The course of my future lectures, including today's, is very simple. Let me remind you that we are facing a dramatic question because if we do not find a satisfactory or reasonable solution to the *demarcation problem,* we have to fall back on either cultural relativism or Polanyi's authoritarianism: and I think none of you will like either. [. . .]

I would therefore like to give you a rundown of the following schools of thought (this is not, however, an exhaustive list): (1) *inductivism.* Its strongest and most influential form is *classical inductivism.* But I shall also present its two weakened versions: i.e., *probabilism* and *verificationism.* (2) *Conventionalism,* (3) *falsificationism,* in all its different forms; and finally, (4) my *methodology of scientific research programmes.* I shall then come back to whether and to what extent we have solved the problem we had ourselves.

Let us begin with *classical inductivism.* We have to go back to the seventeenth century, since science started then. You will of course have come across books with titles such as *History of Medieval Science* in bookshops. I can tell you that I have just discovered from the archives in Rome that there was a secret meeting in 1929 between the representatives of the Pope and Stalin. Just to give you the gist of it, they agreed on the Marxist view that the more developed an economic system, the more developed its science. So therefore we have slavery first, then feudalism, and then capitalism. Let us stop here. We know of course that slavery produced people like Archimedes and Ptolemy, and that capitalism produced people like Newton and Einstein. In between there must therefore have existed scientists who were better than Archimedes but not as good as Einstein or Newton. What about the Catholic Church? The Catholic Church is still smarting under the accusations of people of the eighteenth-century Enlightenment who called the period of the supremacy of the Catholic Church not simply the *Middle* Ages but the *Dark* Ages. People like Voltaire thought that the Church had destroyed science and that until the Renaissance came about there was just darkness, the *Dark* Ages. I think therefore that Catholics and Marxists had a joint interest in inserting something in between. As a matter of fact this was first attempted by Pierre Duhem: he was a eminent French Catholic and also a French patriot who around 1900 began work on a ten-volume history of nonexistent medieval science. In Oxford today, for instance, there is a distinguished man who has actually written a book entitled *The History of Science from Augustine to Galileo.*[5] I am of course referring to Dr Crombie, and Crombie is of course also a Catholic. So you can see that all this medieval scholarship has until recently been in the hands of either Catholics or Marxists, especially in the Soviet Union. There are actually more chairs of Medieval Science in the Soviet Union than anywhere else in the world. Of course, the situation is different nowadays because people are looking for jobs and saying: "Look, there are a few chairs in Medieval Science: why don't we specialize in it?" So you can see how Communism, Catholicism and contemporary Parasitism join to breed Medieval Science. I only wanted to mention this because in England the post was invented by Bernal and Needham, who were actually fellows of the Royal Society. They were quite ingenious. They made out a rather indubitable case: that artisan skills are needed for science and that in fact they can be regarded as primitive science. Much like some agricultural innovations—or alleged innovations—in the Middle Ages. If this is enthroned as science, then continuity is somehow established and nicely democratised.[6] So one cannot complain about this, but I will stick to my guns and maintain that science started in the seventeenth century, and that therefore the philosophy of science could not have started earlier!

This is the main point: what kind of *standards* for science could there

5. Cf. Crombie [1952] 1969.

6. See, for example, Bernal 1954, vol. 1 *(The Emergence of Science)* and vol. 2, chap. 7 *(The Birth of Modern Science).*

have been in the seventeenth century? What did a seventeenth-century man know about it? Most of his knowledge was about God, angels, devils, sorcery, crimes related to punishment in hell. All this vast theological knowledge was clearly *proven* by Revelation and the authority of the Church. Religious knowledge was definitely meant to be certain and indubitable. Therefore *scientific knowledge also had to conform to certain and indubitable standards.* Incidentally, these standards set by the Catholic Church were shared by the Lutherans in the age of the Reformation. We know that Luther said: "A Christian ought to be certain of what he affirms or else he is not a Christian. Mere doubt brings anathema to the Christian who will not be certain of what he is supposed to believe and who does not comprehend it." So if you doubt a proposition, if it isn't completely proven, you have to reject it. Consider, too, the first philosophers of science, like Descartes: his was the famous sentence of 1637 that if you doubt a proposition, you should reject it. Or look at people like Locke, who says that to know and to be certain is the same thing. What I know, I am certain about, and what I am certain about, I know. In other words, as Lord Keynes (an economist) put it, we cannot know a proposition unless it is true.

This attitude, called *justificationism* by William Bartley[7] (who was one of Popper's students), has been extremely widespread among scientists up to the present day. One modification is that in the seventeenth century people were in general more liberal and they admitted three sources of *certain* knowledge. First, of course, is Revelation and the authority of the Church, which the Protestants substituted with the individual reading of the Bible. Second came what was known in the eighteenth century as 'the light of reason', i.e., a priori principles. And, finally, the light of experience—what one actually sees to be true, i.e., factual knowledge.

Empiricists (and scientists were already empiricists in the seventeenth century) thought that such a thing as theological knowledge existed, and they still trusted the Revelation; but as far as science was concerned, they slowly ruled out a priori principles and claimed that what is to be scientific must be proved from facts. You know Newton's famous statement: "I do not play around with hypotheses, I only utter what is true."

The tenacity of this stand can be shown by an anecdote which Hans Reichenbach tells us. Reichenbach is one of the proponents of probabilism, which is simply, as we shall see, a watered-down form

7. See, for example, Bartley 1968.

of justificationism. According to probabilism, we agree to accept statements which are shown to be highly probable, given the factual evidence. It is interesting that Reichenbach, though living in Berlin, belonged to the movement of the Vienna Circle. Reichenbach first met the members of the Vienna Circle at a conference in Prague in 1930. He describes the meeting in his *Modern Philosophy of Science* (1959). Reichenbach gave a lecture on the probability of hypotheses, which was later so well criticised by Popper in his *Logic of Scientific Discovery.* This was in 1930 and Rudolf Carnap, who usually spoke a lot at the discussions, was completely silent and obviously stunned; he then went up to Reichenbach and said: "You know you really shook me completely. Until now I never thought that a scientist could honestly utter a statement which he hadn't proved from fact." Carnap was at the time professor at Prague University, with a very good background in mathematics and physics, and this was his first intellectual shock; the second of course occurred in 1932 during a summer he spent with Karl Popper and Herbert Feigl, but I shall come to that later.

So it is quite clear that these standards have theological origins. Do not forget that when a person makes an error about religion, the punishment is none other than hell, to burn in hell for eternity. Many people in the scientific community today still believe that to publish a paper with an error in it is some sort of damnation. Given continental standards this is, I think, perfectly clear.

How can we criticise the idea that a statement is scientific only if it is provable from facts? We can knock it out completely by logical analysis. You will never be able to criticise conventionalism or falsificationism through logical analysis, as we shall see; logical analysis works only against inductivism. Inductivism claims that a proposition is scientific if provable from facts; what we shall now set out to do is to show that no proposition whatsoever can be proven from facts. And certainly not any scientific proposition.

First of all we have to agree on the fact that a worthwhile scientific law is a universal statement which applies to an infinite number of instances. If we take the law of gravitation, for instance, which states that "all bodies attract each other by the inverse square law," we need two steps, both infallible: (1) from a fact to a factual proposition describing the occurrence; and (2) from a factual proposition which is spatio-temporally singular to a spatio-temporally universal proposition.

Let us take the following fact: this is a tape recorder. Spatio-temporally, that means to say that on this table at the LSE (and we define the spatio-temporal coordinates) there is a tape recorder. This is a sentence, a proposition which I can write up on the blackboard. The tape recorder, on the other hand, is an object. How do we get from this object to this proposition? How can one derive a factual proposition from a fact? One cannot. For instance, knowing Peter Clark[8] and having asked him at twelve o'clock to bring over Sir Karl Popper's tape recorder, without taking any notice of what his assistant may say, I can imagine that he was locked out of the room and brought over this rotating thing, which is not actually a tape recorder—and when I want to tape, it turns out that there is no tape inside. How should I know that it is a tape recorder? Or how should I know that this is chalk, if of course by chalk I mean—my chemical knowledge is rusty—calcium carbonate or something? Well, I do not know, maybe I would have to put the piece of chalk into a liquid and see what dissolves it. All sorts of things can write on a blackboard, but this does not mean they are chalk. And the unpleasant dusty feeling on my hand does not prove it's chalk, either. So between the fact that this is a chalk and the proposition: "I hold a piece of chalk in my hand," there is a big amount of difference.

All this is commonplace for Popperian philosophy, but of course the arguments were first put forward by sceptics. There is a fantastic abyss between a fact and a 'factual' proposition. Nowadays it is much easier to explain this thanks to the intellectual efforts of the New Left. For instance, what a capitalist sees in the North Vietnamese invasion of South Vietnam is "obviously" a civil war. But the occupation of Czechoslovakia is, at least from the point of view of the New Left, the liberation of Czechoslovakia from imperialist agents. So, as you can see, the fact remains. But the interpretation of the fact depends upon your economic interests and your point of view. Nowadays this is commonplace, because epistemological anarchism is becoming fashionable. I do not mind that people understand this relativism. The big question remains: how can we decide what is true and what is false, what has more truth content and what has less? This is taking the point a step further, something the New Left does not do: it means explaining that there is an unhealable gap here which no logic can get to with certainty (or uncertainty). This is one of the basic theses of scepticism. I remember I first came across this statement in fragments of pre-Socratic philosophy in which it was observed that there was a big gap between a fact and a factual statement. The interesting thing is that, at least to my knowledge

8. Peter Clark, research student and then research assistant at the Department of Philosophy, Logic and Scientific Method at LSE, is now senior lecturer at St Andrew's University.

and that of my colleagues, the statement does not occur again until 1831 in the works of Jacob Friederich Fries, who was a Kantian.[9]

So inductivism already breaks down here. But let us imagine for the sake of argument that we can actually obtain from facts certain factual knowledge expressed in absolutely true propositions. Can we still make the next step—from factual propositions to universal propositions? In other words, if I say: "This is a piece of chalk and it writes on the blackboard," can I then say: "All pieces of chalk write on the blackboard"? Or take, for instance, the fact that a planet moves around the Sun according to Kepler's laws; can we *infer* that all planets move around the Sun in this way? Today we call this an *inductive inference.* Most people agree that it is not generally valid—at least *not generally,* which of course means *not* valid. I would just like to point out that this could not have been said before the 1830s, before Bernard Bolzano. I do not want to turn this into a logic lecture, which you can get from Colin Howson[10] (and it would be more worthwhile), but let me just give you a vague outline of what happened.

Aristotelian logic is all about syllogisms. "All cows are dogs." "All

9. Popper too had deferred to Jacob Friederich Fries's work *Neue oder anthropologische Kritik der Vernunft (A New or Anthropological Critique of Judgement)* (1828–1831) on the subject of the acceptance of basic statements. Fries's trilemma brings Popper to argue against the neopositivist view of the 'empirical' basis of science. According to Fries, the statements of science should not be accepted *dogmatically;* but if we demand justification, then we are committed to the view that statements can be justified only by other statements. We cannot avoid *infinite regress;* our only escape route is psychologism: statements can be justified by *perceptual experience.* The latter was Fries's solution, but not Popper's (see Popper [1932–1933] 1979, chap. 5, and Popper [1934] 1959, 93–95). Popper maintained that in order to avoid Fries's trilemma we must admit that basic statements are *accepted* through a convention. Each alternative of the trilemma contains a grain of truth: in order to avoid infinite regress, a dose of dogmatism is called for, though not one considered sufficient to be dangerous. Precisely in the same way that occurs at trials by jury, we can decide on the need for an additional judicial enquiry; while psychological aspects play an explicit role in determining the further decisions of the members of a jury, 'decisions' are not 'justifications'. "Experience can motivate a decision, and hence an acceptance or a rejection of a statement, but a basic statement cannot be *justified* by them, no more than by thumping the table" (Popper [1934] 1959, 105). Basic statements cannot therefore be considered 'basic' in any sense: not *psychologically* since no 'virgin' perceptual base actually exists on which to found universal propositions; and not *logically,* either, since it is not possible to turn general statements into simple experiences, because any description transcends the data it refers to by means of its very terms. They are therefore 'basic' only in a *methodological and pragmatic sense,* and this is why any true Popperian should, according to Lakatos, use single quotation marks to give ironic emphasis to the term.

10. Colin Howson, lecturer in logic in 1973, is currently professor of philosophy of science at the LSE.

dogs bark." "Therefore all cows bark." This kind of logic based on syllogisms came into disrepute in the seventeenth century when science was born. Scientists like Descartes found that all interesting propositions, all interesting inferences are in fact nonsyllogistic. For instance, if you say that both B and C are between A and O, and if G is between B and C, then G is also between B and O. This sort of inference, which you could never get into any Aristotelian form, belongs to relational logic. You simply cannot force inferences like this one into Aristotelian logic form. Thus, seventeenth-century scientists made fun of Aristotelian logic, and they claimed that logic is an intuitive subject and that each inference can be judged to be valid or invalid according to the scientist's intuitions. There are infinitely many valid inferences which are all to be examined a priori, so we can say that this is a logic with infinitely many axioms. In this case we have inferences which increase logical content, because the conclusion says more than the premises. Seventeenth-century thinkers believed that scientific knowledge is possible, and that men have only a finite amount of observation, so that to get to obviously universal statements we need an infallible content-increasing logic. They actually had a psychological theory of logic—namely, that an inference is valid when a sane person feels it to be valid. This was transformed into a linguistic theory of knowledge only in the nineteenth century by Bolzano and Frege and so on. We now have a complete logic (I do not want to go into in what sense it is 'complete', since that was discovered by Gödel in a very technical sense in 1930).[11] A logic which covers nearly all valid mathematical inferences and which is, by definition, content-preserving. Therefore we say that if a conclusion contains more than the premises, it is *invalid.* Thus, *deductive* logic is valid, whereas inductive logic is invalid. If you look at Descartes's original text, it turns out that he used deductive and inductive logic as synonyms. Today we have a sharp demarcation between them. (I shall explain why some other time.)[12] Inductive logic has been degenerating over the centuries, and deductive logic was a progressive programme. We should forget about inductive logic and retain deductive logic, which means that according to the new modern logic the inductive inference is invalid. But I would like to remind you of the historical dimensions of this crossing-out operation: namely, that it is possible only on the basis of the Bolzano-Tarski theory of logical consequence. Unfortunately, I cannot go into

11. For a brief discussion on why Aristotelian syllogistic logic is inadequate, see Howson 1997, 73–76; and, on Gödel's theorem, chap. 11.

12. See Lakatos 1961/1973, especially 79–82, and 1968a, 129–31.

the technical details. But it is good to know why these crossings-out would have been made by the Greek sceptics, and by intelligent New Left-wingers, too. In order to realise that, one has to explain how the Bolzano-Tarski theory of logic superseded Cartesian logic. It is a very technical and difficult topic. Popper takes the Bolzano-Tarski theory as somehow the only possible one—for him, logic is eternal and has no historical dimension; so you will never get this reminder in Popper's criticism of inductivism.

In conclusion, we have returned once again to the point where we prove that by sheer logical means (by the means of modern logic), no scientific proposition can ever be proven from facts, and therefore classical inductivism can be crossed out on merely logical grounds. At this point, we have two ways out. One way out is to look for another theory of demarcation because this sort of Newtonian militant positivism—*Hypotheses non fingo*—has been discarded. The second way out consists in claiming that, having no scientific propositions yet, we have to start science anew from tomorrow and admit only propositions which are deducible from fact. This is utopian, of course, but people still uphold utopian Victorian morality, and utopian standards in science, too.

I will stop here. Next time I will go on by showing other possible patterns of criticism: how to criticise a demarcation criterion. I shall claim that, even without any use of logic, I can cross out classical inductivism with the help of historiographical criticism. In this way, I shall show how one can criticise a demarcation criterion.

Lecture Three

Inductivism and Its Historical Myths

Let us see where we are exactly. If you remember, I gave the following trichotomy in my first lecture: *epistemological anarchism, authoritarianism* and *positivism.* We then went on to talk about the demarcation criteria of positivists. What I promised to do was to deal first with *inductivism,* then with *conventionalism,* and then with *falsificationism,* and finally with the *methodology of scientific research programmes.* I would then come back to the trichotomy which I outlined in my first talk. I have already discussed inductivism to a great extent; you remember this was divided into three: classical inductivism, probabilism, and verificationism. The latter two being watered-down versions of the first. Today I am going to finish with inductivism, and next week comes the really interesting part of the course, when I shall take Popper's approach and start to criticise it.

I would like to remind you that according to the theory of classical inductivism we deduce our theories from facts. Now, to deduce theories from facts, we have to do two tricks in turn. The first is to cross the bridge from facts to factual propositions. Just imagine that there is an elephant in this room, and then consider the proposition: "An elephant is in this room," which is a factual proposition. I mentioned that the unbridgeability of this gap was first pointed out by Fries in 1831. The following is one of my favourite examples. In the seventeenth century, Leibniz thought that one could construct a machine in which we programme a proposition and it will flash out whether it is true or false. Imagine a machine—and Leibniz really devoted a few years of his life thinking about it—which we could feed with an empirical proposition, i.e., not something like $2 + 2 = 4$, but something like: "There is an elephant in this room." Now, we feed this into the machine and we ask the machine: "Is this true or false?" It is clear that the machine not only needs the sentence, but this room as well. So if I want to ask the machine: "Is it true that all planets move in ellipses?" I have to feed into the machine not only the sentence, but also the whole planetary system. The interesting thing is that Leibniz was not a stupid man, but it never occurred to him that, if he wanted to construct such a machine, it had to be bigger than the whole Universe. It is a curious fact in the history of ideas that it took so much time to discover the unbridgeability between facts and factual propositions.

The second jump is from factual propositions to inductive generalisations. Again I only want to remind you, since we discussed this last time, that this argument has now fallen into disrepute with the breakdown of what is known today as inductive logic, as opposed to deductive logic.

These two arguments constitute the logical and epistemological refutation of *classical inductivism.*

Each such school of this kind has a certain *code of intellectual honesty* which is characteristic of the specific school. For instance, the code of intellectual honesty of classical inductivism is: "Reject any proposition that has not been proven from facts." What does this mean, from a *heuristic* point of view? How should one do research if one accepts this proposition? We have to ask these questions because, whether refuted or not, the inductivist code is still very much alive. There are different possibilities. The first is: start your investigation with facts. That means that if, for instance, you want to carry out some research on planetary motion, you should first observe the sky and then write down what you observe. Or, if you want to carry out some research on the ocean, then you should ask for a grant—i.e., you ask the Navy to supply you with a submarine and you measure the ocean's depth at each inch. This provides you with a submarine and with a quiet and pleasant occupation for a lifetime. The interesting thing is that if you ask for such a grant from the Royal Society, I bet you will get it. Because we do actually have such maps of oceans, and somebody must have measured them sometime. The same applies if you want to measure the quality and density of pollution in each cubic inch of air in Regent's Park: you will get a grant. Thus the theory is that you start with the facts and then you can theorise about the problems later.

If you choose to investigate the skies, for instance, you know that there were all sorts of speculations among the pre-Socratics before there was much observation of planetary motion.[1] It would be interesting to have a projector showing planetary motion as observed among the Greeks. What they saw in the sky was that certain stars rotate around the earth in 24 hours, but five star-like objects have a very curious path. These are the *planets,* and the word means "wanderers."[2] Just imagine starting from all the facts, and writing them all down. Actually, the pre-Socratics never did so. What they did first was to *speculate* that all planetary motions are circular. The question is how to construct a complex theory about rotating crystal spheres, so that one rotates around the other—like the Ptolemaic system. Then you get these apparent motions instead of the real circular motions. But it is quite clear that however long you observe these irregular motions, the idea of circular motions will never occur to you on the basis of your observations undisturbed by "bad" speculation. What I want to emphasize is that we do not need to have this sort of fact-collecting heuristics. I think that the sociologists among you would describe this as "bourgeois positivism" as opposed to "Marxist speculation." I am afraid I am on the wrong side in this particular case.

One could also say, as a second possibility: let us speculate *cautiously.*

1. For a further examination of "The Ancient Two-Sphere Universe," "The Problem of the Planet," and "The Two-Spheres Universe in Aristotelian Thought," see Kuhn 1957, chaps. 1, 2, and 3 respectively.

2. See Kuhn 1957, 45: "The term planet . . . was employed until after Copernicus's lifetime to distinguish those celestial bodies that moved or 'wandered' among the stars from those whose relative positions were fixed."

But what does "cautiously" mean? Or—our third possibility—let us consider "bold speculation." Again, what we mean by "cautious" and "bold" is difficult to define; but one could say: let us not go too far beyond facts. William Whewell, for instance, who was the leading philosopher of science in the nineteenth century, thought that there is no need to pay too much attention to facts. We can simply speculate and arrive at some low-level generalisation like: "All planets move in ellipses." Then we should go on to find sufficient evidence to prove the conclusion. After having established stage one, we can ask: "But *why* do planets move in ellipses?" Then you get Newton's theory, according to which all planets are attracted by central forces, the inverse square law, and so on, up to stage two. *Then* you can ask: "Why the inverse square law, and not an inverse cubed law?" Thus an even higher level proposition is reached. This is the so-called *Whewellian pyramid.*[3] This is a cautious way you can go: you are not allowed to build the second floor before the first floor is proven from the facts. This is how, according to Whewell, science should grow. But according to the history of science, science does not grow in this way.

With classical inductivism, in fact, all three procedures are compatible. I only mention this because Popper, who has always been quite influential, messes this up. One can still be a classical inductivist and still encourage bold speculation. The main thing is to be aware of the fact that one is speculating, so that one can store up the facts to prove one's speculation. One can go on with bold speculation as long as one returns to it later to prove it from facts.

Having said this about Popper (and we shall come back to him later on), I have already mentioned that inductivism is still very influential, even though it is logically impossible. I hope to show you that some Nobel Prizes went to people who to my mind did not deserve them, or who were awarded them for the wrong reasons, precisely because of this inductivist tradition.

Just to give you one example: Michelson was the first American to win the Nobel Prize.[4] In Scandinavia, they first decided that he should

3. See Whewell 1840, vol. 2, part 2, chapters 6–9. Whewell himself actually stressed the difficulties of "caution" in inductivism with respect to the problem of scientific change.

4. Albert Abraham Michelson, an American physicist of German origin, was awarded the Nobel Prize in 1907 "for his optical precision instruments and the search which he has carried out with their help in the field of precision metrology and spectroscopy," in the words of the motivation speech given by K. B. Hasselberg at the award ceremony. But no mention was made of his famous experiment that was supposed to have ascertained the nonexistence of "ether wind":

get a Nobel Prize in physics because, as the rumour went, he refuted the so-called ether theory. At that time (1907) some advanced people believed that Einstein's relativity theory could be proved from the Michelson experiment. Here you have inductivism: they decided that there was Einstein's relativity theory, set forward in 1905, on the one hand, and on the other was Michelson's 1887 experiment, from which Einstein proved his theory. This idea was quite widespread, and even today you find it in some textbooks. Einstein's theory was proved from the Michelson-Morley experiment, so they decided to give Michelson the Nobel Prize. And then they started to correspond with him and he believed that the relativity theory was absolute rubbish. Imagine what an embarrassing scene it must have been. The Nobel Prize ceremonies usually go in the following way: a member of the Swedish Academy of Science reads out what a great man is the man about to get the Nobel Prize, and what his achievements are. In the case of Michelson, the speech would have run: "This is the man who conducted the famous experiment from which Einstein deduced his marvellous theory." And then the man gives a reception speech. Michelson's speech would have run like this: "The only embarrassing thing in my career—I am quite proud of everything else—is that some people think that Einstein, this crank, has anything at all to do with my precise experiments." Or he might have said (and we have evidence that he felt this way): "This is the only experiment in my life in which I *simply* do not understand what I observed, because the whole experiment is so confusing."

> As for physics, it has had such remarkable developments as a precision science that we can justifiably claim that the majority of all the greatest discoveries in physics are very largely based on the high degree of accuracy which can now be obtained in measurements made during the study of physical phenomena. [Accuracy of measurements] is the very root, the very condition, of our penetration deep into the laws of physics—our only way to new discoveries. It is an advance of this kind which the Academy wishes to recognize with the Nobel Prize for physics this year.

In the same way, Michelson did not refer to the experiment in his Nobel lecture on "Recent Progress in Spectroscopy." For the whole story, see Holton 1973, chap. 9. Holton goes on to show that in 1907 the theory of relativity was still "far too new, and regarded as too speculative to be mentioned" (276–77).

Michelson is not the only example: in 1922, when the Nobel Committee awarded the Nobel Prize to Albert Einstein, its motivation was not the relativity theory. The official document, dated 10 December 1922, of the Royal Swedish Academy of the Sciences, now kept in the Einstein Archive, stated specifically that the Academy "independent from the value that may be credited the theory of relativity and gravitation after eventual confirmation, bestows the prize . . . to Albert Einstein, being the most highly deserving in the field of theoretical physics, particularly his discovery of the law pertaining to the photoelectric effect" (quoted in Holton 1973, 336).

However, having once decided that Michelson deserved the Nobel Prize, the Academy of Science thought it had better keep quiet about the Michelson experiments. So when he got his Nobel, it was for his precision in constructing instruments to observe certain lengths in spectroscopy. Today nobody anymore remembers that Michelson had anything to do with spectroscopy, but they cooked this up in the last few weeks in order to avoid embarrassment. This is, of course, just a small example of how *historical myths* are developed. In this case the two myths are: on the one side that Michelson's experiment refuted the ether theory, while he actually thought that he had proved it by his experiments (we shall discuss that later, when we come to falsificationism); and, on the other side, that Einstein's relativity theory was derived from the Michelson-Morley experiment.[5]

We see here from these two impossibilities that inductivism is a myth, too. A myth which in turn has a great myth-creating ability. One example which I would like to discuss in some detail, since it really is the classical inductivist myth, is the one which recurs in almost all the textbooks: i.e., that Newton deduced his laws from Kepler's three laws of planetary motion.[6] Just to remind you, the first law states that all planets move in ellipses. We can leave the second one aside for the moment.[7] The third law says that if A is the average distance from the Sun on its elliptic path, and T is the time it takes the planet to orbit the Sun, then A^3/T^2 is a constant for all the planets. Allegedly—and this was the most important result in modern science—Newton derived his theory of mechanics and primarily his theory of gravitation from these three laws. As a matter of fact, if you think about it a little bit, it is easy to see that from Newton's laws it follows that all masses attract each other with an inverse squared force. But then surely planets attract each other, too: Saturn attracts Jupiter, for instance, and therefore there are perturbing forces.[8] One of the most important aspects of the Newtonian theory is the perturbation theory, i.e., the theory about how much planetary orbits differ from ellipses. How can one reasonably claim that Newton

5. See also Lakatos 1970, especially 73–79.

6. The claim that Newton's theory could be deduced from Kepler's laws had already been criticised by Duhem (1906, 190–95). This argument was also used by Neurath (1935) against Popper (1934). Popper in his turn brought the problem up again (1949; 1957b).

7. Kepler's second law states that the orbital speed of each planet varies in such a way that a line joining the planet to the Sun sweeps through equal areas of the ellipse in equal intervals of time.

8. While Kepler's laws are precise when applied to the problem of attraction between two bodies, the solar system involves n bodies, with n greater than 2; and the "perturbing forces" are simply the perturbing effects of other celestial bodies such as planets and satellites. Recommended reading on the subject is Toeplitz [1949] 1963, 150–72.

derived his theory that says that planets do not move in ellipses from the law that says that they actually do? Moreover, one can also easily show that A^3/T^2 is not a constant, but rather a function which in each case depends on the particular mass of the planet. Therefore this number is different for each planet. This follows from Newton's law, but not from Kepler's, according to which it should be a constant for all the planets. Therefore, the claim that Newton deduced this from Kepler's laws is obviously nonsense.

What I am discussing here is how a false methodology, which is false even on elementary logical grounds, can influence historians to the point where they create historical *myths.* I do not want to go into analogies now, but political theories create historical myths in the same way, and so too in the history of science do false demarcation criteria.

These myths are tenacious: it was Newton himself who first talked about having derived his theory from Kepler's laws. Actually, there was a very angry debate. As was quite well known in Newton's time, bodies attract each other with a force inversely proportional to their distance. There was general agreement on the existence of a force radiating from the Sun. The main thing is that if we assume that the force on a sphere is evenly distributed, it is ever weaker as we go to more distant spheres. One can easily calculate from the surface formula of the sphere that the force will be inversely proportional to the distance. It is just like saying that the intensity of light coming from a central source falls in the same proportion. This was common knowledge. Newton had a priority quarrel with Hooke,[9] because when the first edition of *Principia* was published (in 1687), Hooke maintained that he had known this for twenty years before talking about it with Newton, and that Newton had pinched it from him. Newton got very angry, and said: "You only made conjectures, and conjecturing is not a scientific activity." What Newton had done, and he said this himself, was to deduce his theory from Kepler's laws. And funnily enough, this was not challenged in spite of the perturbation theory and all the rest.

It was not challenged because the myth was so fantastically strong that it held, to the best of my knowledge, until 1905 when Duhem published his book *La théorie physique: Son objet et sa structure (The Aim and Structure of the Physical Theory).* I recommend it to you as one of the best books written this century on the philosophy of science. It contains a chapter where he sets out what I have just said. Duhem also

9. With reference to this quarrel between Newton and Hooke, see Westfall 1980, 382–88, and Cohen 1980, 243*ff.*

mentions that there are two examples of what he calls "the Newtonian method and its falsehood," and the second example is very amusing. In 1827 Ampère, the founder of electromagnetism, published *Théorie mathématique des phénomènes électrodynamiques (The Mathematical Theory of Electrodynamical Phenomena).* The subtitle on the cover reads: *Uniquement déduite de l'expérience (In an unequivocal way deduced from experience).* Ampère describes his experiments, the electromagnetic theory, and shows how the theory was deduced from the experiments. The interesting thing is that in the second edition, as Duhem points out, there is a postscript in small print saying: "I think I ought to remark in finishing this memoir that I have not yet had the time to construct the instruments represented in Diagram 4 of the first plate and in Diagram 20 of the second plate. The experiments for which they were intended have not yet been done."[10] Ampère obviously thought that if one wants to be a respectable scientist, one does not publish a theory which has not been deduced from experiments; so he drew all the diagrams until something went wrong. This happens many times when you have utopian standards of morality: in practice you end up in lies. If you have utopian standards of scientific appraisal, you also end up in lies.

What is curious is that, although *The Aim and Structure of the Physical Theory* has been published in twenty languages and many editions since 1905, somehow nobody has ever discussed these painful points.[11] The physics textbooks kept telling us that Newton's laws were derived from Kepler's. In 1947 Popper published a paper in German[12] in which he did not do anything but repeat what Duhem had said—i.e., the criticism of inductivism. Max Born, the great physicist and Nobel Prize winner, published a book in 1948 after having read Popper's paper;[13] Born got so angry that, without mentioning Popper, he wrote an appendix where he showed how Newton actually derived his law of gravitation from Kepler. I need not say that he makes a logical blunder.

10. See Duhem [1914] 1954, part 2, chap. 6, 190–200.

11. In fact, Lakatos does not mention Otto Neurath's criticism of Popper's confidence in the ("pseudo-")rationalist absolutism of falsification that seems not to consider Duhem's warnings; see Neurath 1935. On this specific point, see Zolo 1989 and Rudolf Haller 1991. See Uebel 1991 for a "rediscovery of the forgotten Otto Neurath" and Cartwright et al. 1996, part 2, "On Neurath's Boat," for renewed interest in the subject.

12. Reference here is to Popper 1949; see also Popper 1972, appendix I, 357–58, and especially footnote 10 to p. 358 added in the translation, in which he recognizes his debt to Duhem.

13. See Born 1949, 129–210.

Thus you see that this is the typical reaction of the Royal Society: namely, that scientists cheat when they do not derive major theories from experiment: according to this code of intellectual honesty, this is cheating. This is Born's reaction. Then in 1957 Popper publishes his paper in English (it was later included as chapter 5, "The Aim of Science," in his 1972 book *Objective Knowledge*). In 1962 Feyerabend brings the whole thing up once more, and refers to Duhem and Popper. Feyerabend's article *(Explanation, Reduction and Empiricism)* was published in the *Minnesota Studies,* volume 3, which, like many other publications, has this absolutely idiotic habit of having footnotes at the end of the paper. If the paper is really long, nobody reads them because nobody wants to turn the pages. In 1963, Mary Hesse, a Fellow of the British Academy, reviewed the book and Feyerabend's article especially. The article is entitled "A Revolution in Epistemology,"[14] and she describes "Feyerabend's discovery" that Newton could not have deduced his laws from Kepler's. This, therefore, becomes an epistemological revolution in 1963!

Let me just tell you the story: she sent me the manuscript when she sent it to the publisher. I immediately rang her and said: "You did not look up Feyerabend's footnotes, because he gives the references to Duhem and Popper, so is not it awkward to describe this as Feyerabend's epistemological revolution?" She did not answer and then the paper was published in its original form. So I asked her: "Why did not you correct this?" She replied: "Oh, I misplaced the volume; and anyway, is it important who discovered it?" I just want to add that in 1969 Achinstein, a leading American philosopher of science, published *How Newton deduced his laws. . . .*[15] My friend Jon Dorling, a very able young philosopher of science at Chelsea College, recently published a few papers on *How Einstein deduced his theory from facts,*[16] and so on and so on. Of course what these people always do is introduce additional surreptitious premises which are just as strong as their conclusions.

The most amusing thing is that, while I was out of action last year following an accident, I got a paper from my dear old friend I. B. Cohen, the leading Newtonian scholar who is also the editor of the new three-volume edition of the *Principia.* (Although it is criminally expensive, this

14. In fact, it was published under the title "A New Look at Scientific Explanation"; see Hesse 1963.

15. See Achinstein 1971.

16. See Dorling 1968; 1971.

is the best book ever written about Newton. You will find a comparative collection of the three editions of Newton's work, namely, the ones of 1687, 1713 and 1727. A comparative edition which explains why the changes were introduced: a simply marvellous three-volume book which tells an interesting story.) Anyhow, the curious thing is that a year ago Cohen sent me a paper[17] and asked me to read it fast because he was sending it to the printers and he would have liked my criticism before publication. Since I was ill, I only read it yesterday when I was thinking about this lecture. I asked myself if he had not sent me a paper about how Newton derived his laws from Kepler. So I looked it up and there is this paper fighting against the antiscientism of our time represented by Duhem, Popper and Feyerabend; and it shows once again, in a new way, how Newton derived his laws from Kepler's ellipses. It also holds that Popper and Duhem do not know what the history of science is about, because it is made up of the cumulative efforts of great scientists and scientific theories; that Kepler established that planets move in ellipses and, in fact, they *would,* if they did not have gravitational or inertial mass, and that all this leads to an important continuity. Then, to my surprise, I come to the last footnote, which is two pages long and says that of course Lakatos is quite different, that he understands all of this. Then comes a quotation from me bearing out that I do not share any of the silly ideas of Duhem, Popper and Feyerabend. Of course, not wanting to contradict him, when I described Newton's theory as a scientific research programme, I stressed that—as in classical economics—we have an "ideal" model. First we assume that we have only one planet and that the planets do not have gravitational or inertial mass. From here we can build up a research programme. But of course this does not mean that the subsequent versions of a research programme are always cumulative and consistent. The fact is that a "respectful" historian cannot bear the idea that the rules of science are not cumulative and that there can be revolutions in science. Those have to be explained away!

It is really interesting that Cohen's article will be published in a new volume.[18] Only yesterday, I rang the editor, Elkana: "Can't we somehow stop this nonsense?" This is actually what he said: "What is wrong with it? It is really an excellent work." This is how trivially false logical beliefs can form people's beliefs about what happens in history.

In the last ten minutes I want to run through the two weaker versions of classical inductivism. As I have already pointed out, classical induc-

17. See Cohen 1974.
18. Elkana (ed.) 1974.

tivism is very important and very powerful, but logically it is simply ridiculous. The funny thing is that there are two weakened versions of inductivism, and both were set forward in the twentieth century. Both are difficult to refute logically, unlike classical inductivism: so in fact they are more intelligent and sophisticated. However, they have absolutely no influence. Only a small group of people work on them, and they cannot create scientific news at all. The history of thought often shows that the crude version of an idea is fantastically powerful even though it is inconsistent and idiotic, but the more sophisticated versions, which could be rationally defended, never achieve any acceptance.

I am now coming to probabilism and verificationism. The thesis of *probabilism* is not that theories can be deduced from facts, but that they can at least be probabilified by facts. To go fast: what, then, is the *code of intellectual honesty* which will come from probabilism? A theory can be published in a journal if the probability of the hypothesis, given the factual evidence, is greater than, say, a critical number: 0.934, for instance. If it is highly probable, then it is scientific. (The Newtonian theory, for example, given the evidence, has greater probability than this.) A weaker version of scientific honesty states that one should always calculate the probability of the hypothesis, given the evidence, and (honestly) tell the reader how much this is, even if it is only 0.01.

The only embarrassing thing is that in 1925 Ritchie[19] showed that the probability of *h,* given *e,* equals zero, for any hypothesis and any evidence whatsoever. Intuitively this is crystal clear. What is a scientific hypothesis? A scientific hypothesis says something about the world. For instance, "All bodies attract each other according to the inverse squared law." The information content of such a theory is enormous. A piece of evidence is one piece of factual information about something with a spatio-temporal coordinate, i.e., a little bit of the Universe. So it is quite clear that if we agree that all events are equi-probable, we have infinitely many events which add up to a scientific theory and therefore, since we are capable of producing only a finite amount of evidence, the probability *p* will always be zero. This is what Ritchie showed. In 1935 Popper brought it all a step further in his *Logic of Scientific Discovery.*[20] That was really the end of probabilism. I would like to say—and I think this is the first time this has ever been said, at least at the LSE, and I myself only learned this recently from Colin Howson[21]—that Tarski, the lead-

19. See Ritchie 1926, especially 309–10 and 318.

20. See Popper [1934] 1959, appendix *ix, and Lakatos 1968a, 138–41.

21. Howson (1973) has shown that, given the logical interpretation of probability, Popper's argument is incorrect, for systems of logical probability can be devised so that universal statements do not have zero probability.

ing logician of our time, proved a theorem in which he shows both that this argument is false and how one can construct probability measures which do not fall under the Ritchie-Popper argument.

[. . .] The only problem is, how does one construct such probabilities? One has to formalize scientific languages, formalize the sentences *h* and *e* and then construct these measure functions in these languages. This is a fantastically difficult task, and by now I think that the most complicated measure function which the probabilists have constructed has about twelve predicates in it, because if these were thirteen, it would be too difficult to do. So in the next ten thousand years probabilist theories will be inapplicable to science. [. . .]

There are other arguments against probabilism which I shall not tire you with today and which you can find in my rather technical paper "Changes in the Problem of Inductive Logic." It shows for instance that if we assume that there is a measure function which covers all the available science of today, and gives a probability for each *h* and *e,* it turns out that the same evidence will give quite a different probability to a hypothesis, if we suddenly invent a new predicate; and thereby in each new language all these numbers will change. This is one of the difficulties. In fact, probabilism is useless but not on such simplistic grounds as was believed by Ritchie and Popper.

As editor of the *British Journal for the Philosophy of Science* I received a paper by Professor Swinburne. He proved that Copernicus's heliocentric theory was more probable than Ptolemy's theory, and that it was therefore rational to accept it. I would have loved to publish it because this would have been the first historical myth created by probabilism. My heart broke that I could not.[22] [. . .]

I am afraid that *verificationism* is only a five-minute story, but it is so amusing that I do not want to spoil it by rushing. So I shall come back to it next time. Verificationism is a logically impeccable version of inductivism. It is also totally uninfluential and practically forgotten by now. It was primarily proposed by and advocated at one time by Professor Alfred Jules Ayer. It is interesting to note that a similar theory of appraisal—though logically impeccable and not even suffering from probabilism's limitation in applying only to very small artificial languages—is still absolutely uninfluential and unimportant. Validity alone, therefore, is not sufficient criterion for importance.

22. The reader can look up a comparison in probabilistic terms of Ptolemy's and Copernicus's theories in Swinburne 1973, 60–62.

Lecture Four

Comparing Demarcation Criteria: Verificationism and Conventionalism

I start today with *verificationism.* Since I get bored with the negative aspects of the subject, today I will try to run through both *conventionalism* and *falsificationism,* which is a cruder version of the former. Next time we should be able to get into a more positive frame of mind.

In order for demarcation criteria to work, one of the prime requirements is that it should cut all theories into two: the baddies on one side, the goodies on the other. The main trouble with verificationism is that it never manages to reach this aim. I will now explain why and how.

Verificationism originally came from Ludwig Wittgenstein and Moritz Schlick in the Vienna Circle in the early 1920s. According to Wittgenstein and Schlick, only those statements whose truth value is decidable by experiment are valid. Even if one could decide the truth or falsity of factual propositions by experiments (and we know from Fries that this is not the case), what about inductive generalizations? It is quite clear that for universal statements we cannot perform an infinite number of experiments. Schlick (and not Wittgenstein) found the answer. (Incidentally, Schlick got killed by a mad student in 1936, and I wouldn't say he deserved it.) Schlick had to answer the question concerning the status of propositions such as "All bodies attract each other" or "All swans are white." These cannot be meaningful propositions because their truth value cannot be established. So his idea was that these are in fact truths of inference. We can say: "Swan A is white," and with the inference-ticket: "All swans are white" we already know that if we see swan B it is white as well. This is of course a very crude and primitive idea, and Professor Ryle of Oxford based his career primarily on this.

I would like to switch now from Professor Ryle's career to Professor Ayer's career. Ayer was born in 1910. At the age of twenty-three, an undergraduate attending a beginners' class in German, he decided to improve his German by going to Vienna. And then somehow he picked up the news that the Vienna Circle was there; he got a grant and went to Vienna. As I said, his German was not very good. Anyway, he did not get permission from Schlick to attend the meetings of the Circle. Incidentally, Karl Popper says of himself that he was the main adversary of the Vienna Circle, but he very much wanted to be accepted in it. In fact, Carnap and Feigl recommended him, but Schlick said that he did not want the impertinent idiot in his house, and this is how he became the main adversary of the Vienna Circle. So Ayer went to Vienna and could not get into the Circle, but since its members could always be found in the coffee houses in Vienna after nine o'clock in the evening, he met quite a few of them. With his poor German he could not quite understand the subtleties of the subject. Had he understood the subtleties, he could not have come back to England to publish his little book *Language, Truth and Logic,*[1] which is a sort of journalistic account, at the level of *The Sunday Times* colour supplement, of what was going on in Vienna in those days. If any of you look up the official history of the Vienna Circle written by Professor Victor Kraft, one of its senior members (the book is now translated into English),[2] you will see that when he discusses the historiography regarding the Vienna Circle in the preface, he mentions that Professor Alfred Ayer of Oxford also published a book on the subject.

The Schlick definition of *verificationism* was already in trouble and was losing its interest by the time of Ayer's visit to Vienna. However, Ayer decided that he could in fact correct Schlick's theory—and he thought that this was his main original contribution to philosophy. Roughly his idea was that, in order for a statement h to be meaningful, it does not have to be verified or falsified directly by experience. But the following condition is sufficient: if you take an observation statement, or a factual statement, o, and you conjoin o with h, and if you can derive from them the statement o' such that one cannot deduce o' from o alone, but only with the help of h, then h is meaningful.

This then brings "All swans are white" into the meaningful realm, because with "All swans are white" and "Michael is a swan," you get "Michael is white." So with this construction, "All swans are white" becomes meaningful, which it was not for Schlick. I do not want to go into the details: you can leave aside this tedious scholasticism.

This definition of "verifiability" was published in Ayer's *Language, Truth and Logic* in 1936. I highly recommend this book: you can read it in a single evening and you get the sort of romantic flavour of the Vienna Circle which was there before its members started writing on the subject. You see, if one gets a brilliant idea and then starts writing it

1. Ayer [1936] 1946.
2. See Kraft 1953.

out, the usual effect is that it immediately crumbles and take up a different form, and the romantic simplicity disappears even before it goes to print. In the Circle itself you will never get such a marvellous summary of its obviously false doctrines which received their psychological inauguration in Ayer's *Language, Truth and Logic.* For instance, Ayer dared to claim that ethical statements cannot be verified: they are meaningless, unless perhaps you think that they are poetry because they are not cognitively meaningful. All this brushing aside of all old philosophical problems as sheer meaningless nonsense you will find nowhere else but in Ayer's book (which contains his criterion of verifiability). Of course, it was very successful!

In 1937, exactly 24 hours later, another famous Oxford journalist, Sir Isaiah Berlin,[3] read the book and immediately saw (I do not want to go into the details) that from every sentence *h* such as "The absolute is lazy" (which is a typically famous example of a meaningless statement for the Vienna Circle), some observation statements can be derived such that *h* becomes meaningful.[4] So Isaiah Berlin pointed this out immediately after the publication of the book. Then came the war and Ayer could not answer back. After the war in 1946, *Language, Truth and Logic* was published in its second edition, and Ayer produced an improved definition in the preface: "I propose to say that a statement is *directly* verifiable . . ." Schlick's original criteria was *indirect* verifiability. Ayer had actually called it *direct,* so we could term this new version *super direct.* But then he says that a statement can also be indirectly verifiable, and the actual definition is on page 13 of *Language, Truth and Logic.* I do not want to read it out because it is very long. Actually, I will, so that you get the style of second-generation Vienna Circle books:

> I propose to say that a statement is directly verifiable if it is either itself an observation-statement, or it is such that in conjunction with one or more observation-statements it entails at least one observation-statement which is not deducible from these other premises alone. And I propose to say that a statement is indirectly verifiable if it satisfies the following

3. See Berlin 1939.

4. The argument can be summarised as follows: take *M* as an arbitrary metaphysical statement, e.g.: "The absolute is lazy." Take $O_1, \ldots, O_n$ as observation statements freely chosen, e.g.: "This is white," "This is a pen," etc. *M* then becomes an element of the verifiable systems: $M\&$ (if M, then O_1), $M\&$ (if M, then O_2), . . . , $M\&$ (if M, then O_n). Thus, *M* should be verifiable and therefore meaningful, which in fact it is not.

> conditions: first, in conjunction with certain other premises it entails one or more directly verifiable statements which are not deducible from these other premises alone. . . .[5]

This, of course, confused all the Oxford people because it was so long and dignified. They believed they had a definition of verifiability and in fact no Oxford person was clever enough to show that even under these conditions, *all* statements can be shown to be meaningful.

In 1492, however, Columbus had discovered America, and that led to some trouble because in 1949 an American named Alonzo Church reviewed Ayer's book in the *Journal of Symbolic Logic.* The review is actually twelve lines long.[6] It says that Ayer's philosophy rests upon the following condition of verifiability; the quotation follows and a simple argument is given in five lines which shows that, even according to this long definition, all statements can be shown to be verifiable.

So you see, Ayer's definition does not even divide statements into two: *all* statements are verifiable. You remember that the big philosophical task of all these schools of positivism was: under what criterion should we burn books?[7] Now it turns out that you will never burn any books under the verifiability criterion. Well, some people never give up. Here is an article published in 1968 (it took 12 years). Nidditch, professor of philosophy of science at Sheffield, produced a criterion which I will not read out since it is twenty lines long. Quite frankly, I have never read it. His definition avoids all these pitfalls and divides the statements effectively into two. It was however immediately clear that nobody understood the twenty lines, and quite a lot of Ph.D. theses were written on "How to correctly interpret some of the terms of Professor Nidditch's now false definition." The conclusive criterion was finally published in *Ratio* in 1968 by my friend Professor Wolfgang Yourgrau. His definition is nine pages long and I will just write down a

5. " . . . and secondly," Ayer continues, "that these premises do not include any statement that is not either analytic, or directly verifiable, or capable of being independently established as indirectly verifiable." ([1936] 1946, 181)

6. See Church 1949, 53.

7. Reference to the famous passage which concludes David Hume's *An Enquiry Concerning Human Understanding* (1748):

> When we run over libraries . . . what havoc must we make? If we take in our hand any volume; of divinity or school metaphysics, for instance; let us ask, Does it contain any abstract reasoning concerning quantities or number? No. Does it contain any experimental reasoning concerning matter of fact and existence? No. Commit it then to flames: for it can contain nothing but sophistry and illusion.

few lines. In order for a statement to be verifiable, it has to satisfy the following condition [written on blackboard]:

(1) $D(p)\,(\exists C_p)\,(\forall p_i)\,\{C_p\,(p_i)\text{ ent }(\exists o,o',p,r)\,P(q)\text{ o nent }o')$
$D(oq)\,(oq\text{ ent }o')\,P(r)\,(r\text{ nent }q)\,[D\,(p_ir)\,(p_ir\text{ ent }q)\text{ v}$
$D(\bar{p}_ir)\,(\bar{p}_ir\text{ ent }q)]\}\text{ ent }V(p).$
(2) $(\forall p_i, p_j)\,[D(p_ip_j)\,P(p_i)V(p_j)\text{ ent }V(p_ip_j)].$
(3) $(\forall p_i, p_j)\,[D(p_i\text{ v }p_j)\,P(p_i)V(p_j)\text{ ent }V(p_i\text{ v }p_j)].$[8]

This is a really *"conclusive"* result with regard to verifiability because no one I have come across has ever read it. This is one way of solving an intellectual problem, and a very frequent way too. As you can see, it does not exactly excite people into burning books because they are meaningless. Somehow, the original fervour has gone. What is left is medieval scholasticism.

Later on, in my last lecture, I shall give you the criteria of what I call a degenerating problem shift. But this is already a prime example. You may remember that when I talked about classical inductivism, I said that although it was logically and historiographically refuted, it certainly influenced people throughout the centuries. But this empty scholasticism does not get us anywhere. And the same applies to probabilism, but I do not want to go into that any further.

Let us now move on to a much more important and influential approach to the demarcation problem: *conventionalism.*

Conventionalism is a very important trend in the history of ideas. We have to go back to the Greeks. Inductivism became influential because people believed throughout the centuries that Newton really proved his theories from facts. Conventionalism is just an aesthetic theory. It is uninteresting unless you say: "Look, here is a marvellous masterpiece; look at it and you will find it beautiful, and here is a theory which explains why it is beautiful, and why another painting, say, is not." It follows that, if there is a new artistic masterpiece which does not comply with this aesthetic theory, then of course you think again. Thus, philosophy of science is bound to be very influenced by science itself, just as aesthetics is influenced—almost determined—by the history of art.

What Newton did for inductivism, Ptolemy did for conventionalism. Ptolemy lived in the second century A.D. and with a very complicated and ingenious system he explained the very irregular motion of the planets. With the help of Ptolemy's system, one could predict eclipses, planet movements, etcetera. A whole astrology was based on it for

8. Yourgrau and Chandler 1968, 61.

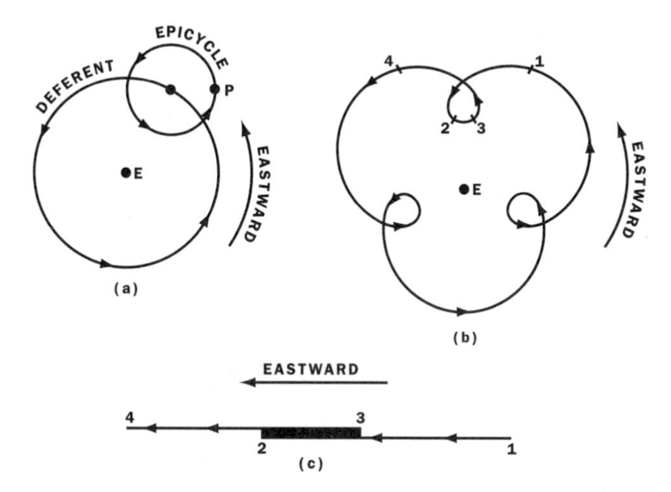

Figure 1 The basic epicycle-deferent system. A typical deferent and epicycle are shown in (a); the looped motion that they generate in the plane of the ecliptic is illustrated in (b); the third diagram (c) shows a portion (1-2-3-4) of the motion in (b) as seen by an observer on the central Earth, E. (From Kuhn 1957, 61.)

many centuries. Astrology, of course, needs a good predictive description of planetary motion.

The interesting thing about Ptolemy was that the Earth was somewhere at the centre of the Universe. There are very complicated things too: a certain number of celestial spheres [. . .] glide on other celestial spheres: many layers of this kind overlap in Ptolemy's system, and new spheres are added whenever a prediction fails. In this way it turns out that by introducing newer and newer circles with the help of Ptolemy's theory, you have a good predictive machine. (These new circles are the famous "epicycles" which now have the abstract meaning of "ad hoc arguments" concocted just in order to take the facts into account.)

But what on earth are these spheres? They are invisible, and like Aristotle, you have to try and make sense out of them. (Aristotle lived before Ptolemy, but the rudiments of the theory were there many centuries B.C.; only Ptolemy, however, developed it into "*near* perfection.") Imagine the spheres are perfect crystal balls, with perfect frictionless rotation: the only problem is that these crystal balls glide through each other! Crystals cannot let other crystals through without friction. So there is something 'idiotic' when we start to explain this in physical terms. In most Hellenistic culture, there was a dichotomy between *astronomy,* which "saved the phenomena" and brought about all the marvellously precise predictive machinery, and *physics,* which was pretty impotent and could not make head nor tail out of it because it could not give a plausible interpretation of the phenomena. Physics was not thought much of, whereas astronomy as a calculating device was. Now: we have "hard facts," the planetary motions, and we have the absolutely impossible idea of these imaginary crystal balls. What, then, is the role of scientific theory in Ptolemy—the first extremely elaborate and precise scientific theory in the history of mankind? The theory is obviously absurd, yet it serves as a marvellous calculating device. If this is so, the theories are not true or false; they only enable us to calculate and predict: they are only a *convention,* or, if you like, *instruments.* This is really the origin of instrumentalism and conventionalism: theories are mere conventions.

What then is the task of science? We have certain facts: F_1, F_2, . . . , and we want to have predictions P_1, P_2, . . . , and never mind what sort of pigeon holes we put them into: the main thing is that we should get as efficient a predictive device as we got in Ptolemy. The idea of conventionalism, therefore, comes from the fantastic success of Ptolemy and at the same time the absolute failure to give a plausible, truth-like interpretation.

When the Copernican system, which held that the Sun was at the centre of the Universe and the Earth moved around it, came along, Copernicus got into trouble with the Church because according to the Bible the Earth is fixed and the Sun moves around it. However, the Copernicans might have argued that their theory was a calculating device just like Ptolemy's. The theory was neither true nor false; it was only an instrument, and an instrument could not be inconsistent with a statement found in the Bible. So in fact, instrumentalism was a protective ideology for anti-Biblical science in the seventeenth century. It is very interesting to note that this type of instrumentalism is *still* a protective device today in the Soviet Union, where quite a lot of what is called "bourgeois economics" (like Marshallian economics and so on) is now applied, and supported by statements like: "These theories are neither true nor false; we can apply them as an instrumental device." Of course some economists may privately have different views, but they fall back on this 'Copernican device', interpreting their theories as instruments and thereby escaping the strictures of the "Bolshevik church."

I only want to say that this situation was changed in a rather funny way, which may make us think about human "rationality." Newton changed our views because he was quite firm about his theories being true and about having derived them from facts. After this consideration, you may start asking what, according to Newton, makes $mass_1$ and $mass_2$

attract each other from a distance and completely instantaneously. In Newton's time, this was just as absurd as Ptolemy's inter-gliding crystal spheres. Indeed, as Leibniz and others pointed out, it was absolutely mad. Newton himself tried to play around, as we now know from his recently published letters, with the following idea: here is a mass ball; here is another one; then there is also some invisible ether which is like a gas and its particles are bombarding and going off in all different directions, just like in a gas container; there is also a shadow effect, so that the overall effect of the ether movement is that the two mass balls are knocked towards each other. Action-at-a-distance is therefore explained in terms of mere push action.

Now, Newton did not really manage to work this out thoroughly, but he tried hard because he did not believe for a moment that action at a distance could be *true*. So his action-at-a-distance theory is actually another support for conventionalism, just as Ptolemy's epicycles were. Quite frankly, I believe that no 'internal history' of science can explain why this 'Ptolemy-like' success of Newton's was turned into a sort of success for realism.

I just want to mention that the twentieth century presents us with quantum mechanics developed primarily by Bohr, Schrödinger, Heisenberg and others, according to which certain types of matter sometimes act as if they are waves and sometimes as if they are particles. The whole thing looks again like the Ptolemaic crystal spheres. Since the 1930s, conventionalism has swept through the scientific community once more. Now the majority of physicists believe that scientific theories from quantum mechanics onwards are very much like Ptolemy's theory. The main thing is their predictive effect, and not their truth value.

All this process is very nicely described in Popper's *Conjectures and Refutations,* the only difference being that curiously, Popper puts a sort of rhetoric on this, as if conventionalism were a vice and realism a "true" philosophical doctrine. If you follow his argument, there is none, but only rhetoric.[9] In fact the Vienna Circle thought that the problem of whether theories are conventions or truth-like was meaningless: what difference does it make? It is not verifiable, and it is not falsifiable! Popper never really comes to grips with this argument. [. . .]

Let me now come back to the demarcation problem, and say something more about conventionalism. There is a stronger version still of conventionalism. It was discovered only at the end of the nineteenth

9. See Popper 1963, especially chap. 3.

century that if we have a theory *h*, and we have any sort of facts, say a set of $F_1, F_2, \ldots, F_n$, then with the suitable addition of auxiliary hypotheses, we can explain any such set. Which means that no fact can compel us to give up any *h*, because we can always devise a protective belt around *h* and rescue it from refutation. I think that this came about as a result of the competition between molecular and heat theory in the late nineteenth century, when it turned out that these are two quite different theories, but that whatever effect you produced, you could explain it by either. The only difference is that one has to add different auxiliary hypotheses.

Let me now mention a very interesting philosophical movement. Bergsonian philosophy was the fashionable philosophy of the French political right in the 1890s. This approach did not lay very strong stress on cognition. (Actually, it is very interesting that what we now read and recognise as the philosophy of the New Left was born from the philosophy of the extreme right in the 1890s.) It asserts that even scientific theories are ideologies that do not mirror the facts and are not the expression of cognition but rather of human creativity, will and determination. In Bergson's philosophy, the key words are "creativity" and "will." That means that we can push the facts into absolutely any framework we choose if we have enough ingenuity, imagination and determination. This was not originally applied to science.

Here is a short but interesting story. Poincaré, the leading French scientist of his time, was not aware of the Bergsonian philosophical movement. In 1891 he published an article on the nature of geometry. Remember that non-Euclidean geometries were discovered in the 1830s. And let me also remind you quickly that in the late eighteenth century, Kant claimed that we are born with certain conceptual frameworks, and we can never demolish this prison inside which we are born. Euclidean geometry, for instance, is not necessarily a characteristic of the outside world, but rather of our physiological brain structure which we just cast out on the world. We catch the world in the conceptual framework with which we are born. It follows, therefore, that non-Euclidean geometry is impossible for Kant. And yet in the 1830s non-Euclidean geometries were discovered independently by a Hungarian artillery officer stationed in a Polish village, J. Bolyai, who, being very lonely and not having read Kant, had no idea how courageous he was; and in 1829 by Lobachevsky, professor of mathematics in Kazan, where I doubt anyone had ever heard of Kant. In the same period (as we now know from his correspondence), Gauss, the leading German mathematician of the time, had a non-Euclidean geometry in mind, but he thought that if he had published it he would have been considered insane by the Kantians, since there would have had to be something wrong with his brain. Non-Euclidean geometries were discussed in 1891 by Poincaré.[10] (Poincaré showed that all empirical data could be accommodated with *either* a Euclidean or a non-Euclidean framework—making, of course, different auxiliary assumptions in the two cases.)

One can therefore 'force' the world into a Euclidean geometry or a non-Euclidean one. Poincaré raised the question of why we prefer to use Euclidean geometry in physics. He claimed that Euclidean geometry was wonderfully simple and economical, whereas non-Euclidean geometry was just not as comfortable for the human mind. So we can *conceive* it, but not really *use* it. It is well known that Einstein based his general relativity theory on non-Euclidean geometry, but this came later (1914). So you see that Poincaré emphasized that we can devise any sort of conceptual framework, but also that some of these are more comfortable than others.

The Bergsonians immediately noticed that by coupling the possibility outlined by Poincaré with the *will* and the *determination* to put the world into some framework (even if it is non-Euclidean), it is possible to fit the facts into it. Take any physical theory, as mad as you like, and the facts can be made to fit. This was elaborated between 1893 and 1901 with fantastically ingenious and interesting examples primarily by two of Bergson's students, Milhaud and Le Roy, along the lines of Poincaré.[11] (Their examples very much remind one of Feyerabend's.)

I forgot to mention one thing: in 1892 Duhem published a paper in which he said how right Poincaré was[12] from a Bergsonian point of view. Duhem was a sort of adorer of Will; the same idea shared by right-wing extremists like Sorel. Milhaud's and Le Roy's philosophy was closely analogous to Sorel's political philosophy. Take any political theory or any idea about the structure of society and if you have enough imagination and Will, you can impose it. Duhem, however, used these ideas for a different reason. Duhem was a Catholic and so he said: "You can take the Bible and you can put anything into a Biblical framework, absolutely anything; science has a million gaps and it is fallible: you can try anything and it works; why should this not apply to the Bible? If

10. See Poincaré's article "Les geométries non-euclidiennes," *Revue générale des sciences pures et appliquées,* 2 (1891), 769–74; reprinted as chapter 3 of Poincaré 1902.

11. Refer to Giedymin 1982, 112–28.

12. See Duhem 1892.

science is not inductively determined, it cannot possibly oppose religion." Duhem was very happy about that. Then came this extremely interesting turn, and both Poincaré and Duhem got cold feet.

The Milhaud-Le Roy business persuaded Poincaré to leave mathematics and physics for a few years. Starting in 1903, Poincaré wrote his famous three books on the value of science[13] and so on, in which after deep thought he explained that both geometry and Newtonian mechanics are a matter of taste. However, Newtonian gravitation and all the rest is deduced from facts. In this way, Poincaré offered a compromise: geometry and the three laws of Newtonian mechanics are a matter of taste and are only accepted because we have got used to them, but the rest is deduced from experiment. Duhem also got cold feet. In 1905 he published his *Aim and Structure of Physical Theory,* which differs from his 1892 article mainly in one thing. He added a new idea, the so-called "natural classification," in which it turns out that somehow, with God's help and the help of induction, we are approximating truth. Why? Because Poincaré and Duhem both thought: "My God, we are now certain of the Devil, namely, scientific obscurantism. So we have no choice but to return to the facts and enthrone them again." This is one of the most interesting periods in the history of science.

Let us sum up. We have these very strong conventionalist ideas and we have to deal with the demarcation criterion once again. Question: how is one conventional framework better than another? Answer: either according to the principle of economy or to the principle of simplicity. However, as we shall see in the next lecture, these principles break down; i.e., they do not provide a criterion to decide how one theory is better than another. Then along comes Popper with falsificationism.

[At the end of the lecture, Peter Clark suggests mentioning Ernst Mach. Lakatos accepts the suggestion and concludes the lecture:]

With his principle of economy of thought, Mach had, in the same years, developed the same ideas [as Duhem], with the difference that in Mach you will not come across this absolutely exciting and direct connection to the lively, vast philosophical movement which has its ramifications in political philosophy. Mach in Vienna simply didn't see these fantastically important *Weltanschauung* implications of this philosophical debate.

13. Lakatos hints at Poincaré 1902, 1905, and 1908.

Lecture Five

THE LIMITS OF CONVENTIONALISM

As I have already mentioned, this lecture was accidentally not recorded.

Regarding the criticism of conventionalism, the reader is recommended to look at Lakatos 1970 and 1971a. Briefly, Lakatos gives conventionalism credit for having emphasised that any theory whatsoever can be saved from refutation by appropriate "ad hoc stratagems." Falsification is then "left to subjective taste or, at best, to scientific fashion, and too much leeway is left for dogmatic adherence to a favourite theory." Its main merit, therefore, is inextricably linked with the disadvantage of "making us unable to get out of our self-imposed prisons." Conventionalism "cannot solve the problem of the elimination of those theories which have been triumphant for a long period. According to conservative conventionalism, experiments may have the power to refute young theories, but not to refute old, established theories: as science grows, the power of empirical evidence diminishes" (1970, 20–22).

The conventionalist can thus *decide*

> to keep the centre of any system of pigeon holes intact as long as possible. When difficulties arise through an invasion of anomalies, he only changes and complicates the peripheral arrangements. The conventionalist, however, does not regard any pigeon hole system as provenly true, but only as "true by convention" (or possibly even as neither true nor false). . . . This version of conventionalism is epistemologically, and especially logically, much simpler than inductivism: it is in no need of valid inductive inferences. Genuine *progress* of science is cumulative and takes place on the ground level of 'proven' facts; the *changes* on the theoretical level are merely instrumental. Theoretical 'progress' is only in convenience (simplicity), and not in truth-content. One may, of course, introduce revolutionary conventionalism also at the level of 'factual' propositions, in which case one would accept 'factual' propositions by decision rather than by experimental 'proofs'. But then, if the conventionalist is to retain the idea that the growth of 'factual' science has anything to do with objective, factual truth, he must devise some metaphysical principle which he then has to superimpose on his rules for the game of science. If he does not, he cannot escape scepticism or, at least, some radical form of instrumentalism. (1971a, 106–7)

Lecture Six

Popper and the Rules of the Game of Science

I would like to start with a few considerations which follow from the recent debate between Shockley and Lord Boyle which you have all heard about.[1] The press is full of it. Consider a letter published in *The Sunday Telegraph* yesterday. I quote from it: "Professor Shockley's thesis was produced from far too limited an inquiry." Let me perhaps put my point forward first so that you see why I am referring to this debate. You may remember I talked about the fact that science has inherited its standards from theology. If we make a mistake about theological truths, we shall burn in hell forever. Committing a mistake is a very serious matter, so we had better take care and mention only the positions which are absolutely proven. I mentioned that the Church banned Copernicus's theory because it was unproven, and it was only taken off the Index in 1820, when it was decided that the theory had

1. The debate had its origin in the publication of an article by Arthur R. Jensen ("How Much Can We Boost IQ and Scholastic Achievement?" 1969) which suggested significant racial differences in intelligence. Jensen's view was based on the assumption that IQ testing was indeed a good way of estimating the different genetic endowments of different racial groups. Jensen concluded genetic inferiority from the fact that black Americans who took the test obtained worse results than white Americans. The 1969 volume of the *Harvard Educational Review* contained about twenty indignant replies to Jensen's theory which aimed to show its lack of grounds. Professor William Shockley, Nobel laureate for physics in 1956, sided with Jensen, bringing into the debate the full weight of his qualifications and his mathematical ability in interpreting statistical data (see Shockley 1971a,b). In his many articles, Shockley supported the theory of the 'involution' of IQ in humans, due to greater reproduction in races which he considered inferior, thus emphasizing the importance of genetic differences over geographical ones (see Shockley 1972a,b). These theories were also shared by H. J. Eysenck, professor of psychology at London, who in that same year wrote *Race, Intelligence and Education.* Soon after the publication of this book, Eysenck was invited to LSE to hold a conference; he was violently attacked by a group of students who maintained that similar views should not be allowed public airing. Reaction to the left-wing students (singled out as enemies of the right to freedom of expression) was similarly violent. Lord Boyle, ex-MP for the Conservatives, and of liberal tendency, as well as one of Popper's admirers, added to the debate by publishing articles in many weeklies against Shockley's 'racist' theories. Lakatos, primarily interested in the methodological aspects of the debate, also sided with Jensen, without coming into the open directly; he did, however, persuade Peter Urbach (then his research student at LSE) to write on the progressive character of the hereditarian research programme versus the degenerating nature of the environmentalist one (see Urbach 1974).

already been proven by facts. The interesting question is, however, another: what was *really* so wrong with the Copernican speculation from the Church's point of view? We now know that all science is *fallible,* and speculative to a large extent. The real problem with Copernicus's theory was that his findings conflicted or seemed to conflict with certain 'truths' propagated by the moral *authority* of the time, the Catholic Church. To claim that the Earth is not the centre of the Universe and that the Sun is, conflicted with the Bible. What would have happened to society if this idea had freely circulated? Just imagine the consequences. It was a totally pragmatic idea: the scientist must bear the social responsibilities linked to his scientific speculations. One can speculate, of course, but until something is proven in a hard way, it is immoral to propagate it if it undermines the basic beliefs of society. (Question: when exactly did the Church react against Copernicanism: in 1543 or 1616? Answer: when *De Revolutionibus Orbium Caelestium* was published in 1543, there was an accompanying preface written by Osiander saying that this was only a calculating device, and that it had nothing to do with the true structure of the planetary system and therefore it could not possibly conflict with the Bible.[2] So the Bible is the truth and Copernicus's theory is just an "efficient device" for social purposes, namely, for calculating the time of Easter more precisely than was done before. It served a constructive social purpose and at the same time it was in no way conflicting with the Bible. Exactly 70 years later, Galileo's crime[3] was to propound Copernicus's system not as a calculating device, but as the *truth about the Universe.* In 1615, in a letter to some prince whose

2. Osiander (Andreas Hosemann) (1498–1552), an Augustinian German reformer, was converted to Luther's ideas while bringing the Gospel to Nuremberg. He sided with the Lutherans at the Marburg Council on the question of the sacraments (1529), although he later abandoned the orthodox Lutheran view on justificationism, maintaining that this depended not solely on the attribution of Christ's merits to the believer, but also on the direct presence of Christ in the Christian's soul. He was a passionate scientist and mathematician, and took Rheticus's place (Rheticus had been directly chosen for the job by Galileo) for the last stages in the printing of *De Revolutionibus Orbium Caelestium,* adding a *Preface* which he left anonymous. In it he claimed that notwithstanding the fact that the new astronomical system was actually an effective measuring device for planetary motion, "these hypotheses need not be true or even probable." Truth has only one source: divine revelation. But "let no one expect anything certain from astronomy, which cannot furnish it, lest he accept as the truth ideas conceived for another purpose, and depart from this study a greater fool than when he entered it." (Quoted in Koestler 1959, 170; this book is recommended reading on the whole matter, in particular, 169–75.)

3. Those interested in pursuing the matter further should see de Santillana 1960 and Geymonat 1957, chap. 8, on the trial, and chaps. 9 and 10 on the long-term consequences.

name I now forget,[4] Galileo began to fiddle with the Bible, reinterpreting it to fit Copernicus, instead of fiddling with Copernicus to adjust him to the Bible, like Osiander did.)

What I would like to draw your attention to is that the idea of social responsibility for science originates with the Catholic Church. It was the Catholic Church which banned Galileo and actually burned Giordano Bruno in 1600: they both propounded truths which undermined the basic beliefs upon which society was founded. The Liberals of today take their cue from the Church when they say: "Look, you cannot speculate irresponsibly: unless you prove your propositions, you have to think very carefully about the social consequences."

This letter about Shockley gives you a very characteristic picture of the whole debate:

> Professor Shockley's thesis was produced from far too limited a field of inquiry. No large selection poll was analysed, and we all know how accurate these analyses can be. In fact they are accurate for the individual questioner examined but not for another person. The Professor may have reported accurately his findings, but they could not *justify* his thesis. Leeds University and Lord Boyle are to be congratulated on them for having destroyed his honour.

You can see reproduced here exactly the same arguments of the Catholic Church: for some strange reason, truth commands (has commanded) respect from mankind. It is a very ancient idea that truth must be respected, but not so speculation. Now we come to the crucial point: only truths fully supported by evidence have to be accepted. Yet we know for logical reasons that we cannot provide full proof for scientific theories.

You may remember that for probabilism the problem was: what is the probability of *h,* given *e?* Let us say that some truths have major

4. Lakatos is probably referring to Galileo's letter to Cristina di Lorena, Grand Duchess of Tuscany, which was completed in mid-1615. The letter reads, at one point: "I would here refer to what I heard from an eminent person of the ecclesiatical hierarchy: that it is the aim of the Holy Spirit to show us the way to the Heavens, and not the way the Heavens work. [. . .] Concerning the other points in the Scripture, which would seem to contradict Copernicus's view, I doubt not that, when his view be true and proved, those same theologians who, while they consider it false, would deem it impossible to make those points agree with it, yet they would find many interpretations in agreement with his view, more so if they should add to their knowledge of the Scripture some notion of the astronomical sciences." (Quotations are from the Italian National Edition of Galileo's works, Florence 1890–1909, vol. 5, 319, 347; see also Drake 1978, 245, 250, 338.)

social consequences. Say also that the probability on the basis of the evidence is greater than 0.3; then perhaps one could argue that, since it could be very dangerous, the probability should be at least 0.35. The trouble with this, as I mentioned, is that $p(h,e) = 0$ for any scientific theory. Therefore, this kind of morality does not help us with this approach.

The problem is: would it be better to subordinate truths, or the search for truth, to social values? In this, of course, and I do not pass any judgement, the Catholic Church and modern liberalism are in the same boat. In actual fact the arguments put forward by Cardinal Bellarmino against Copernicus and Galileo are much more elegant and convincing than those by David Frost against Shockley.

One more red herring before I go back to my subject. I spent the first twenty-two years of my life in Fascist Hungary. That was actually a very curious system; we could call it "*semi*-Fascist." The regime was very keen that truths which could undermine society be kept under control. Instead of adopting this kind of epistemological demand—that one should absolutely prove one's dangerous hypothesis—they had a simple sociological trick. Anything from Marx to Lenin, including a monthly periodical of the Communist Party, could be published on two conditions. The first was that it could not be printed in more than two thousand copies. I remember for instance that when the Hungarian translation of *Capital* by Marx was published, we had to rush to the bookshop because we knew that there could not be more than two thousand copies of it. Xerox machines had not yet been invented. The second condition was that one could not distribute them among the peasants, but only among the workers, who were actually in a rather privileged position. This sociological trick worked quite well, and it led to a sort of closed society of intellectuals among whom everything could be discussed. This is, I think, a rather interesting solution. They certainly did not put the Communists on television to have them interviewed by David Frost!

Since I have started discussing Fascist Hungary, I should also mention that in Communist Hungary, if I wanted to read the "right-wing" bourgeois press of Europe, like *The Guardian,* I simply had to go (and I was in a position to be able to, at that time) to the private library of the Central Committee of the Communist Party. There were about two hundred people who shared this privilege with me. You can therefore compare the liberty of the two societies by these numbers: two thousand people under the Fascists and two hundred under the Communists were allowed to read subversive works, which to my mind means that the semi-Fascist regime was actually ten times better than the present one.

Incidentally, this brings to mind that while I was studying in Moscow in 1949, I asked for a copy of the *Pravda* of 1943. It turned out that I could only have the *Pravda* from January 1949 onwards: the other numbers were on closed shelves because one was not allowed to look at outdated Party lines.

But it is probably better if we go back to our earlier discussion. I somehow think that the epistemological arguments we are dealing with bear some relation to the sociology and policy of science. You remember my imaginary example of how a Newtonian scientist deals with the anomalous motion of a planet.

[The story, which Lakatos probably told in the missing lecture 5,[5] is about an imaginary case of planetary misbehaviour. A physicist of the pre-Einsteinian era takes Newton's mechanics and his law of gravitation, *N,* the accepted initial conditions, *I,* and calculates, with their help, the path of a newly discovered small planet, *p.* But the planet deviates from the calculated path. Does our Newtonian physicist consider that the deviation is forbidden by Newton's theory and therefore that, once established, it refutes the theory *N?* No. He suggests that there must be a hitherto unknown planet p' which perturbs the path of *p.* He calculates the mass orbit, etc., of this hypothetical planet and then asks an experimental astronomer to test his hypothesis. The planet p' is so small that even the biggest available telescopes cannot possibly observe it: the experimental astronomer applies for a research grant to build a yet bigger one. In three years' time the telescope is ready. Were the unknown planet p' to be discovered, it would be hailed as a new victory of Newtonian science. But it is not. Does our scientist abandon Newton's theory and his idea of the perturbing planet? No. He suggests that a cloud of cosmic dust hides the planet from us. He calculates the location and properties of this cloud and asks for a research grant to send up a satellite to test his calculations. Were the satellite's instruments (possibly new ones, based on little-tested theory) to record the existence of the conjectural cloud, the result would be hailed as an outstanding victory for Newtonian science. But the cloud is not found. Does our scientist abandon Newton's theory, together with the idea of the perturbing planet and the idea of the cloud which hides it? No. He suggests that there is some magnetic field in that

5. The example here is given as Lakatos himself describes it in his "Falsification and the Methodology of Scientific Research Programmes" (1970, 16–17). It is likely that he commented on it more fully in the missing lecture 5. The example aims to show the link between falsificationism and the *"Duhem-Quine thesis,"* according to which "given sufficient imagination, any theory . . . can be permanently saved from 'refutation' by some suitable adjustment in the background knowledge in which it is embedded." Briefly, the thesis ("in its weak interpretation") "asserts the impossibility of a direct experimental hit on a narrowly specified theoretical target and the logical possibility of shaping science in indefinitely many different ways" (96).

region of the Universe which disturbed the instruments of the satellite. A new satellite is sent up. Were the magnetic field to be found, Newtonians would celebrate a sensational victory. But it is not. Is this regarded as a refutation of Newtonian science? No. Either yet another ingenious auxiliary hypothesis is proposed, or the whole story is buried in the dusty volumes of periodicals, and it is never mentioned again.][6]

I would like to give you one more example before going on to show you that in the light of Popper's methodology all these "adjustments" are pseudoscientific on a par with Marx and Freud. I discuss the following example in *Criticism and the Growth of Knowledge,*[7] where I deal with a scientific development which is now regarded as one of the most important in twentieth-century physics: namely, the *structure of the atom;* in particular, of the hydrogen atom, which was developed by Bohr in 1913. Let me outline Bohr's arguments and then we shall discuss in what sense they are scientific or pseudoscientific. The great achievement of Rutherford in the first decade of this century was to discover

6. The imaginary story of the little planet is based on many real historical instances. In 1781, for example, Friedrich Wilhelm Herschel spotted a new planet through his telescope: Uranus. Over the following years Bouvard calculated Uranus's trajectory using Newtonian mechanics, the universal gravitational law, and the accepted initial conditions (among which was the assumption that the remaining planets were six). A few years later, somebody signalled that Uranus deviated considerably from its predicted orbit. Some scientists, like George Airy, shifted their concern onto the gravitational law, proposing that it should be modified, if only 'slightly'. This kind of response, however, was not to reap consensus. The winning move was rather to suppose the existence of an eighth, "disturbing" planet—Neptune—and to ask which mass and orbit it should have in order to account for the observed deviation of Uranus from its predicted orbit. In 1846, after three years of busy desk work, John Couch Adams came up with his prediction of where it should have been possible to observe Neptune, and sent it to Airy, who didn't give the calculations much weight since they came from a researcher who was fresh out of college. The question blew up to international scale: when in 1846 the Frenchman Leverrier published his predictions, which were very close to Adams's, the British scientific community finally decided to point its telescopes in the right direction. Neptune was still not identified, even if the calculations were correct, and the Berlin telescopes picked it up soon afterwards. And it was a fantastic piece of good luck, since the approximations which Adams and Leverrier had used to simplify the calculations of Neptune's mass and orbit would not have allowed its identification, had Neptune accidentally not found itself in the immediate vicinity to Uranus at that time. Similar strategies were used in the case of another of those "misbehaving" planets: Mercury. Why shouldn't Mercury's anomalous perihelion be explained with yet another "perturbing" planet? Astronomers had already thought of a name for it: Vulcanus. In this case, however, things didn't work out in the same way as for Uranus and Neptune: the anomaly was to be worked out thanks to a new conceptual framework, that of general relativity; and no doubt this is far from the end of the whole story. (See also Grosser 1962 and Roseveare 1982.)

7. See Lakatos 1970, 55–68, 81–86.

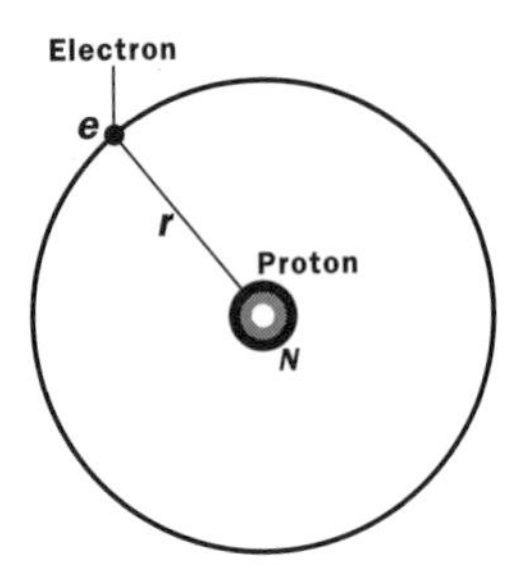

Figure 2 The hydrogen atom according to Ernest Rutherford. *e* indicates the negatively charged electron, and *N* the positively charged nucleus, formed in this case by a single proton. The drawing is not to scale, since the radius of the nucleus would in actual fact be about 1/100,000th of the radius *r* of the atom.

why people, starting from the Greeks, have believed that matter consists of atoms and the void. (You may remember that there were two major rival theories throughout these last two thousand years. The first claimed that matter consists of the void and atoms which move about hitting each other and producing all the objects like, for example, this hard table. The second was the rival theory of the continuity of matter. But I shall concentrate only on the former.) The belief was that atoms are like billiard balls moving around in empty space. Rutherford made the dramatic discovery (the details of which I cannot go into now) that in an atom there is very little matter because an atom is like a small planetary system. It has a very small nucleus and an electron moving around it. The Rutherford model of the atom is called the *nuclear model.*[8] This is a hydrogen atom [explains while drawing Figure 2 on the blackboard].

8. In 1906 J. J. Thomson had received the Nobel Prize for the discovery of the electron. Since 1903 he had been developing an atomic model which supposed the positive charge to be evenly distributed inside a sphere containing the electrons (negative charges). In 1906, however, he changed his initial description. In 1910 Ernest Rutherford, who had been Thomson's student at Cavendish Laboratory in Cambridge and had been awarded the Nobel Prize for chemistry in 1908 for his work on radioactive elements, was ready with "an atom much superior to J. J.'s," to use his words (see Pais 1991, 123). In 1907, after having been appointed to Manchester University, Rutherford started the research that was to bring him to achieve "a feat which I believe to be unique: to make the greatest discovery of his career after having won the Nobel Prize" (Pais 1991, 122). This "revolutionary, highly *creative shift*" (Lakatos 1970, 54), for which Rutherford and his school were responsible, may be thus described:

> In 1908, Geiger had published a paper on the scattering of α-particles by thin foils of gold and aluminium in which he reported that "some of the α-particles . . . were deflected by quite an appreciable angle." Early in 1909 Rutherford suggested to Marsden, a

More complicated atoms have slightly bigger nuclei and have more electrons. In actual fact, 92 different atoms were known at the time, the biggest of which had 92 electrons and a nuclear mass which was much bigger than that of the hydrogen atom. (In our following explanation we shall restrict ourselves mainly to the hydrogen atom.) The interesting thing is that Rutherford's theory implies that an iron atom, which we would expect to have more matter than a hydrogen atom, is relatively empty. Rutherford showed that one can shoot certain particles—alpha particles from the nucleus of helium—onto a thin metallic plate and not hit anything. He showed with a particular instrument that the particles come through in a straight trajectory. Very few are reflected. Rutherford had shown that atoms are mostly void space and consist of a positively charged nucleus and a negatively charged electron orbiting around it. However, although there was an absolutely fantastic amount of evidence bearing out the model, there was one problem. According to the ruling theory of the time, namely, Maxwell's theory of electromagnetism, if any electric charge, for instance an electron, accelerates,[9] it emits energy in the form of light. But if it emits energy, then it loses energy and in the end it is bound to fall into the nucleus. Maxwell's theory had explained practically everything known about light and matter until 1900: it had tremendous confirming evidence on its side. And

---

> twenty-year-old New Zealand-born undergraduate, that he pursue this matter further. In Marsden's words: "One day Rutherford came into the room where [Geiger and I] were counting α-particles . . . turned to me and said, 'See if you can get some effect of α-particles directly reflected from a metal surface'. . . . I do not think he expected any such result . . . To my surprise, I was able to observe the effect looked for . . . I remember well reporting the result to Rutherford a week after, when I met him on the stairs." In May 1909 Geiger and Marsden submitted a paper in which they reported that about 1 in 8000 α-particles were deflected by more than 90°. (Pais 1991, 122–23)

While Geiger maintained that the time was not yet ripe for "discussion of the assumptions which must be put forward in order to understand similar divergences," Rutherford was preparing himself to do exactly that, with his famous paper (1911), in which he described his atomic model and gave "a first (decent) estimate of the radius of a nucleus: about a hundred thousand times smaller than that of an atom. Thus if one imagines an atom blown up to the size of a football field, then the nucleus would be the size of a marble placed at the kick-off point. Nuclei are not only heavy. They are also exceedingly small, even by atomic standards. Matter consists largely of emptiness" (Pais 1991, 124). For further details on the successive relationships between Bohr and Rutherford, see Pais 1991, 128–34.

9. In physics, by "acceleration" we mean a change in velocity, whether in magnitude or direction. If, for example, an electron moves in uniform velocity in a circle, then even though the magnitude of the velocity is the same, the direction changes, and we have a classical case of acceleration.

now we are suddenly faced with the Rutherford model for which again we have tremendous evidence. How can we explain, therefore, that atoms are stable, that they do not collapse into their nuclei all the time? This was actually the central problem in physics in the years between 1905 and 1913.[10]

Bohr solved the problem by following up an idea which had originally been put forward by Planck and Einstein, namely, that energy itself is atomic. This means that, for instance, if the energy of this electron decreases, it can only be decreased by a finite amount and not continuously but in discrete amounts. It is the same idea of photons propounded by Einstein in 1905: light is made of photons with a defined energy $h\nu$, where ν is the frequency of light.[11] Now, if this is what happens, whenever an electron emits a photon, it jumps onto a different orbit like these I am drawing out on the blackboard [see Figure 3].

10. Cf. Bohr 1913a:

> In order to explain the results of experiments on scattering of α rays by matter Prof. Rutherford has given a theory of the structure of atoms. According to this theory, the atoms consist of a positively charged nucleus surrounded by a system of electrons kept together by attractive forces from the nucleus; the total negative charge of the electrons is equal to the positive charge of the nucleus. Further, the nucleus is assumed to be the seat of the essential part of the mass of the atom, and to have linear dimensions exceedingly small compared with the linear dimensions of the whole atom. The number of electrons in an atom is deduced to be approximately equal to half the atomic weight. Great interest is to be attributed to this atom-model; for, as Rutherford has shown, the assumption of the existence of nuclei, as those in question, seems to be necessary in order to account for the results of the experiments on large angle scattering of the α rays. In an attempt to explain some of the properties of matter on the basis of this atom-model we meet, however, with difficulties of a serious nature arising from the apparent instability of the system of electrons: difficulties purposely avoided in atom-models previously considered, for instance, in the one proposed by Sir J. J. Thomson. According to the theory of the latter the atom consists of a sphere of uniform positive electrification, inside which the electrons move in circular orbits. The principal difference between the atom-models proposed by Thomson and Rutherford consists in the circumstances that the forces acting on the electrons in the atom-model of Thomson allow of certain configurations and motions of the electrons for which the system is in a stable equilibrium; such configurations, however, apparently do not exist for the second atom-model. The nature of the difference in question will perhaps be most clearly seen by noticing that among the quantities characterizing the first atom a quantity appears—the radius of the positive sphere—of dimensions of a length and of the same order of magnitude as the linear extension of the atom, while such a length does not appear among the quantities characterizing the second atom, viz. the charges and masses of the electrons and the positive nucleus; nor can it be determined solely by help of the latter quantities.

11. See Planck 1900a,b; Einstein 1905. h is Planck's constant $h = 6.62 \times 10^{-27}$ erg/sec; for a short resume of the subject, the reader is referred to Gillies 1993, 221–28.

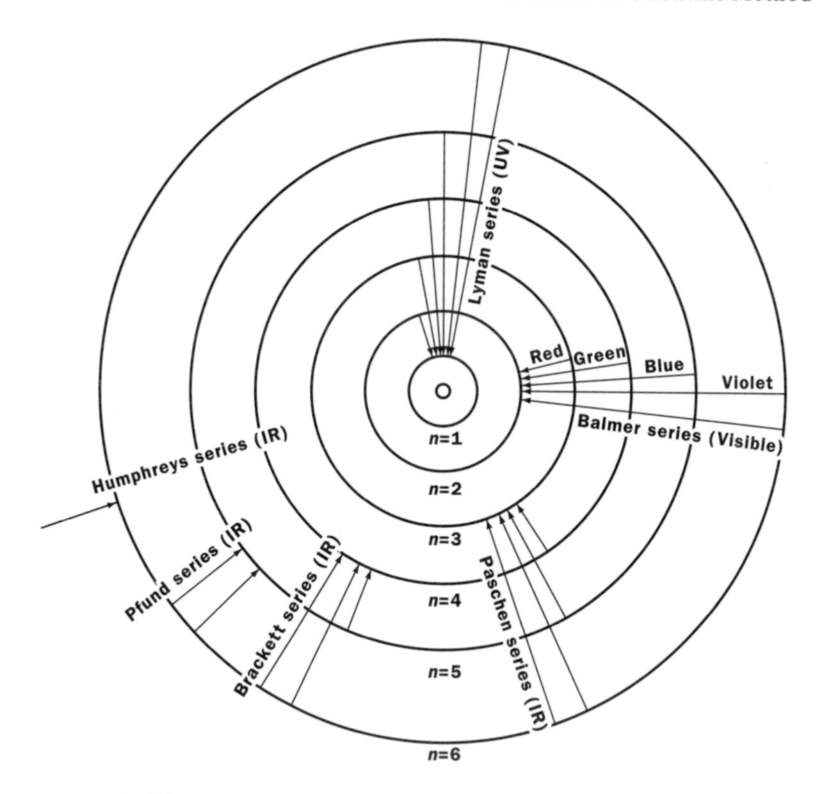

Figure 3 The series of spectrum lines of the hydrogen atom corresponding to successive jumps of the electrons from external to progressively more internal orbits.

This happens every time a photon is emitted.[12] But what Bohr maintained was that in spite of Maxwell's theory, an electron on its circular orbit does not emit light. This means that it does not use energy, and

12. See Bohr 1913a, where he notes how the problem referred to above has "undergone essential alterations in recent years owing to the development of the theory of the energy radiation, and the direct affirmation of the new assumptions introduced in this theory, found by experiments on very different phenomena such as specific heats, photoelectric effect, Röntgen-rays etc." On the problem as a whole, see Kuhn 1978. Further on, Bohr continues:

> Whatever the alteration in the laws of motion of the electrons may be, it seems necessary to introduce in the laws in question a quantity foreign to the classical electrodynamics, i.e., Planck's constant, or as it often is called the elementary quantity of action. By the introduction of this quantity the question of the stable configuration of the electrons in the atom is essentially changed, as this constant is of such dimensions and magnitude that it, together with the mass and charge of the particles, can determine a length of the order of magnitude required.

that it does not obey Maxwell's laws.[13] However, the interesting thing is that if the electron is placed in a level of excited energy, then it decays, emitting a photon. [. . .] So we have a hydrogen atom where the electron starts on an external orbit, then emits a photon and jumps onto a successive internal orbit, etcetera. But we can also *ionize* it (this is a technical term meaning that we can bring the electron to an infinite distance from the nucleus by providing enough energy). There is only a *discrete* range of orbits on which the electron can move around: these orbits are stable and on them the electron does not emit any light.

I shall not go into the technical details. Bohr's aim was to solve the stability problem of the atom: why is the atom stable? Why does not it collapse all the time? We have certain empirical evidence in the colours in the spectrum. For the past 100 years or so, we have known about the hydrogen spectrum, which is shown up by its discrete lines.[14]

In 1885 an ingenious Swiss school teacher called Johann Balmer

13. Cf. Bohr (1913a):

> The inadequacy of the classical electrodynamics in accounting for the properties of atoms from an atom-model as Rutherford's, will appear very clearly if we consider a simple system consisting of a positively charged nucleus of very small dimensions and an electron describing closed orbits around it. For simplicity, let us assume that the mass of the electron is small compared with that of the nucleus, and further, that the velocity of the electron is small compared with that of light. Let us at first assume that there is no energy radiation. In this case the electron will describe stationary elliptical orbits.

14. Cf. Toraldo di Francia 1976, 323–24:

> As soon as the existence of the atoms was ascertained, the problem became that of determining their constitution. Many were the indications taken directly from the physical world that suggested that atoms were not totally respondent to their name, but were rather composite systems. First among these indications was the periodical table [discovered by Mendeleev in 1869] which had convinced many that atoms of different elements were constituted by a single matter present in various proportions. Second was the spectrum lines of the light emitted by the atoms. Gases in an excited state of energy in a flame or an electrical discharge emit characteristic electromagnetic radiations. And the emissions of monoatomic gases are particularly simple since they are made up of discrete systems of *monochromatic* radiations. Each of these radiations has a set wave length . . . and is visible on the spectrum as a *line.* Each element has its peculiar characteristic system of lines which serves to identify it. This was the conclusion reached by G. R. Kirchhoff and R. W. Bunsen, who launched *spectroscopic analysis,* even discovering new elements unheard of up to that point. It is extremely difficult to understand how an indivisible, non-composite atom can give rise to similar phenomena. But there is something utterly wonderful in this.

The reference is to Balmer's series.

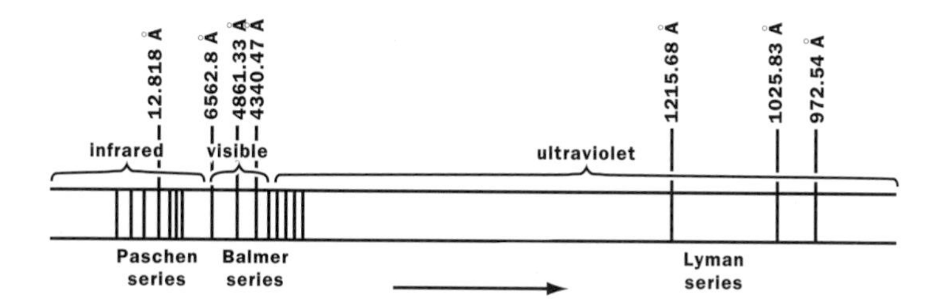

Figure 4 The lines in the emission spectrum of hydrogen.

discovered that the frequencies of these lines in the spectrum can be arranged according to the following formula [refer also to Figure 4]:

$$\lambda = b \frac{n^2}{n^2 - 2^2} = 3645.6 \frac{n^2}{n^2 - 4}$$

[where λ indicates the wave length, from which you get the frequency $\nu = c / \lambda$, c being the speed of light in void; b is an empirical constant, equal to 3645.6 Å; n assumes the values 2, 3, 4, . . .] [15]

This is actually a miracle of patience.[16] But nobody could explain why things worked this way. One must work out the Bohr model theoretically. It is quite simple to do this today and I could put it up on the blackboard for you, but you can find it in any elementary textbook of atomic theory: it is about ten lines long, and very easy to understand.

15. Both the formula and the diagram have been inserted by the editor, who has supposed them to have been written by Lakatos on the blackboard. 1 angstrom (Å) is a unit of measure equal to 10^{-10} m.

16. See also Lakatos 1970:

> Since the Balmer and the Paschen series were known before 1913, some historians present the story as an example of a Baconian "inductive ascent": (1) the chaos of spectrum lines, (2) an "empirical law" (Balmer), (3) the theoretical explanation (Bohr). . . . But the progress of science would hardly have been delayed had we lacked the laudable trials and errors of the ingenious Swiss school teacher: the speculative mainline of science, carried forward by the bold speculations of Planck, Rutherford, Einstein and Bohr would have produced Balmer's results deductively, as test-statements of their theories, without Balmer's so-called "pioneering." (61)

[Here follows the procedure that Bohr used in 1913 to obtain Balmer's formula as a direct consequence of his idea of the hydrogen atom:

If we consider a simple system consisting of a nucleus of Ze charge, with Z its atomic number, and an electron describing closed orbits around it, the line-spectra of the Balmer series are easily deductible from the three basic fundamental assumptions (see Bohr 1913a):

1. The electron revolves in a circular orbit with the centripetal force supplied by the coulomb interaction between the electron and the nucleus.

$$K\frac{Ze^2}{r^2} = \frac{mv^2}{r} \tag{1}$$

where K is a constant assumed for simplicity equal to 1, e is the charge of the electron, r is the radius of the allowed orbit, and m and v are respectively the electron's mass and velocity. The above formula is a special case of Newton's law $F = ma$ where F is the coulomb interaction and $a = v^2/r$ is the centripetal acceleration. These orbits or energy states are discrete, and consequently the corresponding energy values are discrete, too. Moreover, the electron cannot radiate while on them.

2. The angular momentum L of the electron takes on only values which are integer multiples of $h/2\pi$, with h Planck's constant; taking into account that Bohr considers circular orbits exclusively. This can be expressed as follows:

$$L = mvr = n\frac{h}{2\pi} \text{ (with } n \text{ any whole positive number)} \tag{2}$$

3. When an electron makes a transition from one allowed orbit (stationary state) to another, emitting or absorbing radiation, the Bohr frequency condition is satisfied:

$$h\nu = E_i - E_f \tag{3}$$

where ν is the frequency of the radiation emitted (the photon or quantum); and E_i and E_f are the energies of the atom when moving in its initial and final paths, respectively. Eliminating v from equations (1) and (2), we find the allowed radius for the nth state (orbit) to be

$$r = \frac{n^2h^2}{4\pi^2\, Ze^2m} \tag{4}$$

The total energy, E, of the electron is partly kinetic and partly potential, with $K = 1/2mv^2$ and $P = -(Ze^2/r)$. $E = K + P$, so for the nth state:

$$E_n = \frac{1}{2}mv^2 - \frac{Ze^2}{r} = \frac{1}{2}\frac{Ze^2}{r} - \frac{Ze^2}{r} = -\frac{1}{2}\frac{Ze^2}{r} \tag{5}$$

By replacing in (5) the expression for r given in (4), we get

$$E_n = \frac{2\pi^2Z^2e^4m}{n^2h^2} \tag{6}$$

By applying Bohr's formula (3), we get for frequency ν when an electron makes a radiative transition from an initial state n_i to a final state n_f

(7) $$\nu = R_Z \left(\frac{1}{n_f^2} - \frac{1}{n_i^2} \right)$$

where R_Z, which depends on the atomic number Z, is given by

(8) $$R_Z = \frac{2\pi^2 Z^2 e^4 m}{h^3}$$

(7) is none other than Balmer's formula for the hydrogen series, given that n_f = 2($4/b = R$) (R is known as the Rydberg constant).

Bohr's calculation which brought him to formulate the general theory of spectra from a simple formula like (2) was "a triumph over logic. Never mind that discrete orbits and a stable ground state violated laws of physics which up till then were held basic. Nature had told Bohr that he was right anyway, which, of course, was not to say that logic should now be abandoned, but rather that a new logic was now called for. That new logic, quantum mechanics, will make its appearance later" (Pais 1991, 148).]

First of all, Bohr obtained a similar formula to (7):

$$\frac{1}{\lambda} = R \left(\frac{1}{n_f^2} - \frac{1}{n_i^2} \right)$$

Bohr's formula predicts a larger number of line-spectra, since we can take any positive integer for the value of n_f: we can assume $n_f = 1$, or 2, or 3, etcetera. So to each n_f belongs a whole infinite spectrum of n_i. All these series were eventually discovered. They are called the Paschen series and the Lyman series, after their discoverers. Apart from Paschen, who made his discovery for $n_f = 3$ in 1908, all the others were discovered after Balmer had predicted that there must be more. Of course, they did not find them all, since Balmer had predicted there were infinitely many, but they had certainly found them up to the value $n_f = 5$ by 1925. Bohr had actually understood that there should also be lines of helium atoms in the spectra. Now helium is distinguished from hydrogen because instead of a single proton it has two protons and their corresponding orbital electrons. We can perform the following trick: we can ionize helium by shooting out one of its electrons. In this way we end up with a double weight nucleus, and a single electron orbiting around it. The interesting thing is that with Bohr's theory one can predict exactly what its spectrum lines will be like; the constant Rz will be different, but the spectrum will be very similar. The main spectrum of this "crippled" helium atom appears as one of the known spectra of hydrogen.

Bohr made his discovery in 1913. The Balmer series was common knowledge by 1885. Paschen discovered another series with 3 instead of

2 in 1908. Yet another series was known as well, the Evans series, which does not, as it turned out, follow from Bohr's series. It is not among these formulae.[17]

Is this a refutation of Bohr? Bohr said: "It is funny that the hydrogen spectrum comes up in my calculations as a spectrum of ionized helium. Whoever conducted the experiment cannot have done it with hydrogen gas. Systematically present in the experiment there must have been some ionized helium. The remaining ionized helium must be responsible for these alleged additional hydrogen lines." This was Evans's series.[18] It turned out that Bohr was absolutely right. The reason why ionized helium was to be found there in this specific experiment later came out. The anomaly, or refutation, finally turned into a victory for Bohr's programme.

It is interesting that as far back as 1881 Michelson, a great observer, first noticed that some of the spectral lines were doublets or triplets. The Balmer formula simply did not explain a similar phenomenon. According to a much later development of the Bohr programme, since these electrons move with a fantastic velocity (near to that of light), Einsteinian, and not Newtonian, mechanics should apply. If you apply Einsteinian mechanics (and some other things I do not want to go into now) you get the doublets and triplets; these constitute the *anomalous Zeeman effect.*[19]

17. This is the whole story in brief:

> In 1896 Charles Pickering from Harvard had found a series of lines in starlight which he attributed to hydrogen even though this did not fit Balmer. In 1912 these same lines were also found in the laboratory, by Alfred Fowler in London. Bohr pointed out that "we can account naturally for these lines if we ascribe them to helium," singly ionized helium, that is, a one-electron system with $Z = 2$. According to the formula for R_Z this would give a Balmer formula with R replaced by $R_2 = 4R$. Fowler objected: in order to fit the data the 4 ought to be replaced by 4.0016, a difference which lay well outside experimental error. (Pais 1991, 149)

18. See Evans 1913.

19. Pieter Zeeman

> was a young *privaatdocent* in Leiden when in 1897 he discovered that spectral lines split when atoms are placed in a magnetic field. Lorentz at once provided an interpretation in terms of a simple model for an electron moving in an atom. Considering only effects proportional to the first power of the field (the linear Zeeman effect), he showed that a spectral line should split into a doublet or triplet depending on whether the emitted light is parallel or perpendicular to the field direction. For this discovery, Zeeman and Lorentz shared the physics Nobel Prize for 1902. I have often wondered why Lorentz was cited for this particular contribution because it was well known by 1902 that his explanation was incomplete, to say the least. In Lorentz's words (1921): "Unfortunately,

At this stage we can refer back to Popper. I have examined two cases in the history of science: the first was how a Newtonian scientist proceeded, the other was how Bohr proceeded. Two discoveries and two sorts of thought process which Popper's methodology should take into careful account.

Let us sum up: what rules should a scientist observe? As you all know, according to Popper, in order for a theory to be scientific it has to have potential falsifiers. This means that we should be able to specify an experiment which contradicts it. Of course, in order to do that, we have to decide what we would accept as an experimental proposition. For example: you are all familiar with ammeters which measure the strength of electric currents. We attach an ammeter to an electric wire [draws Figure 5 on the blackboard].

An ammeter always looks like this: we have a measuring scale from 0 to 5, and we have a pointer. What is an experimental finding? That there is a current of 0.7 amps. Is this an experimental result? Of course, this is *accepted* as an experimental fact. But Maxwell's theory lies behind it. We actually *need* a theory of the ammeter in order to interpret the movement of the pointer as the *fact* that we have a certain amount of current in the wire: we need an observational theory. Therefore first of all we need a conventionalist decision on what we consider in a certain context to be a basic statement or a potential falsifier and what we consider to be a theoretical statement (first decision). Then, we have to see whether it is true or false (second decision). For instance, in a laboratory, we have to check the device to see if there is any disturbing influence. So in the end we agree *by convention* that the statement is true or false. It is a *decision* because we can never prove that there is no hidden influence which distorts the measurement.

Now: if we have a similar proposition, what then does it refute? Whenever we carry out any experiment, we always have a huge theoretical framework and a *ceteris paribus* clause. This clause claims, for

---

> however, theory could not keep pace with experiment and the joy aroused by [this] first success was but short-lived. In 1898 Cornu discovered—it was hardly credible at first!—that [a sodium line] is decomposed into a quartet. . . . Theory was unable to account [. . .] for the regularities observed . . . to accompany the anomalous splitting of lines." It did not take long to find that this "anomalous Zeeman effect" is the rule, the "normal" effect (Lorentz's prediction) the exception. (Pais 1991, 198).

On the problems arising out of the effect for the "old quantistic theory," and Bohr's reaction, see Pais 1991, 198–99, 201, 207–9, 213–14; and for the explanation of the effect, 241–42, 268

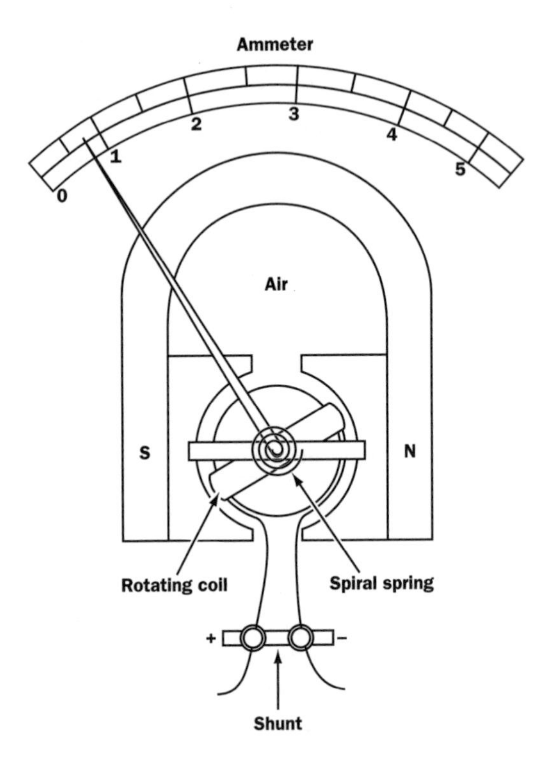

Figure 5 Most electrical instruments commonly used to measure tension and current include a device called an Arsonval mechanism. This mechanism is formed by a rotating coil fixed to an axis around which it turns in a magnetic field, and a spiral spring which is attached to it and tends to return the coil to a point of equilibrium. The figure shows the working of a similar mechanism. When an electrical current passes through the magnetic field, it exerts a momentum directly proportional to the current, which makes the coil rotate by an angle at which the couple is exactly counterbalanced by the responding couple of the spring, proportional in its turn to the angular movement. Consequently, the rotation angle is directly proportional to the current passing through.

instance, that in the case of the planetary system, apart from having six planets and a certain number of stars, no other factor is present. Had there been, all the predictions would have been different (and the potential falsifier would have been falsified). So we have to agree beforehand on the *ceteris paribus* clause and then on what the potential falsifier actually falsifies. Let us imagine, for instance, that I drop this chalk and it does not fall. Do we say that Newton's gravitational theory is refuted? Or if an angel appears and flies upwards, or a bird flies upwards, do we say that Newton's theory is refuted? No. We say that this phenomenon involves air resistance; once we have got the aerodynamics right we shall be able to understand how birds (and angels) fly upwards. This, however, still does not explain the chalk. Suppose I wanted to show with the help of an Atwood machine how Newton's theory exactly measures the downward acceleration of this chalk. I could predict that this chalk will drop in 1.35 seconds. Say I have an exact measurement, and that some of my revolutionary students, knowing that I shall repeat the experiment with a steel ball, put a big magnet up there on the fifth floor. I drop the ball and it goes upwards. Can we say that Newton's theory is refuted? No; we shall say that a *disturbing factor* was present.

Newtonian scientists quite as well as Marxism can get away with their invisible agents! One can always blame a refutation on an invisible factor! This is part and parcel of science. This simply means that refutations are just no bloody good to judge a theory. As you can see, we need a system of decisions. (I am leaving out the problem of statistical hypotheses on purpose since it is another matter.) The main question remains: what *are* Popper's rules if they have to be based on this system of decisions? It is like a game of chess—to use Popper's own analogy—with certain rules which one has to obey.[20]

What are these rules? First of all, one has to put forward a consistent hypothesis. If the hypothesis is inconsistent, throw it out. Bohr's hypothesis was inconsistent. Consider it carefully: it was inconsistent because it contradicted Maxwell's theory. According to Maxwell's theory an electron emits a continuous spectrum, and Bohr never contradicted the corroborating evidence of Maxwell's theory. He never denied, for instance, that if I put an electron into a big magnetron[21] and chase it

20. Cf. Popper [1934] 1959, 53.

21. The magnetron is an electron tube for generating very high frequency electromagnetic waves, especially as used in radar equipment. Schematically, it consists of two cylindrical, coaxial electrodes: one of which, the anode, is positively charged; the other, the

around an orbit and leave it there, when I switch off the magnetic field it will slowly fall into the centre, where there is a positive electrode. If, however, the atom is 10^{-8} cm small, then the electron will not fall into the centre. What sort of inconsistency is this? At what point does Maxwell's law stop and Bohr's law start? Bohr never discusses this. The photon jumps from one orbit to the other, but how long does it take to jump? And where is it *in between* jumps? Does the whole thing happen like this? Bohr's theory actually states that there is a certain time when the electron disappears in one point and re-appears in another point. It cannot possibly be anywhere in the middle for any moment of time. What a mad theory! It is inconsistent with our most elementary geometry. It is inconsistent with Maxwell's equations. If Bohr used an ammeter at any point in his experiments, he used Maxwell's theory which was inconsistent with his own! That is sheer madness.

Bohr actually wrote in another book that he thought that until his theory explained Maxwell's success, there was no point in publishing the paper. But Rutherford advised him to send it to *The Philosophical Magazine* and promised to arrange for its publication. *Popper's first rule of science (that one cannot put forward an inconsistent hypothesis) was violated.*

Secondly, Popper claims that whenever you put forward a theory, you have to specify its potential falsifiers: under what conditions would you give up Bohr's or Newton's axioms? The interesting thing is that Popper later quibbles on this rule. On certain pages he says that it is intellectually immoral and not in the "spirit" of the game to avoid specifying the potential falsifiers of the hard core of a theory. On certain other pages he says that we are always dealing with the refutation of a very complicated system of hypotheses. But even then, he completely forgets about the *ceteris paribus* clause. He never discusses it. In all cases, Popper is wrong anyway because he maintains that when a theory is put forward we have to specify what we would regard as its refutation, and should this refutation take place then it is our moral duty to drop the theory.

---

cathode, is negatively charged. The device is situated inside a magnetic field which is permanently in line with the cylinder axis. The flow of electrons generated by thermoelectric effect from the cathode is subject to both the electrical and the magnetic fields; the electrons tend to group together in beams and circle at high speed around the axis. The circling electrons emit a permanent oscillation which can be extracted by means of a resonant apparatus.

You may remember how many times in the history of science a theory was refuted, and yet we never dropped it. I pointed out how many inconsistencies there were in the first form of Bohr's theory, and yet we did not drop it. Since 1687, when Newton's *Principia* were published, many anomalies had been common knowledge. (I prefer to use the word "anomaly" instead of "refutation.") For instance, Earth's satellite, the Moon, did not move according to Newton's law. If you look at the *Principia* you will discover that the motion of Saturn's satellite is described in detail, but not so the Moon's. It is not discussed because "unfortunately" the Moon did not follow Newton's laws! In 1702 the first textbook on the Newtonian theory was published by one of his students, David Gregory. Newton asked Gregory to allow him to add an appendix to his book on the Moon's motion (this was in 1702, sixteen years after the publication of the *Principia*). However, between 1687 and 1702 Newton had been completely in the dark as to how on earth he could fit the Moon's mad movement into his theory. There are currently big textbooks on the theory of lunar motion, and if you leaf through them you will see that there are dozens of anomalies. In simple planetary motion we have plenty of anomalies which now of course can be explained with theoretical assumptions. We now know, for instance—and I mean "know"—that the Moon's centre of gravity does not coincide with its symmetrical axis. There are various theories about how the Moon came about: whether it was a combination of cosmic dust, or whether it was an ordinary satellite which was hot and cooled down (and this signalled volcanic activity on the Moon), but we now pretty much "know" that somehow the centre of gravity is quite far away from its centre of symmetry. You can explain how this could have happened if, for instance, you imagine a fast rotating body in a fluid state which cools down with big iron masses, big aluminium masses, big chunks which have not reached their hydrodynamic equilibrium, but cool down so fast while rotating that they become rigid before reaching it. The existence of these mad bodies is not in contrast with the Newtonian mechanics because a planet orbiting around the Earth with its axis tilted would produce some very complex dynamic phenomena. But, according to Popper, the Newtonian theory should have been given up in 1686, and the actual lunar motion theory never published!

Incidentally, we learn from science textbooks that Newton's law of gravitation was refuted by Mercury's anomalous perihelion! You all know that planets move in ellipses, but as far as Mercury is concerned, it has a sort of rosetta-like motion (see Figure 6).[22] This movement was explained by Einstein's gravitation theory in 1916. When was Mercury's peculiar movement discovered? In 1816! So, you see, there were a hundred years of illegitimacy. I would not, however, make much of a fuss because there were hundreds of anomalies for the Newtonian theory, and today, after Einstein's theory of gravitation, ninety-nine instead of a hundred remain, simply because Einstein explained one of them, namely, Mercury's perihelion, and even that one very inexactly.[23]

Just to go through Popper's rules, he has a further one: once a theory is rejected, you have to put forward a new theory which complies with certain conditions. The new theory should explain everything that the falsified theory had explained otherwise, without, nonetheless, being a weaker version of it, otherwise it is only ad hoc. To give a very simple-minded example: suppose you put forward that all swans are white, and then produce a black swan. You have to reject "All swans are white." Let us call the black swan "Peter." According to Popper, you

22. Cf. Will 1986: "This precession could be seen in observations of Mercury's motion, and amounted to 574 arcseconds per century. In 2,250 centuries, Mercury's orbit would trace a complete rosetta." Leverrier attempted to account for this precession using standard Newtonian celestial mechanics and to calculate the sum of the perturbing effects of the other planets, but he failed: his calculations gave the precession 38 seconds less than the time actually observed.

> Numerous proposals were put forward to account for the discrepancy. One was the existence of new matter in an orbit between Mercury and the Sun. This matter could be in the form of a planet, which has already been given the name, Vulcano . . . , or it could be in the form of a ring of dust or asteroids. . . . Given enough mass, such matter could generate the required gravitational perturbations to produce the additional perihelion precession of Mercury. Unfortunately, no solid observational evidence for Vulcan or for such a ring was ever found. . . . Another proposal was to modify the Newtonian inverse square law. . . . But for the end of the century this proposal was also dead. . . . From 1959 well into the twentieth century a host of suggestions was made to explain Mercury's perihelion precession, some serious, some outrageous, some simple, some very complicated, and none very successful. In November, 1915, while struggling to put finishing touches on the general theory, Einstein was well aware of the problem of Mercury, and it was one of the first calculations he carried out using the new theory. To his delight, he found a precession of 43 arcseconds per century! He later wrote, "for a few days, I was beside myself with joyous excitement." (91–93)

Further technical details can be looked up in Pais 1982, 275–80; on the method for calculating Mercury's retrograde motion, see Schwinger 1986.

23. As a matter of fact, observations carried out by Robert Dicke and H. Mark Goldenberg in 1960, which showed the sun's slightly flattened appearance, raised doubts—which have not as yet been clarified—regarding the validity of Einstein's prediction of Mercury's retrogression (see Will 1986; Roseveare 1982).

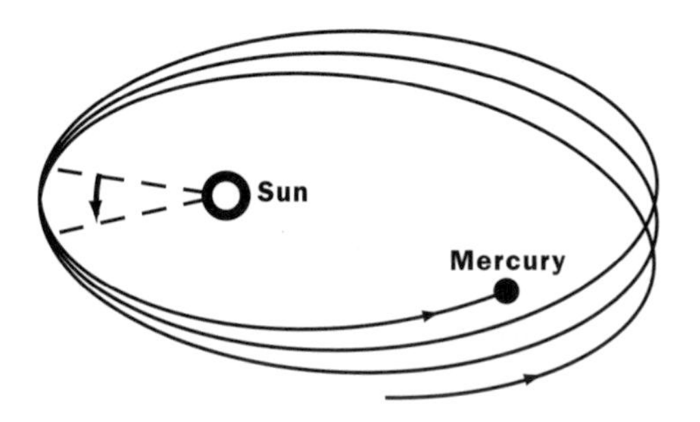

Figure 6 Mercury's elliptical orbit rotates on its plane, so that the planet's perihelion (point of maximum proximity to the Sun) also rotates: the planet's actual orbit thus describes a kind of rosetta. The direction of the perihelion is never fixed, but rather subject to a rotatory movement, a "precession." (From Will 1986.)

are not allowed to put forward the new theory "All swans are white except for Peter," because this is an ad hoc manoeuvre, and this kind of exception-barring is not allowed. You are not allowed to put forward the theory that all socialist regimes are marvellous except for Russia, China, Chile,[24] Cuba and all the existing ones. Popper is great when he makes fun of Marxism or Freudism, when he shows that whenever there is a refutation the two theories diminish their content and reassert themselves in a poorer version. But just remember this Newtonian business: after all, we had a hundred anomalies in 1686. And what the scientists actually did was to shelve these anomalies and say: "The Newtonian theory is all right except for these hundred or so anomalies." It is just the same today when Socialists say that Socialism is all right except for the facts. *Of course, there is a difference, but where does it lie?*

What I tried to demonstrate today is that if there is a difference between the Newtonian theory and Marxism, it certainly does not lie in Marxism's disobedience, and Newtonianism's obedience, to Popper's falsifiability criterion or *anti-ad hoc-ness* rules. So the difference must lie somewhere else. This is what I shall come to next time.

24. Lakatos is referring to the Socialist presidency of Allende in Chile between 1970 and 1973, before the violent military coup d'état perpetrated by Pinochet.

Lecture Seven

Falsification and Intellectual Honesty

I have already criticised Popper's demarcation criterion by using the example of the imaginary Newtonian scientist who, following empirical refutation of the theory which predicted a given planet's supposed orbit, predicts the existence of a disturbing planet, which he does not find. He then predicts the existence of cosmic dust which would hide the perturbing planet, and does not find it, so he predicts a magnetic field, and so on. Mr Bellamy[1] pointed out that this argument is *for* Popper and not *against* him. Indeed, according to Popper, our imaginary scientist is first of all a Newtonian who puts forward the completely falsifiable conjecture that the planets move in a particular way. Thus, a potential falsifier can be easily specified in that at a given time the planet will be somewhere else from the prediction of the Newtonian theory. So this is a falsifiable and also consistent conjecture. Nothing is wrong with this step. The conjecture is refuted and the refutation follows the conjecture: everything is all right. Then the Newtonian scientist abandons this conjecture and puts forward another one, namely, that there is a disturbing planet which we have not yet discovered. This is a non-ad hoc manoeuvre in the sense that it increases the content of the theory: we had a theory in which the Newtonian laws were applied in conjunction with the initial conditions; we now replace these with a different set of initial conditions in which this new planet figures, and we end up with a new theory. This is a non-ad hoc manoeuvre which is completely permissible in Popper's sense. We then try to look at the planet through a telescope, but we do not find it. A new conjecture is followed by a new refutation. *Conjectures and refutations!* We put forward a new—the third—conjecture, this one concerning cosmic dust. Again, it is not ad hoc, because it can be tested, for example, by sending up a satellite. But if we do not find any cosmic dust, then refutation. What is wrong? Again we abandon our whole conjunction of elaborate hypotheses and we replace it by another one which contains the hypothesis of a magnetic field, which is independently testable in Popper's sense (i.e., non-ad hoc). There is nothing wrong with it. So the whole procedure

1. Alex Bellamy was a postgraduate student reading Logic and Scientific Method at LSE in 1973. Of particular interest are some of his letters addressed to Lakatos concerning the difficulties that the methodology of scientific research programmes faces in solving the demarcation problem (Archive, section 12, item 92).

which I have described can be shown to be a paradigmatic example of Popperian rational behaviour on the part of the Newtonian scientist.

I give my friend Bellamy this much credit: I was not clear enough in showing that this is in fact a deadly counterexample to the Popperian approach. So let us just go into a little bit more detail.

First of all, and this is an exegetic red herring, what I have just described is the Popper of 1935, i.e., the author of the *Logik der Forschung*. As a matter of fact, in 1960, Popper claimed in his "Truth, Rationality and Growth of Knowledge"[2] that we need an additional requirement: namely, that a conjecture can really be saved if it is non-ad hoc in two senses.[3] This means that we have to abandon it if at least one of its empirical consequences is not confirmed. [. . .] Popper is systematically ambiguous in his demarcation criteria, and for good reason. In a long footnote in chapter 1 of *Conjectures and Refutations,* he says: "I am asking the Marxist or the psychoanalyst to specify a potential falsifier not just to a specific version of his theory, but to the 'hard core' of it" (actually this is my expression: Popper says "the main axioms of it"). This is Popper's *moral* requirement. I mentioned that most demarcation criteria contain a sort of code of intellectual honesty. If you put forward a conjecture, you have to specify a potential falsifier. If it is falsified, you are morally bound to abandon it and publicly admit that your assumption was false. Whenever Popper refers to the 'hard core' of a theory, he is saying that you cannot fiddle around with these initial conditions.

All this is on page 38 of *Conjectures and Refutations,* and I will read it out:

> real support can be obtained only from observations undertaken as tests (by "attempted refutations"); and for this purpose *criteria of refutation* have to be laid down beforehand: it must be agreed which observable situations, if actually observed, mean that the theory is refuted. But what kind of clinical responses would refute to the satisfaction of the analyst, *not merely a particular analytic diagnosis but psychoanalysis itself? And have such criteria ever been discussed or agreed upon by analysts?* [Popper, 1963, chap. 1, 38, footnote 3, emphasis added]

This is one of the many quotations to this effect. If applied to Marxism, the Marxist is morally bound to specify a falsifying event and if it takes

2. This paper was later published as chapter 10 of Popper 1963.

3. Any scientific theory, according to Popper (1963), can be ad hoc in two senses: (1) if it simply explains all the *explicanda* for which it was originally designed, without implying further testable consequences; (2) if it does not pass independent tests, meaning that none of its new consequences is verified (240–44).

place, he must not only abandon a specific variant with its initial conditions, but Marxism itself. How on earth can we interpret Marxism if it is only a sort of specific version with twenty-five initial conditions and auxiliary hypotheses? If Marxism sums up to this, it can never be refuted. Incidentally, the weaker version of Popper's theory, i.e., the one in the *Logic of Scientific Discovery* (I'm afraid I cannot give you the page number now, but I shall give it to you next week),[4] states the exact opposite: what can be refuted is not just one hypothesis but a *conjunction of hypotheses.* Popper never goes into this deeply and he never discusses the *ceteris paribus* clause problem seriously. Anybody who has read anything about the methodology of economics knows that this clause is always discussed, but exactly the same happens in physics too. After all, magnetic fields can act on planets in the way I have mentioned. If we refer back to our example, I could drop a metal billiard ball and a magnetic field may lift it upstairs. The *ceteris paribus* clause leaves this possibility open.

To sum up: suppose we take Newton's three laws of dynamics, plus the law of gravitation, plus twenty-seven initial conditions, plus thirty-seven observational theories, and we derive an observational statement which is inconsistent with all this, what should we do? Should we cross it all out? Let us imagine the negative case in which our imaginary Newtonian scientist does not find the disturbing planet, does not find the cosmic dust, does not find the magnetic field, and then, for some internal or external reason, abandons the project. By 'internal' reason I mean that he cannot think of anything better; and by 'external' reason I mean that he comes up with some sort of solution, but it is too expensive to carry out in practice. What does the Newtonian theorist then do? Does he give up the idea that there must be a disturbing planet, that there must be cosmic dust, that there must be a magnetic field? No. He says: "I only wish I could see what was wrong with the whole thing"—in other words, what additional elements the *ceteris paribus* clause contains. He does not give it up. In twenty years' time an experimental or theoretical novelty may arise. So, in fact, not even a conjunction of hypotheses will be crossed out, but only *shelved.* Remember that in the

4. Lakatos is probably referring to Popper [1934] 1959, 76, footnote 2: "Thus we cannot at first know which among the various statements of the remaining sub-system t' (of which p is not independent) we are to blame for the falsity of p; which of these statements we have to alter and which we should retain. . . . It is often only the scientific instinct of the investigator (influenced, of course, by the results of testing and re-testing) that makes him guess which statements of t' he should regard as innocuous, and which he should regard as being in need of modification."

case of the Newtonian theory thirty-seven anomalies per year on average emerged.

What we are in fact doing here is an ad hoc manoeuvre: we are saying that the 'new' evidence can be deduced from a disturbing planet, from cosmic dust, from a magnetic field, *plus* something which we cannot specify. It is clearly ad hoc because the hypothesis is untestable. What I am claiming is that the history of science is full of similar cases of tenacity. Without this kind of *tenacity* there would not be any scientific progress at all.

There is a third way of interpreting Popper's falsifiability criterion: namely, by claiming that we should hold as definitely falsified any theory once it has been branded as falsified in specialized periodicals. We would not then be allowed to suppose that something like my auxiliary hypothesis might save it. This idea is too stupid and of course Popper never said anything like it. [. . .]

Most LSE students have read something by Popper. In the first chapter of *Conjectures and Refutations,* which is a sort of autobiography, you will find that Popper says that he really became a full-blooded philosopher by being fascinated with the following problem. At ages sixteen and seventeen he had endless discussions with Leninists, psychoanalysts and Adlerians, the psychologists with whom he was closely associated. Whatever he said, they had a way out. Therefore he decided, and I quote Popper's own words, that: "Scientific theories, if they are to deserve the name 'scientific', should be at least semi-decidable, one-sidedly decidable." This appeared in his famous first publication, a letter in the periodical *Erkenntnis* in 1933: it is only two pages long, and this is the crucial message.[5] Scientific theories must be at least one-sidedly decidable; they might not be proved, but they might be disproved. This is what set him off on his philosophical career.

Allegedly, Popper's three major contributions to philosophy were: (1) his falsifiability criterion—I think this is a step back from Duhem; (2) his solution to the problem of induction—where I think he is a step back from Hume (I cannot argue this in this course: you can look it up in my published writings);[6] and (3) his literary masterpiece "*The Open*

5. *Erkenntnis* 3, 1933; the letter was reprinted as appendix *i to Popper [1934] 1959, 313–14. Lakatos is probably referring to the following passage: "We can, quite consistently, interpret natural laws or theories as genuine statements which are *partially decidable,* i.e., which are, for logical reasons, not verifiable but, *in an asymmetrical way, falsifiable only:* they are statements which are tested by being submitted to systematic attempts to falsify them."

6. See Lakatos 1968a and 1974a.

Society by one of its enemies" . . . what is it called? *The Open Society and Its Enemies.* Sorry; I do not want to plagiarize this joke: you know I like to recognise priority rights, so I will tell you that the source is Professor Maurice Freedman[7] back in 1958. *The Open Society* is frankly a literary masterpiece: not being a political philosopher, I cannot comment on its contents, but I certainly think it is a marvelous book. So, in conclusion, two-thirds of Popper's philosophical fame is based on misjudgement. Perhaps we can come back to this next time.

Bellamy: Can we leave it that what you call *shelving* Popper calls *abandoning?*

Lakatos: No. I will tell you why. Let me say this much: I hope the dramatic difference between *shelving* and *abandoning* will be clear for everybody after what I am about to say. In order to do that, I need to develop some more conceptual machinery. The Popperian framework is too narrow to formulate an answer to this problem.

According to the *Logic of Scientific Discovery* (and, Alex, I think you will find it very difficult to refute this), the rationality of the scientific enterprise (I quote, and note that it is repeated twelve times so you will recall at least one occurrence) depends on cutting the propositions into two: basic statements and theoretical statements; falsifiable statements and unfalsifiable statements. This is absolutely crucial because if all theories are unfalsifiable—Popper actually uses the word "metaphysical" to describe them—then Newton and Marx are on a par. You remember the table I drew up at the beginning with the goodies and the baddies. Well, the whole enterprise of the demarcation criteria hinges on whether we can have this dichotomy, i.e., whether we can have some criteria to decide about the goodies and the baddies. Now, the Marxists can always say: we "shelve" the Hungarian revolution, we "shelve"the invasion of Czechoslovakia, these facts might be explained later (it is only a matter of imagination and of finding the CIA agent who is responsible for these anomalies). Once we accept the shelving-abandoning equation, Marxism and Freudism are exactly on a par with Newton, and therefore Popper's enterprise flounders completely.

Moreover, I pointed out—and this is a matter of historical fact, not logic—that all theories are anomaly-laden. That means that there are always problems in any theory which have to be shelved. If all theories are born refuted, then it is perfectly clear that all theories are falsified right at the beginning; at this point the dichotomy collapses.

In order to maintain his philosophy, Popper forges history in a most

7. Maurice Freedman, professor of anthropology at LSE between 1951 and 1970.

ridiculous way. I can show you examples taken from *Conjectures and Refutations* where he really seems to believe that in the history of science we have a conjecture which is not refuted when it is put forward, then we have a refutation, then we have a new conjecture which is not yet known to be refuted at the time, followed by a new refutation and so on. As a matter of historical fact, when we put forward any conjecture, we know right from the beginning the anomalies it cannot account for. So what is this refutation, if not some curious anomaly? *How do we decide what is a serious anomaly which is to count as refutation, and what is not?*

This question was first put to Popper by me. It might strike you as funny, but this is the case. Word got around and when Popper had a television interview which has now been published in Bryan Magee's book *(Modern British Philosophy),*[8] Bryan Magee asked him: "If science grows by conjectures and refutations, then why wasn't Newton's theory abandoned with Mercury's anomalous perihelion in 1820?" Popper answered: "Because Mercury's perihelion was not a serious anomaly." In a letter to me written at the same time, he explains that if an object were to move around the Sun in a square, then that would be a *serious* anomaly![9]

If you remember what I said about the élitist, authoritarian position—which is of course completely abhorrent to Popper—it is quite clear that deciding on the demarcation criterion between an unserious anomaly which you can shelve and a serious one which obliges you to give up the theory is obviously *a matter of taste* or of the scientist's *authority.* So at the age of seventy Popper's demarcation criterion degenerates into Polanyiism, which he earlier described as obscurantism. Now, just when his theory crumbles, he arrives at the very same obscurantism. This happens because the demarcation between a serious refutation and a shelvable anomaly turns into a matter of *judgement.* But judgement passed by whom? By scientists? By the Communist Party's Central Committee? Or by which other *Élite?*

One final example will further clarify Popper's complete confusion. In the same way as before when I quoted Maurice Freedman, I do not want to steal this joke from my rather narrow-minded friend Alan Musgrave, who thought up the joke but not its consequences. Musgrave prepared the subject index for *Conjectures and Refutations.* If you look up *Marxism* in this index, you will find the following entry: "Marxism—

8. See Magee 1971.
9. See also Popper's "Replies to My Critics" (1974b), pp. 1004–9.

made irrefutable: pp. 34f, 37, 333f; Marxism—refuted: pp. 37, 333"! Now: Marxism is bound to be *irrefutable* because it is unscientific, that was Popper's original axiom. But unfortunately, Marxism *was refuted:* by the lack of absolute impoverishment in capitalist countries and the first revolution in the least industrially advanced country, and so on. Thus, irrefutable Marxism was refuted! You get both in Popper. And it is a mess.

I think that the fact Popper's philosophy survived for so long is a sociological mystery. Popper's immortality is secured by this idiotic result. Reviewing nowadays happens in two stages: first of all a book is reviewed in weeklies like *The Listener, The Economist, Nature.* When someone like Sir Karl Raimund Popper publishes a book like *Objective Knowledge,* the editors rush to ask another knight to comment on it. So the first reviews are written by great men like Sir Peter Medawar or Sir Hermann Bondi. Since the weeklies have to be up to date, these people have to write reviews on the basis of having read the table of contents. So both Medawar and Bondi say that *Objective Knowledge* is a new classic of the great man. (Actually, I think the first two chapters bear the signs of mental senility.) Then comes the second stage of reviews—and by now, after both Medawar and Bondi have reviewed the book, nobody contests Popper's results on induction and his criterion of falsifiability anyway. By the time professional philosophers agree that it is not acceptable anymore, the weeklies have already started to immortalise it. And, believe me, the history of philosophy shows that the weeklies will have their way. The second generation of reviews comes along two years later in the learned journals. Nowadays reviewing has no social standing. However, I remember, as a young man, with what fascination I read Whewell's review of Herschel's book of 1831: it was forty pages long and it was thorough.[10] At that time everyone thought highly of reviews. Nowadays, reviewing is carried out—I hope none of you will take offence—by research students. I had better not comment further.

Since the statement I made about Sir Karl's senility, although true, can be libellous under English law, as Michael Azander[11] will testify, I had better add another piece of truth.

Zander: It's all right, because if it is true it is not libellous.

Lakatos: Oh God.

10. Lakatos is referring to Whewell's extended review of John F. W. Herschel's *Preliminary Discourse* (1830), which appeared in the *London Quarterly Review,* vol. XLV, no. 90, July 1831, 374–407.

11. Michael Zander, professor of law at the London School of Economics.

Zander: The only trouble is that you have to be able to *prove* it.

Lakatos: The law, of course, is based on sixteenth-century methodology! I do not want to defend my statement now; I am simply embarrassed to have repeated Maurice Freedman's libellous remark—"The Open Society by One of Its Enemies"—in public. This is actually a rather dramatic prediction which was rather dramatically confirmed by fact. I am sorry that what I am about to say is a rather personal remark, but first I want to amuse you, then I shall go on to research programmes. If you look up the paperback third edition of *Conjectures and Refutations* and you go to the name index, you will come across: "Lakatos, I., page ix." If you turn to page ix, you will find the "Preface to the Third Edition," but my name is not there. The background of this is that he had originally written in that preface that "Lakatos's criticism of my falsifiability criteria and my theory of induction come too late to take into account in this volume," but he subsequently changed his mind and decided not to give publicity to this lousy criticism.

This was in 1968. In 1972 Popper published his book *Objective Knowledge,* and if you look up the name index, the name Lakatos appears again and the reference is to "p. 38f," which means that it is a lengthy passage. Once again, you do not find my name on those pages, but it says: "I disagree with *those critics* of my views who assert, for example, that Newton's theory is no more refutable then Freud's," and so on.[12]

Then I remember that it says something that I definitely never claimed, but that is how Popper misunderstood it. This, of course, has nothing to do with the world of ideas!

I shall now introduce the concepts of *progressive* and *degenerating problem shifts,* and point out a very interesting sociological question which is, however, completely outside the realm of this intellectual course. I am talking about the problem of whether the representatives of degenerating problem shifts can maintain (or usually maintain) standards of elementary personal honesty. That is a very interesting question, and perhaps I shall go into it next year when I retire and turn to the sociology of science.

Let me sum up where we now stand. First of all I introduced a trichotomy: (1) *epistemological anarchism,* which claims that any belief is as good as any another; i.e., that science is a set of beliefs on a par with Azande beliefs. Feyerabend is the main advocate of this theory. I gave you the reference so that you could look up his famous paper "Against Method" in volume 4 of the *Minnesota Studies,* and oddly enough

12. This entry in the name index was cancelled in the 1979 edition.

none of you complained that the library forgot to buy it. (2) *Élitism,* or *authoritarianism:* it is primarily represented by Polanyi and Kuhn. According to this view there is a demarcation criterion between science and non-science; however, only great scientists can see it, and only they can judge if a scientific theory is better than another. Then I said that I did not like this view because I am a democrat in favour of some balance between *statute law* and *case law.* (3) *Demarcationism.* If you remember, I said that we can divide it into *inductivism, simplicism,* and *falsificationism* which can be placed a step back from the others.

Each of these stands determines a code of conduct and a code of intellectual integrity. Feyerabend, for instance, claims that intellectual integrity does not exist and you can "do your own thing." Polanyi, on the other hand, says: "Do your master's thing." Kuhn: "Do your master's thing, or your own if you can convert others, but not before then!" (Élitism). Each version of demarcationism holds a different code of conduct. Inductivism holds, roughly: "Never utter anything which is unproven" (this is codified in the libel law). Simplicism, followed by probabilism, says: "Never utter anything without at least specifying the probability of *h* given the evidence *e*".

Remember that both these positions collapse on logical grounds. Simplicism, on the other hand, demands you "give up a theory when it becomes too cumbersome, and replace it with a simpler one." I cannot remember if I have already gone into this in detail, but it is possible to show that the paradigm cases which support simplicism, for instance that Ptolemy's theory was abandoned because Copernicus's was simpler, are false. Another example is that scientists have abandoned Newton's theory because Einstein's is simpler. Exactly the opposite applies: the new theories were not simpler. Finally, falsificationism claims that under well-defined conditions we have to give up our theories. If we applied this code of honesty, it would appear that Newton collapses along with Marx!

I also mentioned that each of the three classes [inductivism, simplicism, falsificationism] determines a different historiographical outlook, and if they are false they create historical myths. Epistemological anarchism is actually the least prolific in this sense, because one can describe anything as it happens and not attach any normative values to it. Élitism, on the other hand, has some historical evidence: remember the genealogical trees of the great scientists. Demarcationism is the most influential historiographical view. According to inductivism, unproven theories are replaced by proven ones. According to simplicism, complex theories are replaced by simpler ones. According to falsificationism, falsified theories are replaced by unfalsified ones: this is, as you know, the great myth of negative crucial experiments. This is a huge industry in the history of science textbooks, where you always read lengthy essays on some famous negative crucial experiment or other. You can read that Michelson-Morley refuted the ether theory, or that Lavoisier refuted the phlogiston theory, and so on. If you have a look at contemporary periodicals, you find out that the "refuted" theories survived and were very lively for decades after, and that the *real* crucial experiment was actually invented at least forty years later. This big historical myth-making is alive and well.

Does this imply that we should give up the demarcation criterion altogether, or practice science without putting too much thought into it? We are about to reach the last lecture, and we seem to be lost in considerable chaos. [. . .]

Before I go on to my tentative solution, I would like to point out that in order to separate the goodies from the baddies in Popper's frame of reference, I have to differ from him on at least two absolutely vital points. *Firstly,* I have to abandon refutation as something which causes us to decide the rejection of theories. As I shall show you next time, falsifications are somehow irrelevant: the greatest victories in science were verifications and not falsifications. We have turned the clock back forty years, back to the time when most people agreed on this: how Popper managed to fool them, I shall never know. We therefore come back to verificationism. *Secondly,* if we want to separate the goodies from the baddies, for instance if we want Einstein and Max Planck to be goodies, then somehow we have to explain why nobody, Planck included, believed them for at least ten years. Planck made a great discovery in 1900, and from that date until 1908 he did nothing else but try to disprove his own theory because he thought it was idiotic: he had invented the quantum theory! In 1908 he abandoned his theory completely because he simply could not accept it, and even gave up the idea of refuting it in order to take his mind off the whole problem. He then joined Einstein's relativity programme, which at the time had nothing at all in common with quantum theory. He just wanted to change the subject, and forget about it.

It is quite clear that if I want to make the goodies "goodies," I hold dear both *dogmatism* and *tenacity,* at least under certain circumstances. So I have to distinguish good dogmatism and good tenacity from bad dogmatism and bad tenacity. And we have to forget about refutation because all theories are born in an ocean of anomalies and inconsistencies. I quite agree with Popper that to try to verify Galileo's theory of free fall by dropping the chalk and measuring the time it takes to reach the table (even by repeating the experiment a million times instead of just a thousand, with a dropping machine which records the time automatically while you are asleep) does not add anything to the value of the theory. My problem is therefore how to distinguish between important verifications and unimportant verifications, and not, like Popper, between serious and unserious anomalies. Popper once wrote to me: "If you understood anything about physics, you would see what is a serious refutation, and what is not, for example Mercury's rosetta motion." *If you understood anything about physics . . .* I am quoting Popper, not Polanyi!

In short, I am giving you criteria to distinguish between good and bad *dogmatism,* between good and bad *verifications,* so that science may be tested by the layman, whom I treat with a certain respect. I shall come back to this in my last lecture.

Lecture Eight

The Methodology of Scientific Research Programmes

I started this course with the problem that we are currently facing both an academic and a publication explosion. We therefore need a demarcation criterion in order to be able to 'burn' the right people and the right books. (I only discussed one aspect of the problem in this short course; namely, how to burn books, not how to burn people. That will be left for the next academic year.)

I then hinted, as you remember, that there were three schools of thought: *epistemological anarchism, authoritarianism* and *conventionalism. Epistemological anarchism* is currently represented by Paul Feyerabend. As from next week the school's lending library will have copies of Feyerabend's *Theses on Epistemological Anarchism,*[1] which could be considered a sort of Luther-like manifesto in twenty pages, and a conclusive refutation of everything I have said in the last seven lectures. According to this paper we should not burn anything: let all weeds grow.

Authoritarianism is primarily represented by Polanyi, Kuhn and

1. See Feyerabend, "Theses on Anarchism," this volume.

Merton. They claim that we must identify the scientific community, abide by its judgements and follow its religious conversions. This school reduces the problem of burning books to the problem of burning people: you have to find the "right" people first, and they happen to be the scientific community. Once you have defined that set of people, you just follow them in their views. This school, in short, *does not detach the views from their holders.*

I would, on the other hand, like to face the two problems—of burning *people* and burning works—separately. And this is precisely where authoritarianism draws the demarcation line between scientific communities and other communities: in my view, a hopeless exercise. From the demarcationist point of view, we distinguish between the baddies and the goodies as printed *works,* irrespective of their producers.

We have also discussed several variants of this school of thought: *inductivism,* for example, which we have concluded to be logically impossible: if science were really inductivist, there would be no such thing. It is logically impossible to arrive at valid theoretical inferences from facts or factual propositions.

On the contrary, the second variant, *conventionalism,* is logically possible. Conventionalism puts facts into convenient pigeon holes, and replaces clumsy pigeon holes with simpler ones. But two things are wrong with this: firstly, 'simpler' is a matter of taste, and the choice is therefore too subjectivist; secondly, if you consider the history of science, you will never come across any conception of *simplicity.* In other words, you cannot possibly define 'simplicity' in such a way as to make Einstein simpler than Newton, or Newton simpler than Copernicus, or Copernicus simpler than Ptolemy. Such a history of science is simply not possible, it would be all myths. Conventionalism can be criticised by historical investigation. This means that if conventionalism indeed selects the better principles, then science has up to this moment grown in a completely irrational way.

I then discussed *falsificationism* and Popper. I would like to go into this in more detail because it will lead us to my solution of the problem, i.e., *the methodology of scientific research programmes.*

What is wrong with falsificationism? First of all, let us go back to our table of goodies and baddies. These are boundary conditions necessary for any theory of demarcation: we would, for instance, prefer to include Galileo, Newton and Einstein among the goodies, and certain other doctrines which we shall not name, among the baddies. The problem is that with falsificationism, absolutely everything is listed as a baddy! I have already claimed as a matter of historical fact that any theory is born anomaly-ridden. Contrary to Popper, these anomalies are shelved instead of used towards rejecting the theory. It follows that falsificationism considers the growth of science as a whole and the behaviour of individual scientists as *irrational.* In other words, Popper cannot explain in rational terms the presence of anomalies.

Secondly, consider the role played by scientific *dogmatism.* Scientists are dogmatic and scientific theories are tenacious. You may remember my favourite example: Newton's theory. However, according to Popper, the tenacity of theories is immoral. It is primarily a characteristic of Marxism and Freudism, whose lengthy survival is immoral. The point I am driving at is that what applies to them, *applies to any great scientific theory!* Consider that shelved anomalies and the tenacity of a theory are equivalent.

These are *historiographical* ways of criticising Popper's demarcation criterion. From a logical point of view, it is quite possible to 'play the game of science' according to Popper's rules: a consistent theory which is experimentally falsifiable is produced, the experiment performed, the theory rejected, the next theory produced, then rejected and so on. This is all logically possible; the only problem is that it has never happened in this way. So if we accept Popperian philosophy, then we also have to accept that the history of science is irrational, and that scientists have always behaved irrationally and immorally.

There are other troubles with regard to Popperian philosophy. According to Popper, anomalies should have a shattering effect on scientists. This is a psychological corollary to an objective appraisal. In Popperian terminology, if a theory is produced along with its potential falsifiers, then one should see a problem in every anomaly, and one cannot just go ahead with the 'core' of the theory, simply ignoring the problem, without first trying to solve it. It is one's *moral duty* to face the negative crucial experiment. I have already claimed that instead of acting in this way scientists just ignore the problem. But scientists also do something much worse.

According to Popper, verifications do not count, only falsifications do. So we should not think that a theory is verified by a confirming instance; we should simply shrug our shoulders since this instance does not add anything to the value of a theory. The only added (moral) value of a theory is to be the best on the battlefield. Only the refutation of a theory proves that it was scientific. No theory can be scientific *before* it is refuted: it might be unfalsifiable, how is one supposed to know? If we look at the history of science, we see that certain theories, though not all, stand victorious only after having passed through dramatic confirming instances. Theories grow in a sea of anomalies, and counter-examples are merrily ignored, but certain dramatic confirmations are given great attention.

Just to give you two examples: for about a hundred years after Newton had proposed his theory (the *Principia* was published in 1687) the French Academy offered an annual prize for whoever refuted the theory. About a dozen prizes were actually awarded, and then something happened. Newton had a disciple called Halley. He extrapolated data from a short interval observation, calculated certain initial conditions, applied Newton's four laws along with other theories, and finally calculated that a certain comet he had been observing, which is now known as Halley's comet, would disappear and reappear again after 72 years in exactly the same place in the sky. He gave the time of return within the exactness of a minute. And he was, in fact, wrong by half a minute. Let me just fill you in on the background: in the 1680s there were two major rival theories regarding comets. According to the first, when God is angry he sends out a comet, i.e., a celestial body characterised by a tail which makes it very different from any other, to show that he is in a bad mood and that disaster will come. (In 1755, for instance, a month before the Lisbon earthquake, a comet had appeared in the sky and it was duly predicted that some big disaster would follow.) The other theory came from what I call Kepler's fourth law, which is not mentioned in any textbook nowadays. Let me remind you that Kepler's first law states that planets move in ellipses; this law is followed by two others regarding the planets, and by the fourth, according to which comets move along straight paths. Predicting to the minute and against both these theories that a particular comet moving in an ellipse would return after 72 years, was indeed a stunning achievement. After that, the French Academy did not put out any more prizes for the refutation of Newton's theory.

Now, according to Popper, this example shows the irrationality of the French Academy, because one should not be impressed by a comet and the Lisbon earthquake, or by the verification of Halley's prediction.

Let us move on to the second example, which is just as famous. Einstein's general relativity theory, proposed in 1916, was a rather earth-shattering theory, in the sense that it completely replaced Newton's law of gravitation. Had you asked physicists their opinion in around 1918, you would have discovered that very few people had even heard about the theory, but things had changed by 1919. One of the consequences of Einstein's theory was this: suppose you select two stars in such a way that during a solar eclipse when the Moon obscures the Sun the two stars are visible for a few minutes, and are again visible during the day; you then measure their distances during the night and again during the day, when the solar eclipse occurs. That distance will be significantly different! No one would even dream of observing such a thing without a theory to predict it, since it is so incredible. Then came the Eddington expedition to South America in 1919 organised to test the Einsteinian prediction during an eclipse, and, lo and behold, that is exactly what they found!

This verification tilted the balance from Newton to Einstein. According to Popper, scientists should choose to completely ignore any confirmation. If we accept his theory of appraisal, we have to accept that scientists are irrational all the time. Firstly, because they ignore anomalies, and secondly because they are shattered by verifications rather than crucial experiments!

Bellamy: But this is a crucial falsifying experiment of Newton's theory, not simply a verifying experiment for Einstein's.

Lakatos: But why should it be any worse than Mercury's perihelion, or the Moon's anomalous motion? There had already been one hundred and twenty negative experiments. The only thing that differentiates this refutation of Newton from the other one hundred and twenty refutations is that the South American expedition is at the same time a verification of Einstein's theory, whereas the others are not. Curiously enough, Einstein's theory inherits 99% of those refutations,

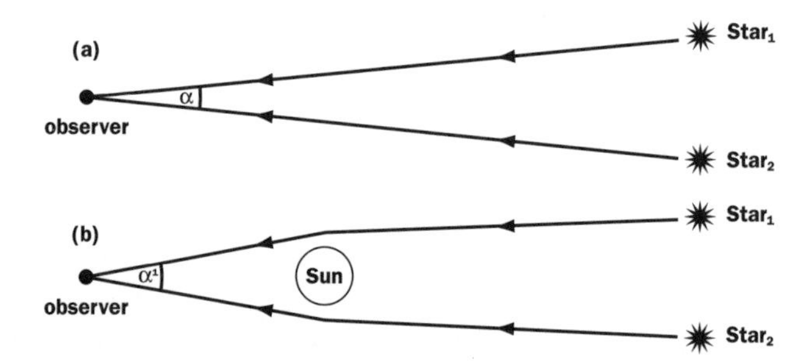

Figure 7 Gravitational deflection of light rays in proximity to the Sun's mass: $\alpha' > \alpha$. If light is gravitationally attracted by massive bodies such as the Sun, then the effect may be detected at times of total solar eclipse by examining the pattern of an appropriate stellar configuration in the sky. Suppose two stars normally subtend the angle α at the surface of the Earth. The angle will be modified to α' on occasions when the light has to pass the Sun to reach the Earth. Under normal circumstances, it is hopeless to try to measure α', for the starlight is negligible compared with the light emitted by the Sun. But on the rare occasions of solar eclipse, α' may be measured. This is what Eddington did. (From Oldroyd 1986.)

eliminates only 1% and that is precisely what counts in its favour! As a matter of fact, Einstein's theory increases the number of anomalies; the sea of anomalies does not diminish, it only shifts. In other words, certain anomalies are eliminated, and some new (Einsteinian) ones are acquired.

Then we come to the, at least according to Popper, completely irrational behaviour of scientists, which consists in the positive appraisal given to "dramatic" verifications.

Finally, I would like to introduce a new historiographical notion which also comes in as a sort of explication of Popper's ideas. This is the notion of *problem shift.* I call "problem shift" a series of scientific theories which replace each other with the passing of time. The term "hypothesis shift" would have been better, but it sounds very awkward. I call such a problem shift *theoretically progressive* if each H in a sequence $H_1, H_2, \ldots, H_n$ predicts everything that the preceding hypothesis predicted, and something else in addition. A theoretically progressive problem shift is an example of growth of knowledge, of growth of predictive conjectures. We get an *empirically progressive* problem shift if at least some of the predictions are corroborated. One good thing about Popper's theory is the ad hocness business: progressive problem shifts would be described by Popper as non-ad hoc. So if a theory grows in a non-ad hoc way, I call it progressive. If it never predicts any new facts but only eliminates anomalies through verbal tricks, it is regressive and content-reducing.

Suppose H_1 is put forward and some anomalies appear, then we can say that H_2 equals H_1 minus the anomalies: in this case the problem shift is not progressive. Another example of regression is when you add an untestable auxiliary hypothesis. Consider Marx's political theory. Some people claim that Marx's theory can explain everything. Let us apply it to the Hungarian revolution. According to Marxist theory, no other revolutions are possible after the Socialist one; then you confront a Marxist with the fact of the Hungarian revolution. And he says: "Well, I can explain that: it was actually a *counterrevolution*" (you will find the same explanation in some Soviet texts). This is interesting because the classics of Marxism-Leninism do not discuss the possibility of a counterrevolution breaking out in a Socialist country surrounded by Capitalist powers thanks to the infiltration of imperialist agents. This was never discussed before, but now you can account for this case too by introducing an ad hoc auxiliary hypothesis. *Everything* can be explained in this way, but there is never any prediction of a corroborated *novel fact.*

Actually, Marxists predicted things all the time. Even today, for instance, one of the most recent predictions is inconsistent with some of the earlier predictions. And the accumulation of these ad hoc inconsistencies is another sign of degeneration. A funny modern prediction of Marxist theory is that a peaceful parliamentary transition from a bourgeois to a socialist system is possible! We are "testing" this theory in Chile at the moment. I am sorry for the Chilean people, but I have no doubt that this conjecture will be refuted and I wonder what the next ad hoc hypothesis will be. I shall show in what sense according to my MSRP the whole Marxian political theory constitutes a theoretically progressive, but empirically degenerating, problem shift.

The interesting thing is that, according to Popper, if there is a progressive problem shift, then everything is all right (this would, for instance, be the case if Trotsky is really as commonly described).[2] The funny thing is that there are similar progressive problem shifts in science all the time, and yet scientists do not always like them. We have a prediction, the prediction is corroborated, and the scientists in some cases say: "This is no good, let's forget about it!" I give an example in *Criticism and Growth of Knowledge* which I cannot go into now for lack of time. You will find it on page 153 where I give the formula for diatomic molecules predicted by Bohr's theory of molecular (band) spectra at the origin of an important research programme.[3] Subsequently, somebody pointed out that the formula was anomaly-ridden, and somebody else said: "If you replace a given part of the formula with something similar but slightly different, you will get the right results in the theory of spec-

2. Lakatos does not give any further explanation of what he means. He may be referring to the claims Trotsky made according to which the Bolshevik revolution made sense only if considered as the starting point of a vaster international revolution. Only when the latter was brought about could there be any chance of rendering vain all the aggressive attacks of capitalism surrounding the USSR, and avoiding a Soviet Thermidore. In this way, Trotsky "creatively" opposed Stalin's idea of socialism spread in a single country; and he also criticised as a poor ad hoc idea Stalin's claim that Russia was in fact peculiar (the most backward among capitalist states, which had nonetheless been host to a proletarian revolution!). Last but not least, Trotsky gave an alternative explanation to the Stalinist orthodoxy with which he predicted the failures of the Third International's policy outside the USSR.

3. Bohr's influential article dated July 1913 ("The Spectra of Helium and Hydrogen") reports the direction in which the Danish physicist hoped to continue his research. In September of that same year Bohr participated in the discussion with an article on the structure of atoms heavier than hydrogen and on Mendeleev's periodic table; in November he made his contribution to molecular structure. This "annus mirabilis" (1913) signals the first step in an articulated research programme which Bohr carried out and which was to become a pivotal reference point for the old quantum theory.

tra." The scientists immediately realised that this was an ad hoc adjustment. And yet, something new had been predicted and then corroborated. It was found out that this adjustment holds for *all* cases, and not just for the two or three observed. Scientists still maintain that this is an ad hoc adjustment, and degenerating, but not in the two senses I have described. I shall now refer to my published papers where I introduce the idea of ad hoc_3 as a sort of ingenious "patching up" of theories which might even bring you to true predictions, if you are lucky enough. Somehow, though, the theory is out of touch with the "spirit" of the problem shift.[4]

I am obliged to use this word "spirit" for the moment, but I shall explain it in a few minutes. Stalinists, for instance, would argue that the problem shift by Trotsky is not *in the spirit* of Marxism. Let us leave aside the discussion as to whether it is right or wrong. My answer is given in the theory of scientific research programmes, of which I shall give you a summary, and then I shall draw some consequences from it and turn back to consider the question of Trotsky.

First of all, if one considers the history of science, it turns out that what we appraise is *not* in fact a long synchronic conjunction of hypotheses $H_1, \ldots, H_n$ (with $n = 17$, for example), any one of which you can choose to change, as has been suggested by Duhem or Quine. Usually, we are faced with what I call a *hard core* of two, three, four or maximum five, postulates. Consider Newton's theory: its *hard core* is made up of three laws of dynamics plus his law of gravitation. Along with the core, we have some auxiliary hypotheses which I call its *protective belt.*

A historical example will clarify my choice of the term: Flamsteed, the Astronomer Royal of Newton's time, sent him a table of data on the Moon's motion which showed that Newton's theory was no good. Having received the table, Newton took a coach to Greenwich to meet Flamsteed, and said: "Suppose Earth is here and the Moon here; how can you calculate the Moon's true position from your observations, since you must know that light rays travel so as to break when they reach the atmosphere? If we observe the Moon from this point, we see it as if it were here because we have to take this phenomenon into account. My suspicion is that your theory of atmospheric refraction is wrong, and not

4. See Lakatos 1970, 88, footnotes 1 and 2, and 80. A typical example of an ad hoc hypothesis is Max Planck's first formula of the energy quantum (see Planck 1900a,b). See also Lakatos 1971a, 112, footnote 2, and 115. Another of Lakatos's examples of an ad hoc_3 hypothesis is Ptolemy's introduction of equants, which would have gone *against the heuristic* of the same programme. And this justifies Copernicus's dissatisfaction with it.

my four laws." "What is *your* theory of atmospheric refraction?" asked Flamsteed. And Newton: "I shall send you my theory of atmospheric refraction, as long as you recalculate your table on it; do not bother to send me the new one, because I am sure it will be all right." And it was! This sort of thing happens all the time.

When Evans refuted Bohr's theory with a new hydrogen series, Bohr, without even looking or asking for the details of the experiment, wrote a letter in which he claimed that it was quite clear that there should have been some ionized helium in the tube of the experiment that Evans had carried out. Bohr suggested that Evans should have been more careful in taking account of the ionized helium and how it had penetrated systematically into the tube of the experiment. And this was exactly the case.

Instead of directing the *modus tollens*[5] of refutation to the hard core, one leads it against these *auxiliary hypotheses* which, in this sense, create a protective belt around the hard core. But there is much more to the question than this: in all major research programmes you have a hard core and a protective belt which hold for centuries. Using the degree of tenacity as a unit of appraisal is more interesting than deciding whether a single theory is falsifiable. In Popper's view, on the contrary, the acme of a scientific theory is "All swans are white," because one single observation of a black swan refutes it. This is absolute nonsense.

Take some of the major theories in physics, like Newton's, or Maxwell's, or Einstein's, or the quantum theory: they are all terribly complicated and growing entities, not static entities. They are growing entities with a fixed hard core but an ever-increasing protective belt and mathematical techniques used for problem solving. You may ask, for instance: Newton invented the hard core of his theory in the *Principia* of 1687; how did his programme develop between 1687 and 1900? The answer is that in the eighteenth century there were no means, given the velocities

5. To say it in Popper's words ([1934] 1959): "Given the relation of deducibility, $t \rightarrow p$, and the assumption p, we can then infer t (read 'not-t'); that is, we regard t as falsified." Popper goes on immediately to add: "By means of this mode of inference we falsify *the whole system* . . . which was required for the deduction of the statement p . . . Thus it cannot be asserted of any one statement of the system that it is, or is not, specifically upset by the falsification." And he winds the argument up in the footnote: "It is often only the scientific instinct of the investigator . . . that makes him guess which statements of t he should regard as innocuous, and which he should regard as being in need of modification." Neither does Popper ignore the fact that it is often the change in what we consider to be "obviously innocuous (because of its complete agreement with our normal habits of thought) which may produce a decisive advance. A notable example of this is Einstein's modification of the concept of simultaneity" (76).

and the respective positions, to calculate from Newton's equation the movement of a planet involving other bodies in the system because the many-body problem was still an unsolvable mathematical problem.[6] Furthermore, scientists of the time had to contend with the idea that there were actually no mass points, but rather some sort of strangely-shaped bodies whose centre of symmetry does not coincide with their gravitational centre.

This was the case for one of Saturn's two moons,[7] which are not at all ball-like. How can we calculate the gravitational effect on an irregular body? We need integration and partial differential equations. It takes about a century or two in order to produce all the mathematical techniques required to get predictions from this hard core. And of course, in the meanwhile, there were anomalies and unsolved problems. Observations are not actually very important, since we know that they are wrong because the adequate mathematical techniques have not yet been developed. So it takes a hundred years fully to develop a good theory. And anomalies can only be expected.

So I have replaced the descriptive unit in science. I shall not ask anymore whether a hypothesis is true or false anymore, or even whether a problem shift is good or bad; what I *shall* ask is whether a research pro-

6. The article from which all further argument on the question stems is Poincaré 1890. This is the hub of the question, as set forth by Poincaré (1891):

> Newton's law is the simplest among all physical laws: its mathematical expression, however, is a differential equation, and in order to calculate the coordinates of astronomical bodies one has to integrate this equation. This problem, which is one of the most difficult in analysis to date, is far from finding a solution, notwithstanding tenacious geometrical research. . . . Which is the movement of n material points which attract each other with a force directly proportional to the square of their distance? If $n = 2$, in the case where we have a single isolated planet and the Sun, for instance, leaving aside other planet perturbations, the integration is quite simple: the two bodies describe ellipses, and therefore obey Kepler's laws. The question turns complicated when the number n of the interacting bodies is equal to 3: *the three body problem* has so far escaped all the analysts' attempts at solving it.

See also Poincaré 1898:

> The mathematician considers only celestial bodies as fictitious, reducing them to simple material points, and subject *exclusively* to the action of their mutual gravitational attraction, which strictly obeys Newton's law. How will a similar system behave? Is it stable? The analyst faces a problem as difficult as it is interesting. And yet it is not the same problem which is present in the natural context. Real stars are not material points, and are also subject to forces other than Newtonian attraction.

7. Saturn actually has eighteen satellites. The furthest among them is Pan, discovered in 1990 thanks to the computer analysis of the images obtained from *Voyager.* It is also probable that many other small satellites orbit around the planet.

gramme is progressive or degenerating. It is degenerating if—during its growth when we add layers of mathematical techniques to its protective belt—(1) it does not lead to stunning new predictions (at least occasionally, like Newton's theory did over two and a half centuries); (2) if all its bold predictions are falsified; and (3) if it does not grow in steps which follow the spirit of the programme. In this context I would like to introduce what I call *positive heuristics,* which is a problem-solving, anomaly-digesting technique. If you do this in a funny way like replacing a formula out of the blue without any integration with the programme, then the scientists will call it ad hoc [ad hoc_3]. I must have given at least a dozen examples of this already. Popper does not have this notion, which is akin to Duhem's concept of simplicity, with the difference that Duhem worked with conjunctions of hypotheses from which he was able to choose the simpler. In my case, on the other hand, we are dealing with a dynamic, *dialectical* picture (a characteristic that explains why some Marxists like my approach). Thus a theory is always judged by its growth, and this growth must be marked. Progress is made by stunning novel facts, like the return of Halley's comet or Einstein's prediction of the curvature of light. It is interesting to note that you do not have similar predictions in the social sciences.

I only have time to make two further points. One is that most of what we know with the name philosophy of the social sciences is to *explain away* the weakness of social theories. When evaluated as research programmes, even the most respectable theories turn out to be degenerating! Unfortunately, it is very difficult to have real case studies. For instance, nobody has yet applied the MSRP appraisal, because it is not as simple as saying that if "All swans are white" is falsifiable by one black swan, it is scientific. It is not as easy as that. It is not a matter of pure logic. One has to have a historical investigation, to identify the hard core, to describe the theory's growth over the decades (nobody has yet done this with Marxism),[8] to check what is ad hoc and what is not, as well as what is really in the spirit of a research programme and what is only a patching-up procedure.

Marxism is certainly a research programme in the sense that it has a defined major apparatus whose function is to deal with facts. This is also the case with Freud. No doubt these are research programmes and intellectual achievements. The question is: are they *empirically* progressive or degenerating? I think it will take a good three or four years to find out to what extent Marxism actually had a progressive stage, and

8. This is not true anymore; see, for example, Blaug 1980.

when it actually became regressive, if there were any creative shifts and so on. All this simply has not been carried out. The only case study on research programmes in the social sciences was done by Spiro Latsis, who showed in his paper on situational determinism (*BJPS,* June 1972) that allegedly the most respectable social theory, neo-classical economics, is actually degenerating, and does not seem to have had a progressive stage at any point in its lifetime.

As I said, most of the philosophy of social sciences boils down to explaining away the lack of growth in empirical content. An explanation is given in the sense that social sciences cannot be predictive simply because of the nature of human beings. However, we can predict all sorts of things about rats! So why shouldn't this apply to human beings, unless, of course, they are God-created and this is their privilege? This is where you have to see the revolutionary character of logical positivism: one of its slogans was: "Unity of science, unity of human knowledge." According to this, you cannot let social scientists get away with standards like: "Our purpose is to *understand* human behaviour, while it is not possible to predict it" (this is where all the important theories of "understanding" [*Verstehen*], especially in the German tradition, originate). There are all sorts of standards designed to make the social sciences respectable, and yet they are not. The social sciences are on a par with astrology, it is no use beating about the bush. (Funny that I should be teaching at the London School of Economics!) But this claim has to be shown in detail. Peter Clark has shown why the kinetic theory was successful so late.[9] The first person to show why Einstein's theory succeeded Lorentz's (a rival relativity theory now forgotten) was Elie Zahar in our department.[10] But I cannot get students to do research programmes in the *social sciences* because they obviously do not like to antagonise their professors by showing that the whole subject is nonsense, even though the papers would be well received in our department.[11]

9. See Clark 1976.

10. See Zahar 1973.

11. Following the conference held in Nauplion (Greece) in 1974, many case studies have revolved around the connections between the MSRP standards and actual scientific practice. Together with the proceedings of this conference edited by Howson (1976) and Latsis (1976), see Worrall 1990, 1991 and Zahar 1989. In particular, regarding the social sciences, see de Marchi and Blaug 1991; Hausman 1992, chap. 11; and Hands 1993. For a comprehensive summary of the history and methodology of economics according to Lakatosian standards, see Blaug [1980] 1992. Refer to Fisher 1986 for an analysis of the marginalist revolution. For the general equilibrium theory, see Weintraub 1985; Rosenberg 1986; Salanti 1991; Backhouse 1993. For the debate on the theory of capital,

Let me make my last point. I cannot go into the details of how to appraise a scientific research programme: you can look them up in my publications. One important aspect of my approach is that I start with *singular judgements.* I start with the goodies versus baddies lists, about which I am however prepared to be flexible. I would, for instance, like to agree finally that Newton is better than Ptolemy, that Einstein is better than Newton. If according to a theory of appraisal Ptolemy is better than Einstein, I become suspicious: I shall say that something is wrong with the standards of appraisal. I have shown in a paper, which was originally intended as an appendix to *History of Science and Its Rational Reconstructions,* that Copernicus did not present any progress over Ptolemy. I withdrew it in the galley stage because I was not happy. I would like to see Copernicus emerge at least as a little bit better than Ptolemy. I would not like to abolish the Copernican revolution.

I just want to mention that Elie Zahar in our department came up with a new definition of a *novel fact* which I cannot go into in detail, but it is very interesting. I shall read out the crucial sentence of his paper:

> A fact will be considered *novel* not just if it is temporarily novel, but, with respect to a given research programme, if it did not belong to the problem situation which governed the construction of the research programme or of the hypothesis in question.[12]

In order to apply this criterion, we have to know in great detail how the theory is actually constructed. For instance, nobody did this with Marxism nor with Freudism, because people committed to Marxism or Freudism are usually hero-worshipping, and they would either not undertake any such critical analysis, or think a priori that it is rubbish and it is not worth spending three years of one's life on it.

---

see Birner 1990. Refer to the discussion between Blaug (1976, 1990, 1991), Hands (1985, 1990), and Ahonen (1989, 1990) for the Keynesian revolution. Look to Blaug 1980 for an appraisal of Marxist economics, and Moseley 1995 for a criticism. For an analysis of the basis of microeconomics, see Latsis 1972 and Rosenberg 1976. On the decision theory, see Mongin 1988. With reference to the research programme known as Bayesian statistics, and its development around the core of subjectivist probability as described by de Finetti, see Mondadori 1989. On the wage fund doctrine, see Vint 1993. For an up-to-date general view of methodological estimates of economic theories, see Blaug 1994 and Backhouse 1994 and 1997, especially chaps. 8–13; see also Backhouse et al. 1997. Regarding research programmes in mathematics, see Hallet 1979; Giorello 1981, 1992; and Koetsier 1991, chaps. 5–8.

12. See Zahar 1973, Lakatos and Zahar 1976, and "Editor's Note: The Value of Novelty," this volume.

Nobody, therefore, has ever undertaken a critical history of Marxism with the help of better methodological and historiographical instruments. Nobody has ever tried to find an answer to questions like: were Trotsky's unorthodox predictions simply patching up a badly degenerating programme, or did they represent a creative development of Marx's programme? To answer similar questions, we would really need a detailed analysis which takes years of work. So I simply do not know the answer, even if I am very interested in it. There is lots of work to be done in this line, and I think it is not out of order to refer you to my papers for further details.

In the transcript of this lecture there follow a few lines of comment in which Lakatos announces that Feyerabend is working on a detailed criticism of MSRP, and adds that he thinks his students "will, after all these considerations, see that he is wrong."

Editor's Note

The Value of Novelty

The second chapter of *Proofs and Refutations* ends with a remark which neatly sums up the critical approach involved in Lakatos's sophisticated falsificationism: "If you have *serious* doubts about a dominant theory then scrap it, and replace it by another dominant theory."[1] This is also his recommendation for *methodologies* once they are tested against the history of science. From this point of view, the changes undergone by the concept of a *'novel' fact* in the passage from Popper to Lakatos (and after) are representative of the fruitful interaction between history of science and philosophy of science. Since Lakatos deals with it only very sketchily in the last lecture, the main points of this debate should be mentioned here.

Both Popper and Lakatos regard empirical support *not* as a simple two-place relation between theory and evidence, but rather as a three-place relation between *theory, evidence,* and *background knowledge.*

1. Lakatos 1976a, 125.

Hence, a theory is supported by a fact when this fact is correctly described by the theory and is not part of the background knowledge. According to Popper, 'background knowledge' consists of "all those things [we accept for the time being as] unproblematic."[2] The *time* element is thus relevant in view of any appraisal of the theory's empirical merits: any theory whose empirical consequences are *facts* belonging to background knowledge (that is to say, "known and accepted by science" at the moment the theory is being tested) will receive little or no confirmation from those facts; however, should those same facts result from the theory's predictions and not from the current background knowledge, then they would confirm the theory.

Popper's strictly *temporal* (or historical) approach to confirmation, however, fails to capture scientists' intuitions about empirical support in particular cases. It implies, for example, against any intuitive judgement, that the 'anomalous' motions of the planets (stations and retrogressions) which Copernicus boldly explained did not support the heliostatic theory, simply because those same facts were already known to the Ptolemaics long before the advent of the Copernican revolution. Similarly, and again contrary to scientists' "basic value judgements," Galileo's experiments on projectiles cannot be taken as evidence in favour of Newton's theory of gravitation. Moreover, we would also have to claim that Michelson's results, known and accepted since 1887, could not support either Lorentz's 1904 theory of corresponding states or Einstein's 1905 theory of special relativity. Finally, the explanation of Mercury's anomalous perihelion, a major critical point for astronomers *before* the common acceptance of the general relativity theory, and heralded by the scientific community as evidence in favour of Einstein's theory at least as strong as the discovery of (real) new facts such as the gravitational red shift, would not be credited in favour of Einstein simply because this fact was recorded long before general relativity was proposed.[3]

Several important episodes in the growth of knowledge suggest, therefore, that although new and unexpected facts may support a theory, *not all empirical results which support a theory are necessarily novel.* The temporal order of theory and evidence cannot therefore constitute in its own right the discriminating factor on which to base the distinction between genuine and spurious confirmations. To paraphrase Lakatos,

2. Popper 1963, 238.
3. See Worrall 1978a, 47, and Zahar 1989, 13–17.

"if the rationality of science is Popperian, actual science is not rational; if it is rational, it is not Popperian."[4]

On the contrary, the idea implicit in the methodology of scientific research programmes is that a fact may not be used twice, first in the construction of the theory and subsequently in support of it. This suggests the following definition of a 'novel' fact: *"A fact will be considered novel with respect to a given hypothesis if it was not used in the construction of the hypothesis."*[5] Thus any further evidence which was not involved in the process of "arriving at" the theory and which is therefore, in this specific sense, *independent* of the theory, supports it.

With reference to the famous controversy between supporters of Descartes and of Newton, we should note that facts predicted by Newton's theory and subsequently accommodated within the Cartesian programme did not provide the latter with any empirical support. This is so not because (as is suggested by Popper's view) these facts were *already known* before the subsequent developments of Cartesian theory which 'explained' them, but rather because these facts themselves were used by the Cartesians in the construction of their theory.[6]

The "neo-Lakatosian" approach to confirmation thus captures scientists' intuitions by considering, for example, why Copernicus's explanation of the retrogressions and stations of planets is better credited than Ptolemy's. Indeed, although these facts were previously known by Ptolemy, "they lend much more support to Copernicus than to Ptolemy within whose system they were dealt with only in an *ad hoc* manner, by parameter adjustments."[7] We may also understand why the explanation of the anomalous precession of Mercury's perihelion "gave crucial support" to Einstein's theory, even though, as a low-level empirical proposition, it had been known for almost a hundred years. The *known* facts regarding Mercury's orbit played no role in Einstein's original design: "Its exact solution . . . was an unintended by-product of Einstein's programme."[8]

4. Cf. Lakatos 1971a, 123–36.

5. Zahar 1989, 16. As is clear from lecture 8, Lakatos's initial development of MSRP was in terms of temporal novelty. It was Zahar's influence which led to the change from temporal novelty to heuristic novelty. This was accepted by Lakatos and developed further by Zahar and Worrall, and christened the "Zahar-Worrall view" (see also Worrall 1978a, 1985; Zahar 1973, 1983).

6. See Worrall 1978a, 45–51.

7. Lakatos and Zahar 1976, 185.

8. Lakatos and Zahar 1976, 185.

It follows that: (1) *Temporal novelty* is a sufficient but not necessary condition for confirmation: an old fact can be new with respect to a theory if the latter has been conceived independently of that specific fact. (2) *Background knowledge* should not be taken to denote the general unproblematic knowledge inherited at a given moment in time by the scientific community, but rather the whole set of *facts and parameters used in the construction of a given theory.* (3) Likewise, *evidence* is always *in support* of a certain theory which has been arrived at in a certain way (i.e., *evidence in support of a theory along-with-its-heuristics*). (4) Lastly, if the way the theory has been arrived at is relevant in appraising its worth, then the very notion of empirical support is *heuristics-dependent,* and the distinction held dear by Reichenbach and again taken up by Popper between "context of discovery" and "context of justification" should be thought over once more.[9]

9. With regard to a critical revision of the heuristic approach to confirmation in a Lakatosian framework, see Musgrave 1974; Gardner 1982; Gillies 1988; Murphy 1989. For a criticism of the notion of "novelty of facts" from a Bayesian perspective, see Dorling 1979, 1982 and Urbach 1989; on the problem of ad hoc hypotheses, see Redhead 1978; Howson and Urbach 1989, 110–12 (and for a criticism, Gillies 1990). The question of the 'novelty' of evidence is also the focal point of several case studies in various fields, ranging from the theory of continental drift (Frankel 1979 and Nunan 1984) to the Keynesian revolution (Hands 1985; Blaug 1976, 1991; Ahonen 1989).

TWO

THESES ON ANARCHISM

Paul Feyerabend

Feyerabend formulated these theses in rough draft for a conference planned for 20 March 1973, where he was to criticise Lakatos's defence of "Law and Order" from an anarchist point of view. The theses were enclosed with a letter to Lakatos dated February 1973; the letter reads: "Look what I have done for you: [. . .] I have written THESES ON ANARCHISM (enclosed), and I have also made an hour tape, complete with arguments and music in case my feet get bad again and I cannot come. [. . .] At any rate, you will be *threefold prepared* and should be able to make *mincemeat* of me (usually I am very bad when I prepare too much—like a boxer who overtrains before a fight). Has any opponent ever made your task that easy?" A slightly revised form of these theses has already appeared as Feyerabend 1975d, 176–81, and Feyerabend 1996, chapter 1.

Anarchism is opposed to the existing order, it wants to destroy this order, or escape from it. Political anarchists oppose political institutions, religious anarchists may oppose the entire physical world, they may regard it as a lower realm of being, and they may want to eliminate its influence on their lives. Both hold dogmatic opinions about what is true, good, valuable for man.

For example, post-Enlightenment political anarchism believes in science and in the natural reason of man. Remove all boundaries, and natural reason will find the right way. Remove educational methods, and man will educate himself. Remove political institutions, and he will form associations that express his natural tendencies and may thus become part of a harmonious (un-'alienated') life.

The faith in science is partially justified by the revolutionary role which science played in the seventeenth and eighteenth centuries. While the anarchists *preached* destruction, the scientists *smashed* the harmonious cosmos of earlier ages, they eliminated fruitless 'knowledge', changed social relations and slowly assembled the elements of a new type of knowledge that was both true and beneficial for man.

Today this naive and childlike acceptance of science (which is found even among 'progressive' leftists such as Althusser) is endangered by two developments, viz. (1) by the change of science from philosophical inquiry to a business enterprise, and (2) by certain discoveries concerning the status of scientific facts and theories.

Twentieth-century science has given up all philosophical pretensions and has become big business. It no longer threatens society, it is one of its most powerful supporters. Humanitarian considerations are at a minimum, and so is any form of progressiveness that goes beyond local improvements. Good payment, good relations with the boss and colleagues in their unit are the chief aims of these human ants who excel in the solution of tiny problems but who cannot make sense of anything transcending their domain of competence. Let someone make a great step forward—and the profession is bound to turn it into a club for beating people into submission.

We have also discovered that science has no solid results, that its theories as well as its factual statements are *hypotheses* which often are not just locally incorrect but entirely false, making assertions about things that never existed. According to this view, which was introduced by John Stuart Mill *(On Liberty)* and whose most vociferous contemporary propagandists are Karl Popper and Helmut Spinner, science is a collection of competing alternatives. The 'accepted' view is the view that has a temporary advantage, either because of some quirk, or because of some real merits. There are no revolutions that leave no stone unturned, no principle unchanged, no fact untouched.

Unpleasant in appearance, untrustworthy in its results, science has ceased to be an ally for the anarchist. It has become a problem. Epistemological anarchism solves the problem by removing the dogmatic elements of earlier forms of anarchism.

Epistemological anarchism differs both from scepticism, and from political (religious) anarchism. While the sceptic either regards every view as equally good, or equally bad, or desists from making such judgements altogether, the epistemological anarchist has no compunction in defending the most trite, or the most outrageous statement. While the political anarchist wants to remove a certain form of life, the epistemological anarchist may want to defend it, for he has no everlasting loyalty to, and no everlasting aversion against, any institution and any ideology. Like the Dadaist (whom he resembles in many respects) he 'not only has no programme, he is against all programmes' (Hans Richter, *Dada: Art and Anti Art*—an excellent textbook for a dadaistic science), though he will on occasions be the most vociferous defender of the *status quo,* or of its opponents: "To be a true Dadaist, one must also be an Anti-Dadaist." His aims remain stable, or change, as a result of argument, or of boredom, or of a conversion experience, or because he wants to impress some people, and so on. Given some aim, he may try to reach it with the help of organised groups, or alone. He may appeal to reason, or to emotion. He may decide to proceed violently, or in a peaceful manner. His favourite pastime is to confuse rationalists by inventing compelling reasons for unreasonable doctrines. There is no view, however 'absurd' or 'immoral', he refuses to consider or to act upon, and no method he regards as indispensable. The only thing he opposes positively and absolutely are universal standards, universal laws, universal ideas such as 'Truth', 'Justice', 'Honesty', 'Reason' and the behaviour they engender, though he does not deny that it is often good policy to act as if such laws (such standards, such ideas) existed and as if he believed in them. He may approach the religious anarchist in his opposition to science, common sense, and the material world that is examined by both; he may outdo any Nobel prize winner in his vigorous defence of scientific purity. Behind all this outrage lies his conviction that man will cease to be a slave and gain a dignity that is more than an exercise in cautious conformism, only when he becomes capable of stepping outside the most fundamental convictions, including those convictions which allegedly make him human. "The realisation that reason and anti-reason, sense and nonsense, design and chance, consciousness and unconsciousness [and, I would add, humanitarianism and anti-humanitarianism] belong together as a necessary part of the whole—this was the central message of Dada," writes Hans Richter. The epistemological anarchist might agree, though he would not express himself in such a constipated manner.

Having *stated* his doctrine, the epistemological anarchist may try to sell it (alternatively, he may keep it to himself, considering that even the most beautiful ideas get worn and shabby when they start circulating). His methods of selling depend on the audience. If he faces an audience of scientists and philosophers of science he will produce sequences of statements liable to convince them that the things in science they appreciate most have been brought about in an anarchistic manner. Using propagandistic moves most likely to succeed with this type of audience, i.e., using argument, he will demonstrate from history that there is not a single methodological rule that does not occasionally inhibit science and not a single 'irrational' move that may not further it, given the right circumstances. People and nature are very whimsical entities which cannot be conquered and understood if one decides to restrict oneself in advance. He will rely heavily on anarchistic utterances by revered scientists such as the following utterance by Einstein: "The external conditions which are set for [the scientist] by the facts of experience do not permit him to let himself be too much restricted, in the construction of his conceptual world, by the adherence to an epistemological system. He, therefore, must appear to the systematic epistemologist as a type of unscrupulous opportunist . . . ." Using all these bits and pieces of propaganda to maximum effect, he will try to convince his audience that the only *universal* rule that can safely be in agreement with the moves the scientist must make to advance his subject is *anything goes.*

Imre Lakatos disagrees. He admits that the existing methodologies clash with scientific practice, but he believes that there are standards which are liberal enough to permit science and yet substantial enough to let reason survive. The standards apply to research programmes, not to individual theories; they judge the evolution of a programme over a period of time, not its shape at a particular time; and they judge this evolution in comparison with the evolution of rivals, not by itself. A research programme is called 'progressive' if it makes predictions that are confirmed by subsequent research and thus lead to the discovery of new facts. It is called 'degenerating' if it makes no such predictions but is reduced to absorbing the material discovered with the help of its rival. The standards judge research programmes, they do not advise the scientist what to do. For example, there is no rule that tells the scientist to remove a degenerating programme—and rightly so, for a degenerating programme may recover and come out on top. (Such developments occurred in the case of atomism, the temporal finitude of the world, the moving earth. All these research programmes advanced and degenerated numerous times, and all of them are now a solid part of science.) It is 'rational' to pursue a research programme on its degenerating branch even after it has been overtaken by its rival. There is therefore no 'rational' difference between the methodology of Lakatos and the 'anything goes' of the anarchist. But there is considerable difference in *rhetorics.*

For example, Imre Lakatos criticises research programmes which are in their degenerating phase and he demands that support be withdrawn from them. His standards *permit* the criticism, and they *permit* the action. However, they do not *encourage* it, for they also permit the opposite: they permit us to praise such programmes and to support them with everything at our disposal. Lakatos often calls such praise 'irrational'. Doing this he uses standards different from his own, for example he uses commonsense standards. Combining commonsense standards (which are independent of his standards) with the methodology of research programmes, he utilises the intuitive plausibility of the former to support the latter and to smuggle the anarchism into the brain of the most dedicated rationalist. In this he is much more efficient than I, for rationalists are constitutionally unable to accept anarchism when it is offered to them undisguised. One day, of course, they will discover that they have been had. This will be the day when they are ready for anarchism, pure and simple.

Nor has Lakatos succeeded in showing 'rational change' where Kuhn, according to him, has recourse to 'mob psychology'. Revolutions lead to quarrels between opposing schools. The one school wants to *abandon* the orthodox programme, the other school wants to *retain* it. The standards recommended by the methodology of research programmes permit either move, as we have seen. Hence, the fight between the opposing schools is a power struggle, pure and simple. Kuhn, as described by Lakatos, is right after all.

Finally, Lakatos has not shown that Aristotelian science, magic, witchcraft are inferior to modern science. Criticising Aristotelian science (and other 'pseudo' subjects) Lakatos uses his standards. How does he obtain his standards? He obtains them via a rational reconstruction of modern science 'of the last two centuries'. Measuring Aristotelian science by his standards therefore means comparing Aristotelian science with modern science 'of the last two centuries'. The comparison leads to a condemnation only if it has been shown that modern science is better than Aristotelian science, that is, if it has been shown (a) that it has better aims, and (b) that it reaches its aims more efficiently than its rival. Nowhere has Lakatos shown that the aims of modern science (progress with the help of 'anticipations of the mind') are better than the aims of Aristotelian science (absorption of facts into a stable body of basic theory; 'saving' of phenomena), and that they are reached more efficiently. Hence, taking Lakatos as our guide, the case of science vs. witchcraft (for example) is still entirely open.

Conclusion: neither science nor the methodology of research programmes provide arguments against anarchism. Neither Lakatos nor anybody else has shown that science is better than witchcraft and that science proceeds in a rational way. Taste, not argument, guides our choice of science; taste, not argument, makes us carry out certain moves within science (which does not mean that decisions on the basis of taste are not surrounded by and entirely covered by arguments, just as a tasty piece of meat may be surrounded and entirely covered by flies). There is no reason to be depressed by this result. Science, after all, is our creature, not our sovereign; *ergo,* it should be the slave of our whims, and not the tyrant of our wishes.

THREE

The Lakatos-Feyerabend Correspondence (1968–1974)

Introduction

In the Preface to *Against Method,* Feyerabend explains the origin and the scope of his and Lakatos's joint project in the following terms:

> I was to attack the rationalist position, Imre was to defend it, making mincemeat of me in the process. Taken together, the two parts were supposed to give an account of our long debate concerning the matters that had started in 1964, had continued, in letters, lectures, telephone calls, papers, almost to the last day of Imre's life and had become a natural part of my daily routine. The origin explains the style of the essay: *it is a long and rather personal letter to Imre and every wicked phrase it contains was written in anticipation of an even more wicked reply from the recipient.* It is also clear that as it stands the book is sadly incomplete. It lacks the most important part: *the reply of the person to which it is addressed.* I still publish it as a testimony to the strong and exhilarating influence Imre Lakatos has had on all of us. (1975a, 15)

The letters chosen have been published here for the first time with the intention of filling that gap. They do not, however, make up the whole correspondence. Feyerabend recalls that

> Imre and I exchanged many letters about our affairs, ailments, aggravations and most of all the recent idiocies of our colleagues. [. . .] Cambridge University Press wanted to publish our letters, but could not: as usual I had thrown away Imre's part of the correspondence. Only a few postcards survived as bookmarks, or to cover holes in the walls of my house. (1995, 130)

Lakatos, however, kept and filed most of the letters he received and a copy of his outgoing responses. It is therefore thanks to him that the correspondence survived. (It is deposited in the Archive together with other interesting letters such as those he exchanged with George Pólya, Victor Kraft, Rudolf Carnap, Richard Popkin, Paul Bernays, Alan Musgrave, Adolf Grünbaum, Thomas Kuhn, and Karl Popper.)

I have selected those letters which appeared most relevant to the debate surrounding the ideas of *Against Method* and those most useful in suggesting Lakatos's possible response "For Method." Also included are letters dealing specifically with the student revolt of 1968 and important international political affairs of the time, since they reflect the strong (and often contrasting) opinions that the authors held with regard to social and political matters. Finally, special attention was given to those letters in which the authors comment on books, papers, conferences, and works of general and specific interest. There is necessarily a subjective element involved in this choice.

By and large, the letters appear in their original form and in full. Square brackets signal cuts and omissions that I have deemed necessary in reference to strictly private matters, or simply in order to avoid repetition and improve the readability of the text. My editorial notes provide the main references to books, papers, people, friends, colleagues, scholars, politicians, and events to which the letters refer.

The reader should not expect a reply to every letter. The exchange is in fact often interrupted at times when the two interlocutors had the chance to meet regularly and to continue their discussion face to face. I would therefore suggest looking at these letters as a patchwork, rather than a continuously woven, fully articulated, and coherent tapestry.

1968

17 December 1967[1]
University of California, Berkeley[2]

Dear Imre,

It is settled: the best medicine for my present anti-theoretical mood is history; and so I have decided, as you have advised me to do, to start work on the two *historical* books I am supposed to write, viz. history of philosophy of nature and history of 20th century physics. However, the *philosophical* attitude behind this change of interest will come out in my Kuhn paper[3] which I shall *completely rewrite.** The new version of which I already have an outline will criticise *all* theoreticians of science, and by "all" I mean of course only those who count, viz. Popper, Kuhn, Lakatos (the anti-Kuhn Lakatos[4] and not the anti-Carnap + Karl [Popper] Lakatos[5]). So, prepare yourself for a strong criticism of your Kuhn paper, and please send me your Kuhn paper as soon as you can (you will of course get my paper so that you can have the last word and annihilate me).

My "position" now is thoroughly Popperian: Karl always starts his lectures on scientific method (he did so in 1952, when I listened to him in London; he did so again in 1962 in Berkeley) with the remark that "there is no scientific method."[6] It is *this* remark which I accept—and

I reject everything which comes afterwards. The only theoretical restriction (or "definition") of science which I am prepared to tolerate is what follows from a principle of *general hedonism:* all those elements of science which are inconsistent with hedonism *must go* (which, of course, does not mean that people will be forbidden to be masochists; only that they should exercise their masochism *privately* and not advertise it as a principle of truth, or of professional integrity, thus misleading themselves and everyone else; they can even be sadists; but again they should choose their friends not by misleading propaganda—"you are now going to do the most important thing that man has invented," but honestly (not in the "professional" sense): "I am a sadist; you are a masochist; so let us have some fun together"). Here Karl has stopped midway between Plato and his own Open Society.

Essentially my paper will contain the criticism of Kuhn already in existence (with historical material added from the period of the early quantum theory, that is of Planck and after, until 1905) together with the material of my lectures (which you know) and some not yet existing criticism of your approach ("research programmes" etc.). On the whole the tenor will be: "A plague on all your houses." *This I must say* (this is *my* Carnap paper), and then I can go on to something else such as case studies and aesthetics.

In some way you are much better off than I and I envy you for it. You believe in something such as the truth, you have some ideas how to reach it; and apart from minor fluctuations (such as the Carnap paper)[7] you proceed steadily. Right now, I am lost, and it is largely your doing. Listening to your lectures on Popperianism meant the end of my dogmatic slumber. *Now* I am awake, and at first I don't see anything at all *except* that the reality I now imagine I am seeing has nothing whatever to do with the preceding Popperian dreams. It is of course an advantage to have been a Popperian. For the other philosophies are not only inadequate, they are also quite definitely sickening (having returned to Berkeley I just got a little inkling of how bad philosophy really can be). Still, I must admit that I am somehow lost and the best one can get in such a situation is criticism. So, can you let me have some criticism of what you heard in my lecture? What did you make of it? Was there anything in it? etc. etc.

Apropos lectures. I think I was a little impolite not to talk to the lady—the blond Helena—who was supposed to be my assistant. Can you point out to her that this was not due to rudeness but simply to the inconveniences caused by a very strong attraction (inconveniences: because she is married). And I would really like to know what *she*

Dear Paul,
Two nights ago I had terrible pains for 3
hours. Gillian got me a doctor who found me
unconscious in the bathroom, where I fainted and
~~I~~ knocked my forehead into two, and now at
least I know that ^the^ two hemispheres, contrary to
recent journalism, are equal. ~~I was~~ taken in
an ambulance into hospital with a (heartattack)
('cardiac arrest').
In these modern hospitals one is first seen
by a sociologist & a psychologist. They decided that
I am completely Anglicised & ~~the~~ ^my^ heart has
~~I~~ therefore vanished. Then another chap took
over who was an industrial psychologist. He
discovered that for 20 years I ~~have~~ ^had^ no
time to breathe, therefore no lung. (I forgot
to say that their second hypothesis was pleurisy.)
Now I am here as a Great Lump of
^= muscular^ Myalgic Pain and I scream all the time. I
asked for Bleuler's & Freud's early works
to read but I only got Trollope.

Houghton Street,
London, WC2A 2AE
Telephone: 01-405 7686

The London School of Economics and Political Science
(University of London)

Figure 3.1a Imre's last letter to Paul.

Didn't say that when I fainted Gillian
fainted too & so she too was taken ~~[illegible]~~ to
the hospital with the ambulance. But she was
let out as unworthy of attention. Walked home. (NO TAXI)
I am now waiting for a Top Conference
to decide whether a patient who, although
clearly dying, terrorizes the ward, should
be discharged for Reasons Unknown, or whether
they should put me under glass & Observation (!)
and [illegible] *
Little do they know that this is a free country
and I shall leave tomorrow on foot. But
tonight I want to see three more chaps die.
Very funny how people come in & die.
All for now. I don't know whether
Wendy has typed out my long letter to ye.
I instructed Greg that in case of my death your
MS should be sent to Hannay with a POSTCRI
FROM THE THIRD WORLD BY IMRE LAKAT
I shall send it from there.
This is my first letter since I
fainted away. You cannot complain. However, don'
spread the news because should I recover I do not
want a Sick IMAGE.
Tell John Searle that I am asking for
a salary rise since I now stop poisoning the
[illegible] All my lectures have been cancellec
[illegible]

BERKELEY: DEPARTMENT OF PH

30.7.

Dear Imre,

I am now slowly starting to rewrite AM, and by this I mean sections 2, 31, 32. Everything else will remain the same (except for some stylistic changes). In sections 2, 31, 32 I shall change the arrangement of the assertions made, not the assertions themselves, and I am also going to do some cutting down. My criticism of you is twofold. First, that your standards do not exclude anything. Secondly that if this problem could be fixed, they would be too restrictive. The standards contain a reference to content and increase of content is a virtue. But (and this is a historical assertion) theories which succeed each other in some domain are hardly ever in the relation

but much more often in the relation

T T′

and the second case occures especially when the successor is the result of an exciting revolution (example: Galileo, Einstein, as described in my part of AM). And (this is an assertion concerning standards) if we want our theories to be interesting, coherent, simple etc. etc. we shall have to give up on increase of content. Secondly, there is the problem of incommensurability. This is all I have to say about you. I also say that you are the last stronghold of mechanical rationalism, so if you fall, mechanical rationalism falls to be replaced by dialectical rationalism, or anarchism (this bit about mechanical rationalism vs. dialectical rationalism will be included in the new version). I am sorry to be constantly running away from you in this way, but our book should really be good.

In two weeks another girl, Janet, is going to turn up at your door with a letter from me in her hand. She is entirely different from Joan, takes Quine seriously etc. etc.

As a matter of fact, I am constantly debating with you, as Bohr (= J) with Einstein (= you). And Bohr also constantly reformulates his views.

UNIVERSITY OF CALIFORNIA (Letterhead for Interdepartmental Use)

Figure 3.1b An original typewritten letter.

thought about the whole jazz: about content, manner of presentation etc., of the lectures. I ask her this not just because I want to start a flirtation (which I do not, considering the circumstances) but because I have a bold hypothesis that she is intelligent and, moreover, that she will say what she thinks. And in my present state (see above) the more I hear about what I said, the better.

Part of my lectures, by the way, is written up in a not too efficient way in my paper "Counterinduction" which you have.[8] My other paper "Classical Empiricism"[9] contains a detailed analysis of Newton and may be of interest for your anti-Kuhn paper.

Here it was stormy and cold—trees have been uprooted, my fence is blown to pieces, and the last part of the flight, from LA to SF was a veritable nightmare. The nightmare continues when one enters the dept. of philosophy and sees the interesting topics taught by such nuts [. . .]. Popper may be wrong, but he is still the best of all. And those grumbling idiots at LSE don't know how well off they are.

So—and now, four months of work and solitude. I am a little afraid it will be hard for me—but we shall see.

Be well, and all the best.

Paul

*P.S. My paper, of course, will *no longer be an anti-Kuhn paper,* but it will be *an independent piece* that will touch on Kuhn only *as a side issue* (though it will contain almost all the material from my original Kuhn paper). There will be a longish utilisation of Havas on relativity (eternal thanks to you for this paper!)[10] and the title will be "Against Method" (this in analogy to Susan Sontag's *Against Interpretation*). If you could lure Karl [Popper] into commenting upon it I would be eternally grateful. (This is pure curiosity on my part). But I wonder how *you* will get out of my onslaught. I haven't got a clue. Have you?

All the best again. (There is a hailstorm here now and the roof is leaking.)

1. Feyerabend was not in the habit of putting dates on his letters. The dates found in the top margin were reconstructed and penned in by Lakatos himself, who kept his files tidy in this way.

2. Feyerabend became professor of philosophy at the University of California at Berkeley in 1958. In *Killing Time,* he recalls that "comments such as 'I have to leave Berkeley'; 'I can't stay here any longer'; 'this is a cultural desert—I am suffocating' occur on almost every page [of my notebooks], especially after visits to London. Very, very slowly I adapted to the Californian way of life. Indeed, I might still be in California and

might have been buried, burned, or eaten there had I not been chased away by the October 1989 earthquake" (1995, 112–13).

3. Feyerabend is referring to his 1970c.
4. Feyerabend is referring to Lakatos 1968b.
5. Feyerabend is referring to Lakatos 1968a.
6. Cf. Popper ([1956] 1983): "As a rule, I begin my lectures on Scientific Method by telling my students that scientific method does not exist. . . . (1) There is no method of discovering a scientific theory. (2) . . . There is no method of verification. (3) There is no method of ascertaining whether a hypothesis is 'probable'." (5–6)
7. See Lakatos 1968a.
8. Feyerabend is probably referring to his 1962a.
9. Feyerabend 1970b.
10. See Havas 1964.

3 January 1968
London School of Economics

My dear Paul,

Many thanks for your Christmas card, charming letters etc. etc. I had a lonely Christmas which I spent with the flu and nothing else. Now that I have sent off the Carnap-Popper paper, my resistance has collapsed and I feel worse than ever. But the doctor says there is nothing wrong with me and within a fortnight, if I don't work, I shall be healthy and merry. However, he suggested that (a) I should never work too hard and (b) I should sack all my present set of girls and adopt a more relaxing set.

Incidentally, my relations with Karl will have a dangerous turn in the coming few weeks (as from report).

All for now. I miss you, but now I shall read your papers instead of enjoying your company.

Gillian, Eva, John and Alan send their love. So do I.

Till soon,

Imre

28 January 1968
University of California, Berkeley

Dear Imre,

A letter from you always cheers me up and it does so especially now when my work proceeds at snail's pace. I write, I write again, but nothing reasonable comes out of my pen. Lecturing at LSE was so simple (comparatively, that is, though I was constantly in a state of fear of disaster), but now that I want to put the damn thing onto paper as an addition to the Kuhn paper, I do not seem to be able to know how to start. But first things first: many thanks for your remarks on, and your criticism of, my Bohr eulogy.[11] They have been most helpful and I shall change the presentation in various places, trying to make it clearer and to take into account your objections. My point is as follows: Karl may score against a few bum-physicists who have also written on philosophical matters, but he has not scored a single point against Bohr. Nowhere does Bohr *commit* the mistakes which Karl criticises, he knows that one could commit them and warns against them, and the ghost which Karl wants to exorcise is *nowhere* to be found in his writing. Quite the contrary: Bohr can almost be regarded as a materialist (as Fock[12] has pointed out in his defence of him against his Russian opponents, in the *Allunionskonferenz* whose Proceedings you have in your office). Moreover, the propensity interpretation is *part* of the idea of complementarity, so that Bohr has anticipated Karl. He has also examined the consequences of such an interpretation much more carefully than Karl has. So, if we take Bohr as the opponent, then Karl neither exceeds him (he has nothing new to say—except erroneous things), nor has he found any critical point. This is, I think, all I want to say.[13]

The Copenhagen interpretation, however, is an altogether different matter. It is a mixed bag, consisting of the errors and misunderstandings and the superficialities of many people, and Bohr has always had a hard time putting things right. Hence, putting your hand into this bag you may come up with almost anything you want. Now I would say, first of all, that criticism of this kind is too comfortable and should not be attempted. Secondly, even here the "grave logical errors" etc. are entirely absent (I guess Karl got them from Landé who invented them in order to have something to rave against). Of course, one can criticise collective views—but then this would have to be done in a very different manner, and paying due attention to the positive points also. But I am not really interested in the Copenhagen interpretation—all I am interested in is Bohr and here Karl has not said a single enlightening thing. Others too have misrepresented Bohr, as a matter of fact almost all physicists misrepresent him, but I would think that a philosopher should go a little further. This is the background of my paper. I see it has not come out too clearly, so I have had to add a word here and there. And lots of Bohr quotations—which, incidentally, are very clear, *crystal clear.* The idea of Bohr's obscurity is nothing but a myth created, among other things, by Einstein's unwillingness to pay attention to him: it is Einstein whom Karl leans on when he says that he has failed to understand complementarity, "just as Einstein has." This is not worthy of a philosopher. Einstein does not comprehend; is this reason for Karl to be proud of his own incomprehension? But I wander—all these things will be added, very briefly, to my paper which you will see before it goes out into the world, to *Philosophy of Science,* that is (they have accepted and even praised it—I don't know who the referee was).

[. . .] Your final suspicion, viz. that "the interpretation [I] put on Bohr's complementarity principle [is] an arbitrarily selected and improved, consistent subset of his inconsistent position" will, I hope, be mollified by a few more quotations which I shall add. I don't think that Bohr's position is inconsistent (if we have reasonable standards of consistency, that is). I don't think that I have added anything to it—all I have done is to approach it in a somewhat sympathetic spirit and if one does that one will be surprised at every turn by the richness of his ideas. Just one example. Bohr says that all our measurements must finally be related to *classical concepts* because it is in terms of classical concepts that we see the world. Now this entails that only magnitudes such as position, or momentum, or angular momentum which have a classical analogue can be measured independently. There is no possibility of an independent measurement of a *spin* which has no classical analogue. This was rejected by most physicists—but Bohr succeeded in refuting all the measuring arrangements they dreamt up (for this cf. the Proceedings of the VIth Solvay Conference, Pauli's paper—217ff—as well as the final discussion—267ff—). This is a property of almost all "philosophical" statements made by Bohr. At first sight they seem to be very general and obviously incorrect, but then they have concrete applications in places where no one would have suspected them. This combination between abstract speculation and concrete physics is the most attractive feature of Bohr's philosophy. But it is also much more realistic than for example Einstein's. Our ideas are "free creations" etc., says Einstein and, of course, Karl (by the way: it seems to me now that the best way to criticise Karl would be to criticise Einstein. When Einstein falls, then Karl falls with him. I am working on that). But this is just empty talk; a child that has been kept in darkness and not taught a language invents nothing. And if it is capable of invention, then his ability is certainly restricted by the structure of his brain and especially by the structure of the brain *stem* which contains the results of millions of years of evolution. Is it asserted that these results can be "freely changed"? This is certainly not the case. Some changes are not possible without mutations and certain basic structures of our perceptions are of exactly this kind. Now, it seems to me that Bohr, who makes a bold hypothesis concerning the nature of these structures and who tries to adapt his physics to it, is a much more profound philosopher than Einstein, who simply talks about the free creativity of the mind as if no limits existed for it. Bohr may of course be wrong. But his hypothesis, being more detailed and also more highly refutable (it will be refuted by the first theory that fundamentally changes our perception of space and time) is the more interesting conjecture.

Now: as regards the papers in *Criticism and the Growth of Knowledge.* I shall leave my latest anti-Parmenidean paper[14] pretty much unchanged (unless you think drastic changes are indicated) except that I shall add some details, especially in connection with the general theory of relativity and with the final sections on statistics. I shall, however, add about 20 pages to the anti-Kuhn paper. The structure of the paper, as I intend to write it, will be roughly as follows. I shall start with some remarks concerning the need to criticise the idea that our knowledge changes by conjectures and refutations and that it approaches the truth. I shall point out that this can be shown *historically,* that is, it can be shown historically that the idea gives a wrong description of the actual development of science; and that it would also have given the wrong *advice* to those bringing about certain changes of our scientific knowledge (Karl as Galileo's house philosopher would have been a disaster). I then say that history does not easily give us the critical material we need, as the historians have just escaped the clutches of inductivism and are now doing research confirming the Popperian account. The only exception is *Kuhn*—and then comes my Kuhn paper, first still criticising him from a Popperian point of view, as I do in the paper which you have, but then turning to his positive side: incommensurability; gains always accompanied by losses etc.—and here I shall start the new section against critical rationalism, with a few historical examples. The end is supposed to be a criticism of Einstein's point of view and his confrontation with Galileo. Galileo the greater humanitarian, Galilean science being more entertaining, more interesting, more artistic. I would be most grateful if you could let me have your views on these two papers which, after all, you are going to edit. I hope to send you the new Kuhn (which will be much less anti) in about a month. But any comments you have to make at any time are very helpful. A simple line from you often suffices to put things into perspective for me.

Three more months—and I shall be back in England. However, it seems that I shall have hardly any time for LSE on account of having to go back and forth between London and Berlin (unless of course I decide to give up Berlin altogether). [. . .]

Best wishes for your Russian *Proofs and Refutations.*[15] Your return to case studies in the spirit of critical materialism should now be imminent, n'est-ce pas?

I have resumed my translation of Galileo's tract on the sphere. It should be finished shortly. Don't buy Sacrobosco—Galileo is much better.

Give my best regards to Hintikka and tell him that formalising girls is a hopeless enterprise. Even informally they hardly satisfy the simplest conditions of adequacy.

[. . .] All the best

Paul

11. See Feyerabend 1968–1969.

12. See Fock 1962, 192, 210.

13. For Popper's criticism of Bohr, see Popper 1963, chap. 3, and 1967. On Feyerabend's judgement of Popper's ideas on quantum physics, see Feyerabend 1968–1969. "'I am not going to read your diatribe!' Popper had shouted when he saw my comments on his diatribe against Bohr. (He calmed down when I told him that many people had complained about my aggressive style and had ascribed it to his influence. 'Is that so?' he said, smiling, and walked away.)" (Feyerabend 1995, 145–46)

14. See Feyerabend 1970e.

15. In 1966 "Proofs and Refutations" (in its 1963–1964 *BJPS* version) was translated into Russian as *Dokatatelstva i Oprovershenia.* Lakatos himself was very proud of it for the success it had met.

16 February 1968
London School of Economics

Dear Paul,

Many thanks for your wonderful letters. I am looking forward to your *Galilean Studies* and your final MS on Bohr rehabilitated.

Karl, incidentally, had a bad knee which kept him in bed for 4 weeks, now he is OK. He still has not seen my Carnap paper: he will only see it in print. He is now working on quantum mechanics and may publish two papers in some physics journal.

I am now reading some mathematics to keep my mind fresh. But in a couple of weeks I'll return to my Kuhn paper, polish it up a bit and send it to you for annihilation (I shall leave the last word to you).

The financial situation at LSE is catastrophical; but I am trying to get some money for Ernest Adams. I already wrote to him.

Joske [Agassi][16] sent a copy of his *Science in Flux*[17] to everybody, but not to me. However, I read it, and did not find a single interesting line. He repeats himself, he tells in detail about his (alleged) intellectual autobiography, he announces priority claims—he is becoming a caricature of Karl: all the worst of Karl, none of his greatness. I hope that this characterisation will apply only to this particular paper. But Judith Thomson's comments on Agassi show that even Agassi *at his worst* is better than Harvard-MIT average at their best.

I liked your "Counterinduction" very much. Could I have the exact reference? I may have to quote it.

My students, girlfriends etc. keep asking when is the wonderful Paul coming back? Please let me know the date. In April I shall be off for two weeks, I hope to see you before my holiday.

Much love

Imre

16. Joske (Joseph Agassi) was a member of Popper's group of students and was Popper's assistant at LSE in the sixties. Agassi and Feyerabend became "more or less" friends notwithstanding some doubts on Agassi's part: "[Joske] did not trust a former Nazi officer" (Feyerabend 1995, 96).

17. Agassi 1975.

20 February 1968
University of California, Berkeley

Dear Imre,

How lucky you are! You have assistants (*attractive* assistants!), secretaries working for you and you can discuss every new idea that occurs to you with a whole seminar. I have neither the one, nor the other, I hardly get out of the house, for I am now rewriting all my papers. I have started on the quantum paper[18] and on "In Defence of Classical Physics."[19] Why did I ever write these bloody papers! [. . .] The only relaxation I have now—and I need it, believe me—is cooking: I spend hours in the kitchen, preparing a new dish, and then I move up the stairs again to the typewriter to throw away the last ten pages I have written and to start all over again. And I still have to write my papers on general philosophy of science (the two old men supporting each other, as you expressed it so well)—and Galileo is so much more interesting, for my quantum paper is really not so much *argument,* but just putting the record straight.

As I told you they want me in Yale and they invited me to go there to talk to them. I declined, having too much work to do. So the chairman of Yale is now coming here to talk to me. I had also to decline participation to a small conference in Hawaii—all expenses paid. All Karl's fault! If I have not finished my papers by the end of April, I think I shall cancel London plus Berlin and stay here.

So, having got this off my chest, and hoping that *you* at least are well (and please send me just a line) I remain, miserably,

Paul

18. See Feyerabend 1961b.
19. See Feyerabend 1970e.

2 March 1968
London School of Economics

Dear Paul,

How I envy you! In the marvellous privacy of 6041 Harwood Avenue you can think, read, write, cook, watch television without being harassed by students, administration, committees etc. etc. When I get an idea, I have to get to the LSE and when I get back, the idea has disappeared.

Anyway, let me speak to you solemnly as *Editor.* I want you to deliver the *final version* of "In Defence of Classical Physics" on April 29th. It must not be more than 100 printed pages. At the same time, you will receive Karl's final version which will differ only in long footnotes concerning Greek texts from the original, so I don't think you will want to modify your text. But if you wish to do so, you will have 1 week.

On May 5th you will get my Kuhn paper (final title: *Rationality versus Paul*). You will have five weeks to finish your contribution. But I can tell you a secret. I think that your ideas about incommensurability are either very important or they are not; it seems to me, however, that this may be the crux of many matters (incidentally, Hattiangadi, in Princeton, works on incommensurability, Kuhn, Feyerabend as Ph.D.

thesis).[20] So, *hic Rhodus, hic salta:* elaborate it as sharply as you can for the Kuhn volume; whatever my paper contains it will neither anticipate nor criticise your thesis.

I have to confess that I have only now read your paper on the Bunge volume[21] to check whether you anticipated me in my views on Karl's paper "Three Views."[22] You bastard, you did! It is a *marvellous* paper. I talked about it for an hour and have now made it compulsory reading after Karl's "Three Views." But now you say that you elaborated your points further in your Bohr paper. You must send it to me immediately, I shall quote it at length in my paper. Also I *urgently* need the final version of your "Classical Empiricism"[23] for similar reasons (where will it be published?).

In a rush, with much love,

Imre

20. Starting with his "Explanation, Reduction, and Empiricism" (1962a), Feyerabend developed his thesis of *variance of meaning,* which held that two theories T and T′ may contain homophonous terms, but in more than a few cases where a transition is made from one to the other, what happens is "a *replacement* of the ontology (and perhaps even of the formalism) of T by the ontology (and formalism) of T′, and a corresponding change of the meanings of the descriptive elements of the formalism of T." Along with this, one must take into account both the argument that observation is theory-laden (which states that one does not encounter brute facts, but facts which are mediated both linguistically and theoretically), and that of the relativity of methodological standards, which can change from one programme to another. While similar ideas were being put forward by Kuhn ([1962] 1970), Feyerabend was responsible for turning the *incommensurability* between theories, programmes, or paradigms into both the pivotal point for his criticism of Lakatos (see especially his 1975a) and the starting point for a more general cultural relativism (cf. his 1980a).

21. Feyerabend 1964c.

22. See Popper 1963, chap. 3.

23. Lakatos refers to Feyerabend 1970b.

10 March 1968
University of California, Berkeley

Dear Imre,

No reason for envying my "marvellous privacy" as you call it—the very same is *at your fingertips.* All you need to do is disconnect your telephone, lock your door and show enough self-discipline not to run to the LSE every day. I have not been on campus for over three weeks now, I have not read my mail, even the mailbox in front of my house gets voided only once a week (this is why I found your letter rather late). And I do the very same thing when I am teaching. I was disturbed only once, by a physics student who had to write a paper on the quantum theory of measurement and told me about the "most recent" research. Well—nothing has changed since 1957 when I left the field, so one does not even read the *journals.* What I *did* read, however, as the result of your letter, was my "Realism and Instrumentalism"[24] and I agree with you it is a rather good paper. What I am adding now to the ideas in there are only certain historical items. For example, on page 98 line 5 of *Conjectures and Refutations* (this is the *Three Views* paper) Karl quotes Osiander as saying: "There is no need for these hypotheses to be true, or even to be at all like the truth; rather, one thing is sufficient for them—that they should yield calculations which agree with the observations," and from the context of Karl's paper it follows that this is supposed to be a kind of *philosophical* attitude which tries to debunk a beautiful theory (Copernicus). *But Karl stopped quoting Osiander in the middle of a sentence,* for Osiander immediately continues, *even without a period:* "except someone should be so ignorant in geometry and optics as to believe that the celestial path of Venus be like the truth and the cause of its motion in the neighbourhood of the sun [. . .] For who does not see that this assumption necessarily implies that the diameter of this planet when close to earth must be more than four times as large and its body more than sixteen times as large as when it is in its most remote position *and this is contradicted by the experience of all ages?*" What Osiander means is that the actual variation of the brightness of Venus is not as it should be according to Copernicus—*hence* Copernicus must not be assumed to be true or like the truth (it is this discrepancy that Galileo alludes to in praising Copernicus for having trusted reason more than the senses). As I said, this reason of Osiander's follows immediately after the passage quoted by Popper and is not separated from it by a period, neither in Mezzner's German edition nor in the original Latin. By omitting it, Karl has given an entirely *false* account of the historical situation. About Bellarmino, Karl says (footnote 2 on that page) that he "saw in instrumentalism one of the possible ways of dealing with scientific hypotheses." Instead of *inconvenient,* one should read, after what I have just said, *refuted,* for Bellarmine knew the *whole* Preface to *De Revolutionibus.* This is the *first* remark I add to the account in my *Realism and Instrumentalism.* Secondly, Karl accuses the Copenhagen people of simply having been instrumentalists by temperament almost. "The view of Osiander, etc. etc. has won the battle without another shot being fired. Without any further debate about the philosophical issue, without producing any new argument etc. etc." (*Conj. Ref.,* page 99, from bottom lines over to page 100). Now, this remark shows that Karl *cannot have read a single line* of Bohr's scientific papers. For Bohr makes it clear, *in every paper between 1913 and 1925,* that a new *"theory"* of atoms (by which he means a realistic account) is not yet available, that all he was able to provide by his own work were *instruments* of prediction (which he calls "formal considerations leading to correct predictions") but that he hopes by an improvement of these instruments finally to arrive at a real theory. *He constantly says this.* Thus he says, in an address to the Copenhagen Academy of December 1913 (!) that his "present objective" was not "to propose an explanation of the spectral laws," but rather "to indicate a way in which it appears possible to bring the spectral laws into close connection with other properties of the elements, which appear to be equally inexplicable," and he emphasizes again and again that this restriction was necessary only "at the present state of research" (a phrase which recurs *ad nauseam*) and until new knowledge arrived concerning the "nature of radiation." The distinction between realism and instrumentalism is therefore seen very clearly by him; he is aware that he has not yet got a theory, but only an instrument, but he *hopes* at some time to get a theory. Now it seems that he gave up this hope after the refutation of Bohr-Kramers and Slater. This may have been premature. BUT it was done on the basis of physical difficulties and in full awareness of the philosophical issues involved. Now I do not mind if Karl, who *knows nothing* about all that, also says that there *has been nothing.* But I do mind the new distorted history of the quantum theory which the readers of Karl (such as Bunge and Landé who should know better) now present almost as an obvious fact, and if new myths are introduced in this way. Incidentally, seeing the way in which Karl quotes Osiander and describes Bohr I no longer doubt those classical scholars who accuse him of having done the same with Plato.[25]

Now—I have *at last* finished the second version of my Bohr-eulogy and it is exactly 80 pages again, twice as long as the first version (mainly *historical* material, to set the record straight). Today I finished (except for minor corrections) the *Defence of Classical Physics.* The title (which I copied from the style of 17th century titles) is now: "An Essay in Defence of Parmenidean Methods and Parmenidean Apologies and especially of Classical Physics, containing also various Problems and Paradoxes from the History of Science as well as some benevolently critical Glances at the Method of Conjectures and Refutations"[26] and it is exactly 70 pages long, the increase mainly due to more on the quantum theory and on relativity and to a 20 page appendix concerning Galileo and the telescope (which is more footnote than text). All this material will be relevant to my Kuhn paper and I shall have to ask Kuhn to read it also. You will get a copy as soon as I have finished the final corrections and have had it typed (I think I shall let the secretary here type it—but that may take *another two weeks*). At any rate, only minor changes will be made after that. I think I shall now turn at once to the Kuhn paper and I shall comment on your objections in an *appendix* to be added on as soon as I get *them.* I hope that Karl will read the pro-Parmenides. (Is the *appendix* idea agreeable to you?) I also hope to be able to send you the pro-Parmenides with Galileo etc. *before* you mobilise rationality against me (and Galileo, and Kepler, and Einstein etc.) for the appendix to that paper contains a brief account of all I have said in my sermons at LSE.

Last week I was invaded by the Chairman of Yale. I was supposed to come to Yale in January, but I cancelled it because of lack of time (after all, I have to fight rationality!). So he now came to me and said that if I come to Yale they would offer me at once $26,000 (I now have $18,000) for sure, and $28,000 very likely, and in two years I would get a Chair of my Own (the something-something Huggins Chair—whoever that is) with $30,000. *I almost fainted* and I have a new quandary in addition to Berlin, London and Berkeley. What do you think? I do not need to make up my mind at once, so I shall be able to talk all this over with you in London.

Incommensurability: you, the chief inventor of clean-cut examples of incommensurability, ask such questions? "My paper will not anticipate your thesis" you write. But you *have* anticipated it, at least in clear presentation, in your *Proofs and Refutations!* Today my question to the opponents of incommensurability would be as follows: "Friends! Do you know of Piaget's research concerning the development of perception in children? Do you remember what Piaget says about the development of object perception? He says here that objects at first look to a child *like after-images.* They become *substantial* only much later. *But at the same time we cease to be aware of our after-images* and must bring them to our attention by rather complicated methods. Now, I would say that after-images and objects are certainly incommensurable categories of things. Or should we perhaps try to base all our judgements of objects on whatever after-images they create? You shake your head? Well, my battle is almost won. For *what makes you so sure that the perception of grown-up people cannot undergo similar changes?* I for one would think that the change of our perception (and thought) concerning celestial objects as a result of the use of the telescope is of a very comparable kind (for this cf. my appendix for pro-Parmenides). And, friends, remember that Dr Imre Lakatos has shown the *conceptual changes* which will accompany such changes of perception. So, why do you continue being obstinate? Because you are afraid that "rationality" will be lost? Why should you be afraid? Unless you think that your reason as it is now is already *finished* and that any further change will only be deterioration. This, however, I must say, is rather a dogmatic point of view. I much prefer to believe, as Hegel has conjectured in his *Phaenomenologie des Geistes* that "further changes and further progress are possible." So much for what I would say. And if somebody says: "But what of argument? What if we cannot argue from one stage to the next?" The reply is, of course, the truism, the really *trite* statement that man lives not by argument alone (or would you like to maintain that one should use only *argument* to get a woman into one's bed? And why should what is good for Casanova not also be good for Einstein and Galileo? Or are they supposed to use only *part* of their faculties? And why insist on such self castration in the sciences? Because of "rationality"? Not a good enough reason—not by a long shot).

I cannot send you the final version of "Classical Empiricism" as I have sent it to the publisher (I have no copy). It will be published in a book with articles by Hanson, Laudan etc. under the title *The Methodological Heritage of Newton,* ed. by Robert E. Butts, John W. Davis, University of Toronto Press (I shall try to write to Butts to send you a Xerox copy).

So—and now I go for my long walk. All the best!

Paul

24. Feyerabend 1964c.

25. For Popper's criticism of Plato, see Popper [1945] 1966. For his criticism of Bohr, see Popper 1963, 1967.

26. Feyerabend 1970e.

25 March 1968
University of California, Berkeley

Dear Imre,

Your letter comes precisely at the right time. I am sitting here, full of pills, dizzy, and the pain will not go away, so I just have to wait and to hope for the best. It is not a very pleasant state to be in. In addition, I have some temperature which is due to my kidneys, which ailment is in turn due to a weakness in the defensive mechanisms of my blood. Well, we shall see . . .

Congratulations for your success! And it is well deserved. For there is nothing in contemporary philosophical writing that is so full of ideas and at the same time so charmingly written.[27]

You are quite right in criticising me on Osiander. I am getting too bloody righteous—which is quite contrary to my temperament and to my ideology. Only Karl occasionally strikes me the wrong way. No doubt this is due to the fact that I followed him slavishly in many ways and am now somewhat sore at having been taken in (through my own fault; for why was I so uncritical?). Well, growing up is a difficult business and with some people it takes longer than with others.

As regards the *substance* of my remark I think, however, that I am correct. Karl speaks of a "New Betrayal of Galilean Science" (subtitle to the *Three Views*). My point is: there was no "Betrayal"; and there was also no "New Betrayal." As regards Osiander, *Karl insinuates* that he is a philosophical dogmatist for whom Ptolemy and Aristotle are the truth and everything else an instrument. *Actually,* he simply says that Copernicus, taken as a description of the real world, is refuted, but may still be used as an instrument for prediction. Galileo's whole effort, his whole life work is to remove the refuting instances which he does with the help of propaganda tricks, *ad hoc* hypotheses and so on. From a Popperian point of view, it is Galileo who is the dogmatist and Osiander who is the critical scientist. All this interesting tension is buried by Karl who acts as if there were only a philosophical issue at stake, and supports his procedure by partial quotations. As regards Bohr, Karl says that instrumentalism won the day "without another shot being fired." "Without another shot"? *The whole period between 1913 and 1925* is a discussion of the issue between realism and instrumentalism. All this time, Bohr says that so far he can only offer an instrument, but that he hopes at some time to arrive at a theory in the full sense of the word (cf. my quotations on section 6 of my pro-Bohr, especially the footnotes). Karl's "without any debate about the philosophical issue," therefore, is simply not true—because this issue played a central role in the older quantum theory. The situation changed in 1925/26 as a result of the breakdown of Bohr-Kramers-Slater and the invention of wave mechanics: these were the "shots" which seemed to convince Bohr that an instrument was all he could have. And even then he did not *slide* into instrumentalism, but accepted it *regretfully,* as a result of refuting instances he accepted. Now, what is one to think when Karl says "without another shot" or "without debate"? Must one not assume that he does not know the literature (especially as he himself philosophizes as if the refuting instances which Bohr accepts did not exist)? And if he *does* know the literature, then why does he write such a thing? Moreover, Bohr showed how one must on occasion speculate against the evidence and he is in this respect very similar to Galileo. (I found a marvellous quotation in Bohr which asserts precisely the story of the two weak men supporting each other, as you concisely expressed the content of my ramblings at LSE.) Now, if one is as dramatic as Karl, one must say, in the light of all this, that it is *he* who is guilty of a "New Betrayal of Galilean Science"—as I also say in a footnote of my pro-Parmenides (note, also, that Galileo used Parmenidean apologies rather freely, whereas Karl wants to see them eliminated, another sign of a "New Betrayal" etc. etc.). Well, enough of that, for I am feeling the evil waves of righteousness already rising in me again. But I think I have at least shown that my "inductive basis" for generalisations to Plato is not false. You point out that Karl never said anything against Bohr *between 1913 and 1925.* But he does not restrict his comments "without another shot"; "without any debate" etc. to a specific period either: there are *no* shots, and there is *no* debate, according to him. There *are* lots of shots, and there *is* lots of debate, as a matter of fact. And, to repeat it, this debate and these shots are not buried in some footnotes, but they are *omnipresent,* so that it needs a little more than superficiality to say that they don't exist.

OK. Incommensurability. Well, I hope I shall be able to write something sensible about it. Right now, I feel quite written out and wish I could forget all philosophy and withdraw to a mountain top for a year or so. But there is no rest for the wicked. As regards private property: I like money, but as regards girls and ideas I think I am not so particular.

Alas—there will be no time for me even to *appear* at the LSE—half of the week I am in Berlin anyway. Besides, you and your "students" have already heard *everything* I have to say, and it will take me 10 years to invent something new, but by the time the 10 years are over I shall most likely be too senile, or too indolent to invent anything new, so I have nothing more to say to you.

I was mistaken as regards the quantum-conference: it is in *July.* I originally thought it would be a *small* conference and I was somewhat inclined to come. Now it turns out that it is going to be a monster conference—and that is the end as far as my attendance is concerned. I am lecturing in July in Berkeley, but they are willing to let me go for a week, and to pay for my journey. They also want me to give a paper in Cambridge. But I think I am now definitely not coming.

Please, if you have a little time, have a look at my quantum-paper.[28] I am occasionally quite blunt in it, but I don't want to be nasty, or self-righteous. So tell me where I err. I have sent the paper to *Philosophy of Science* who have already accepted it, but I can send corrections, I think, for another month (there is some doubt as to publication: the original paper which they were prepared to publish was only half the size and the present size is forbidding as far as their rules are concerned).

Finally, apologies when on occasion I sounded a little incoherent, but in my present state, with all the pills and the pain, I cannot think very straight.

All the best

Paul

27. Feyerabend is probably referring to Lakatos 1963–1964a.
28. Feyerabend is always referring to his 1968–1969.

26 March 1968
London School of Economics

Dear Paul,

Many thanks for your pro-Bohr paper. I cannot judge it as a whole; I shall need another few weeks to do so. A few odd remarks after a first very cursory reading:

(1) You should say that you were (and where you were) anti-Copenhagen earlier and why you changed your mind; the introduction would have a nice personal touch.

(2) "The myth of Bohr's dogmatism which Popper accepts . . ." (p. 45 and p. 79). This is very bad. You do not bring any evidence of the existence of the myth (is it your strawman?) and to quote Popper as asserting it is falsification. If I say that "there is a basic dogma in X's thought," this does not mean that "X is dogmatic." The latter is only true if X ignored explicit criticism. Don't be naughty. Popper always criticises (in his writings) people's *ideas* (third world), not their *character* (second world). You seem to say that he (Popper) criticised Bohr's intellectual character. On a dozen pages you defend Bohr's personality against Popper. This is nonsense, and ruins your paper.

(3) I disagree slightly with pp. 56–7, but it is very interesting.

(4) I do not really know Bohr's writings. But Popper is concerned with the (certainly *dominant*) subjectivistic interpretation of quantum theory and it is of very little interest (except for Bohr's biographers) what Bohr personally thought. Your misquotation of Fock against Popper (your p. 60) is simply dishonest. Fock in fact says that Bohr "hat seitweilig positivistische Ansichten vertreten"[29] (Fock [1962], p. 193), and his *polite* separation of the "realist Bohr" and the "subjectivist Copenhagen interpretation" is based on an unpublished personal letter *only* (since published in Russian), which, I take, you have never read.

(5) Your *Conclusion* is incredibly rude and equally incredibly uninteresting. It is an "orgy of name-calling, dogmatic pronouncements, inane lyricisms . . . ." Scrap it. (Incidentally: "Back to Bohr"? Back to *which* of Bohr's ideas? Or back to his *intellectual attitude?* Or back to a *higher appreciation* of Bohr?)

I am sorry, my dear Paul, for this letter. I don't know what the devil is driving you. This paper is subjective, emotional; its valuable material and very important arguments are submerged in irrelevant and, on many points, distorting anti-Popperianisms. You know how fond I am of you and how much I admire you; so I hope you will not be too angry for my *Odd Remarks.* For a reasoned appreciation of the whole essay, I am afraid you will have to wait. I am very slow.

Affectionately,

Imre

29. "[Bohr] has temporarily held positivistic views" (Fock 1962).

Saturday, end of March 1968
University of California, Berkeley

Dear Imre,

[. . .] The main issue between us is now (after the matter of rudeness treated in my last letter) the matter of the *third world*[30] and the importance and *self-sufficiency of argument.* I don't believe in the third world and even think that it might be a step in the direction of a dull totalitarianism. Nor do I think that argument is sufficient for introducing a new point of view. Mill, in a wonderful passage of his autobiography says that he did not understand half of his father's explanations of logic, but that these ununderstood explanations acted as a *nucleus* (his word) in terms of which *new* things were interpreted giving content in turn to the "nucleus." This means that you may often have to run around with an ununderstood and mystical phrase which, as the years go on, will help you better understand what you are constantly learning and which will become understood in the process. Being brutally rationalistic may in this case help one to be always clear about what one believes and says, but one's ideas will be much impoverished. It is for this reason that I think "irrationalist" moves are absolutely essential, though of course not sufficient. Karl is too much of a rationalist. Put Karl and a poet (or a Galileo) together, and you might have the right balance. But this is just rambling. Incommensurability is so important because, once accepted, it reveals the need for more "irrational" moves.

I am rereading Kuhn's *The Copernican Revolution.* A marvellous book—and I wish he were still writing in the style he was using then. His more recent papers sound much more pompous—but, maybe, that happens to all of us.

Be well

Paul

30. Feyerabend is here referring to Popper's theory of the three worlds: i.e., the first world of physical objects or physical states, the second world of states of consciousness or mental states, and the third world of objective contents of thought such as scientific theories and works of art (see Popper 1972, chap. 3). This distinction is discussed in many of the letters that follow.

30 March 1968
London School of Economics

My dear Paul,

I just received your very nice letter which crossed my Second Critical letter. I feel awful because I hate to hurt people of whom I am so very fond and whom I respect so tremendously; especially since you write that you have physical pains too.

Please don't be hurt. Of course I realise that accepting my basic criticism *(i.e. the conflation of second world and third world arguments)* would involve you in a great deal of rewriting, but let me just bully you into it; let others sort it out.

[. . .] From April 5th to April 20th I am in Majorca with Eva [Pap].[31] God save me.

And don't be angry with me. Our clashes are all in the third world, never in the first or second.

Very fondly and with warmest best wishes,

Imre

31. Lakatos married Eva Pap on 7 February 1956 in Budapest. Their marriage lasted for about three years, but, as this letter indicates, they continued to see each other occasionally.

19 April 1968
University of California, Berkeley

Dear Imre,

Thank you so much for your nice postcard. How are things with you and Eva now?

[. . .] On your desk you will find (if the postal service has not lost it) three copies of my pro-Parmenides: one for you, one for Karl [Popper], one for Alan [Musgrave][32] (John [Watkins] has his own copy). The difference between Karl and me can now be expressed thus: we *both* admit that one has to deal in some way with imprecision, irreversibility, chance and so on. Karl wants to take care of those things by changing the *laws* (or the kind of hypothesis introduced) but keeping the method unchanged (he still wants to take refutations seriously, eliminate *ad hoc* hypotheses, and so on). I, however, want to keep deterministic etc. laws and make the method more liberal (take a more liberal attitude towards refuting instances; admit *ad hoc* hypotheses, and so on). Briefly, *Karl keeps the method clean and muddies up the laws, I keep the laws clean and muddy up the method.* And the historical material I present tries to show that the Copernican revolution was done "my" way.

Enclosed you will find a further essay which uses some of the material on the pro-Parmenides, but makes it independent of Karl's anti-Parmenides. I have also added two long sections on Galileo's theory of motion, made the non-historical part clearer, and added a short history of the origin of this point of view of mine which is of course the fact that I listened to your lectures. All this is said on the *first three pages* of the paper as well as on page 84, which is the first page of the footnotes.

There you will find yourself in company with the blond Helena. Now, please, can you do me a great favour? I do not know Helena's second name. Please insert it into the MS, make a Xerox copy of the MS and give it to Helena with my regards. Can you do that? I shall be eternally grateful.

I shall now arrive on May 6th. I have been improving the last two weeks, but it is a precarious matter, I seem to have a wandering infection, my feet were swollen and I looked like Frankenstein's monster and I am going to have a thorough check-up at the Mayo Clinic before starting on my insane chase between London and Berlin.

In this connection: you said you *might* be able to teach in Berlin in spring 1969 (end of April to end of July). You could bring your own assistant who will get his own salary, you will have full use, with no one else interfering, of a two language secretary (and I shall try to find an attractive one for you)—so why don't you come? I have mentioned you (for this period) in my acceptance letter which does not commit you to anything, but keeps the place open. I know from myself that one cannot make up one's mind years in advance, or even weeks in advance, but I just want to let you know that the possibility is there, that the place is being kept open, and I would also like you to *occasionally* think about it. (Perhaps you can come for a visit this spring, for a few days, expenses paid.)

So that is all for now, and now keep your fingers crossed that I don't fall ill again—I have already become so paranoid—so that I can arrive, when planned, viz. May 6th (I may be there already on the 5th, i.e. Sunday, and shall ring you immediately).

Remember Ravenaugh, the American conductor who briefly visited you at the LSE? His recording of the Busoni Piano Concerto is a runaway bestseller, the reviews are raving about the piece, his conducting, Ogden's piano playing, and some of them even mentioned my "perceptive" essay on Busoni, of which you will soon get a copy. Well, that cheered me up a little.

All the best

Paul

32. Alan Musgrave, after studying under Popper at LSE, became professor of philosophy at Otago University, New Zealand. He was co-editor with Lakatos of the fourth volume of the Proceedings of the 1965 International Colloquium in the Philosophy of Science held at Bedford College, London; see Lakatos and Musgrave (ed.) 1970. See also Musgrave 1993.

25 April 1968

My dear Paul,

I have just arrived back in London from my holiday and used the first couple of days to keep out the students from the decision-making bodies of the School. [. . .] This morning I got down to your mail which now I see will take me two days to get through, and which always gives me great enjoyment and, indeed, is one of the main pleasures of my life. I was fascinated by your letter to the *Dekan* [Dean] because it does not specify exactly what the position is that you are going to take up in Berlin. Does this mean that you are abandoning Berkeley? Anyway, it seems to be a very important and exciting decision in your life, so let me know more about it.

[. . .] I just learned from John [Watkins] that you may come back later than I thought you would. This would be a disappointment, but I see from your letter that you must be here early in May when I am looking forward to the promised supper. By that time I hope I can present you with a copy of "The Problem of Inductive Logic"[33] which should be due about then.

After my absence and my Academic Board excitement, I am now swamped with administration and teaching, so I shall discuss with you your *Problems of Empiricism II*[34] personally: but I looked through it and thought that what I saw was admirable.

I am worried about your health. I hope that your visit to Rochester is only a routine check-up.

With warmest greetings and many thanks,

Imre

33. Lakatos (ed.) 1968.
34. Feyerabend 1970f.

6 July 1968
London School of Economics

My dear Paul,

Your *Classical Empiricism* is a masterpiece; apart from microscopic criticism and one disagreement, I admire it enthusiastically.

Please send a Full Final copy of your Galileo papers; I shall have them both mimeographed and make it compulsory reading.

No doubt these two papers are marvellous and the best you have ever written. I am playing with the idea to rival Joske's offer and edit you with a preface (possibly together with his "Sensationalism" and one or two more (not much more) cases and publish it as a post-Popper *Festschrift* for his 70th birthday and thereby restore Critical Peace). [. . .] Write soon,

Imre

8 August 1968

My dear Paul,

[. . .] I am aghast at your idea of rewriting your paper. Absolutely not! You can improve your section on *Incommensurability* within one week; that is the maximum allowance you get: Kuhn is already working on his reply to you.

However, if within two weeks you send an appendix, entitled *Against Method,* I shall persuade Kuhn to take it into account. Moreover, because I love you, you will be able to write a final note in October about Kuhn's final reply, about my paper, about anything you like—if you are a good boy and *don't tell anybody about it.* So the last word in the volume would be yours (of course Alan the co-editor will have to agree, but he will, I trust).

But my basic condition for all this is that you do not alter your original paper (except for improving your chapter on incommensurability which is *not clear*) and you write a red *appendix* (maybe *"Against Method" or "Unter der roten Fahne"* [Under the red flag]?). Then Kuhn will quickly reply (he may already be writing it!) and then you may toss in a retort when everything is in print! (I assume that will be now rather than December).

I met Grünbaum and family; and liked them immensely.

I am depressed and miss you tremendously. Send me a new girl to cheer me up. When do you arrive in London? What are your plans?

Fondly,

Imre

27 July 1968
University of California, Berkeley

Dear Imre,

Here is the last version of *Problems of Empiricism, part ii.* There is a new introduction, I have changed some footnotes and I have also added some new material. I hope all this meets with your approval. You will let me know, yes?

I have received one of the volumes you edited, on the problem of induction,[35] but nothing else has arrived yet. I am reading your paper on induction[36] slowly and carefully and while I am impressed, very impressed by the way in which you uncover the philosophical shifts, changes, deteriorations in the inductivists' philosophy, I am now convinced that the debate is nothing but shadow boxing, only that in this case the shadows do not imitate the movements of the real things. Right now I am writing a paper on this, for Herbert Feigl, where I shall defend the idea of a methodology as a series of *rules of thumb,* without probability, without quantitative considerations. This, it seems, will be my last paper for a long time. I think you will like it.

You intend to publish a Popperian collection for Karl's 70th birthday? Let me tell you my attitude about this. I would write a paper for a *Festschrift,* but I would not participate in a Popperian collection simply because I do not intend to join any party. I detest uniforms, whether they are worn on the outside, or on the inside. Now, of course, it may be the case that, as a result of having read Popper I have acquired an inner uniform without being aware of it, or without being able to help it (I may be a Popperian). But if that is so, my intention is to get rid of it and not to exhibit it. Writing a contribution to a *Festschrift* is an entirely different affair, for here one simply acknowledges what one owes to a man without signing a party ticket. One may owe a lot to a man and yet be of very different opinion. One may owe a lot to a group of friends and yet not be a member of a party. One may prefer discussion with some people than discussion with others and yet not be a member of a party as long as the affair is not *institutionalised.* A collection of the kind you suggest would be a step towards institutionalisation—and therefore I cannot see any way towards being a part of it (I also refused becoming co-editor, with Joske [Agassi], of a Journal he is editing). Being part of a book which says "look at us Popperians, what we have done" is utterly repulsive to me [. . .]. Finally, read my footnote 2c: I believe, with the sceptics, that, given a sufficient number of intelligent people with sufficient motivation, *any* point of view, however unreasonable, can be made to threaten any other point of view, however reasonable. It is my aim in life to show this by examples. Hence, I shall turn anti-Popperian as soon as Popperians become accepted as "reasonable." The time is very close. Hence, I shall turn anti-Popperian (i.e. irrationalist) very soon. This is another reason why I cannot join the Popperian party.

There is a further problem which became aggravated by my visit to Boston. Joske always urges me to write an autobiography dealing especially with the question as to what I owe Karl. Well, I am thinking more and more about this, viz. (a) what I owe Karl, and (b) what Karl has achieved. As regards (b) I am now convinced that the only advance which Karl has made over and above the Presocratics consists in technicalities, that is, in the very same things I try to get rid of (basic statements vs. theories; initial conditions; corroboration; degrees of falsifiability etc.). All *I* need is Xenophanes; frg. 189: "no man knows, or will ever know, the truth about the gods and about everything I speak of; for even if one chanced to say the truth, yet one knows it not; but seeming wrought over all things." This [asserts] the hypothetical character of all our knowledge. Frg. 190: "Let these things be opined as resembling the truth . . ." That is, although everything is hypothetical, let us not give up altogether (irrationalism) but deal with the hypotheses. Frg. 191: " . . . by seeking, men find out better in time" (against scepticism: the hypotheses can be improved). That is all that is needed, and if we want to take care of *science,* we add *dialectical materialism* to it. Had Popper been content simply to defend this tradition (as he does in *Conjectures and Refutations—without mentioning dialectical materialism*) we would not now be faced with the battle of the technicians, such as Bar-Hillel plus Carnap etc. vs. Popper and (?) Lakatos. As regards (a) I really do not know, but I *do* know that the person who turned me from a positivist into a realist was Walter Hollitscher and that the person who taught me the importance of theoretical principles was Dingler whom I read when I was 16. (Hugo Dingler and *not* Herbert Dingle.) So you see, party-wise I am in a quandary from all points of view. Do you have any helpful comments on the matter?

The meeting with Joske was very peaceful, we talked a lot about various things. Lynn wants to come to University College and get her PhD within a year instead of 3 at LSE (this is what she told me over the telephone—she is now living at Joske's place). Julie Rowntree, who thinks that you are charmingly rude, sends her regards. Grünbaum is going to turn up in London very soon and will visit you. I have rejected altogether six invitations for talks with offers from $500 to $1000 per talk—I just want to be left in peace. Besides, I am not convinced that one can convince people—this is also why I want to stop writing. I do not know your Barbara, but have seen a picture of her. And now, back to the preparation of lectures for the fall (on "Ether and Relativity").

Be well and don't work too hard!

Paul

P.S. Has Karl shown any reaction to my Proclassical Physics? Have you sent the Kuhn material to Berlin (or is it not yet ready?). Could I see the last version of *your* Kuhn paper? And a copy of *my* Kuhn paper (I don't have a single copy of it and would like to see it, so that I can refer to it in other papers). Do you have the Berlin address to which I want you to send the Kuhn material? Again, be well and write soon!

35. See Lakatos (ed.) 1968.
36. See Lakatos 1968a.

29 August 1968
London School of Economics

Dear Paul,

Many thanks for your new Kuhn MS (great improvement on an excellent paper);[37] also for the paper in the Weingartner volume[38] which turned up only today in a battered state; but I bought the volume, and shall read it during the weekend. It looks very brilliant and I am grateful for your kind reference. (Incidentally, I am now starting to return your kindness as you will soon see.)

I have been terribly upset by the Czech events,[39] the brutality of which (I doubt you know as many ghastly details as I do) reminds me of Hitler's and Stalin's worst days. How tenuous civilisation is! The animals in Moscow and the frightening Chicago convention[40] just prevent me from thinking about incommensurability and even about women.

I hope to recover soon, but at the moment I just listen to the radio and watch television all day, and that's all.

Much love

Imre

P.S. I enclose a copy of the first page of my doctoral thesis for your enjoyment.[41]

37. Feyerabend 1970c.
38. Feyerabend 1967b.
39. Lakatos is of course thinking about the events in Prague in 1968.
40. In the spring of 1968 some of the Yippie leaders (the name "Yippie" comes from "YIP," Youth International Party) planned a large gathering in Chicago for the last week of August, when the Democratic Party had scheduled its convention for the official nomination of its presidential candidate. In the programme for their "Festival of Life," the Yippies had included rock concerts with the most famous groups of the time, an 'alternative' candidate-naming ceremony, public distribution of psychedelic drugs (accompanied by the threat to spill LSD into local water reservoirs), and liberal, free-thinking, generally "shocking" actions ("Make love, not war" was the motto of the Yippie leader, Jerry Rubin). The mayor of Chicago, Richard Daley, alarmed by the Yippies' plans, called in the entire city police force and asked for special protection from both the Army and the National Guard. The Yippie festival opened on 25 August with a psychedelic rock concert; the police moved in quickly. A whole week of street fights between police and protestors ensued, ending with a toll of one dead among the Yippies and many thousands injured. The peak of the clashes occurred midweek in front of the Hilton Hotel, host to the Democratic Convention, when television network cameras filmed the National Guard assault live. The assault was carried out with tear gas, armoured personnel carriers, and helicopters, and involved not only the demonstrators, but also reporters, photographers and simple onlookers.
41. Lakatos is referring to his 1959–1961.

August 1968
London School of Economics

Dear Paul,

I was stunned by your conversion to Marxism. Haven't you read my footnote 1, *Proofs and Refs,* p. 234 (offprint p. 59)? This lays down the Church's position on this matter (last sentence of the footnote).[42] The problem of whether sin is venial or mortal is being investigated.

The Holy Inquisition (Musgrave and myself) recently investigated your life's work and decided that it has been unjustly neglected in the Church's official teachings, and that its Shining Truths should be widely disseminated and honoured. It was also decided that Agassi's second-world excommunication must not affect Our Holy Decision to include his *Metaphysics* paper into the Scriptures.

The lady was graciously readmitted to the LSE; but this is only for the M.Sc. Her bosom qualifies her for her own M.Sc., but her legs rule her out from our Ph.D.

We all miss you. Bring, please, Liz Aschenbrenner with you in October. And send your incommensurability addition immediately.

Love

Imre

42. The footnote Lakatos is referring to reads:

Few mathematicians would confess—like Beta—that reason is too weak to justify itself. Most of them adopt some brand of dogmatism, historicism or confused pragmatism and remain curiously blind to its untenability; for example: "Mathematical truths are in fact *the prototype of the completely incontestable* . . . But the rigour of maths is not absolute; it is in a process of continual development; the principles of maths have not been congealed once and for all, but have a life of their own and may even be the subject of scientific quarrels" (A. D. Aleksandrov). (This quotation may remind us that dialectic tries to account for change without using criticism: truths are in "continual development" but always "completely incontestable.") (Lakatos 1976a, 54–55)

4 September 1968
University of California, Berkeley

Dear Imre,

Your last letter cheered me up considerably. It came in the middle of a period of blackest depression. My only reply is that Popper is of course granted his achievements, but these same achievements, as far as I am concerned, are not big enough to start a new school; or at least they should not be regarded as such lest we are left with a million splinter-groups, and nothing strong enough to advance our political, intellectual (and so on) situation. Dubcek does not pretend to have invented a new philosophy (although he has done more than Sir Karl); his followers do not pretend to have refuted Marxism, or dialectical materialism—they have improved it. Why shouldn't this Western Dubcek, i.e. Popper, have a similar sense of perspective, especially as he is under no pressure at all? To repeat, then: I grant Popper his achievements within the party of dialectical materialism, but I also declare that I belong to this same party. Moreover, I am a little suspicious of the achievements themselves, especially about the criticism of Marx, which I intend to review just as I have reviewed the criticism of Bohr. For, after my *Empiricism II,*[43] and my suspicion concerning the omnipotence of argument, I can fully subscribe to Trotsky: "A political struggle is in its essence a struggle of interests and of forces, *not of arguments*":[44] if *astronomy* can't get on on the basis of argument only, *how should politics be able to?* Therefore—back to Marx, I say to myself, and re-examine what can be preserved of "formal democracy," which plays the same role in politics that argument plays in astronomy etc. I am afraid that not much will remain.

I have discussed all these things with Liz Aschenbrenner who is now my ideological partner (I hope it will not remain there), and she agrees. Even Joske is not averse (though he says that Karl is the head of the 5th International—yet another example of lack of perspective! Where are the concrete programmes?) Anyway, I wish I could discuss these with you personally. Let me add, by the way, that having become more bosom conscious (since my leg appetite has been satisfied by Liz), I no longer criticise Lynn (especially as she addressed me as "beautiful" in her last letter—now what made her say this?).

I sent the Kuhn [paper] about ten days ago. I am still not satisfied with the incommensurability mess, but it will have to stay that way now, for I have no time. I sent the final version of "Empiricism II" long ago, but my secretary sent it surface mail, so it may still take some time to arrive. Be well, and I hope to see you soon if I come. Long live Dubcek and the 5th International, including its rightist troublemaker, Sir Karl.

Have to run to lunch with Liz.

Paul

43. See Feyerabend 1970f.
44. Cf. Leon Trotsky, *The Revolution Betrayed,* 86–87.

18 November 1968
University College
Gower Street, London[45]

Dear Imre,

I have finished Cohn-Bendit,[46] and am *wholly* on his side. He is against theories; so am I. He is against organisations; so am I. He is against "leaders," be they professors who "know," or generals who command; so am I. He is *for* joy and *against* sacrifice; so am I: "The real meaning in revolution is not a change in management, but a change in man. This change we must make in our own lifetime and not for our children's sake, for the revolution must be born of joy and not of sacrifice." He recognises that those whom you call the "monkeys" are not exactly paragons of intelligence, but he has compassion and understanding for them: "It is difficult not to adopt a paternalistic tone when speaking of the struggle of high school boys and girls whose refusal to be cowed often expresses itself in childish ways, *all the more touching for that.*" Students often act stupidly in front of a problem which their elders do not even see and therefore do not treat at all, not even in a stupid fashion.[47] We must indeed change man so that progress, the getting rid of fetters, is not at once turned into an inhibition of a new kind.

Karl was a revolutionary once, in politics as well as in the narrow context of the Vienna circle. What has remained of all that now? Quarrels about footnotes, a bunch of frightened students, and new formalisms (of corroboration). "No new progressive epoch has ever defined itself by its own restrictions and limitations. The Renaissance did not define itself by referring to the limitations of its naive rationalism, but by the rationalistic transgression of all limitations. Romanticism defined itself by transgressing the limits of the classical canon and by the new discoveries it made beyond those limits [. . .] *In our case, however, it is still regarded as virtuous to watch the boundaries rather than transgress them.*" This was said by Milan Kundera at the IV Congress of Czech Authors in Prague, June 1967, and he clearly meant the party hierarchy. But the very same criticism applies to Popperians today (you yourself excepted). But why is it that what starts as the tearing down of boundaries always ends up with the construction of new boundaries? And why are people so meek and gullible that they accept the new boundaries (see what Cohn-Bendit says about the Russian revolution, the role of the party and of professional revolutionaries etc. etc.—here, however, the workers yearning for freedom had to be killed, whereas students give in at the threat of a bad grade, and why? Because they are frightened. Why are they frightened? Because New Man has not yet arrived—but he is more important than any doctrine, for it is only through him that doctrines will be prevented from turning into fetters.)

Anyway, I now see my *Against Method* as a weak and stumbling prologue to what others have done much better: Cohn-Bendit, for example. And when I have finished my "scientific" career, I hope that my "human" career will have some success (first step: a talk in Chicago with the title: "Philosophy of Science—an Expensive Fraud"; I might even write this up as an article for the *New Left Review,* though they are too doctrinaire for my taste—again doctrine, doctrine, doctrine). It is interesting to see, by the way, how many people to whom I have sent my *Against Method* shrink back from it because it contains such names as Lenin etc.

Well, that is all for today. I only wrote you this letter to show you my present state of Romantic Irrationalism etc. etc.

Paul

45. In the late sixties Feyerabend was still "highly marketable." He received offers from London, Berlin, and New Zealand, which he accepted. In London he lectured at University College and at the London School of Economics (see Feyerabend 1995, chap. 11).

46. Daniel Cohn-Bendit was the leader of the French student revolt which led to the Paris events of May 1968. At the end of that same year he wrote *Obsolete Communism: The Left-Wing Alternative,* inspired by important ideas which would recur during the whole student movement, such as libertarian communism, anarchism, and the more general denial of consolidated social conventions. The book contains many ideas which Feyerabend would use in his later articles; for example, the link between theory and politics, the emphasis on action within a libertarian framework, the demand for flexibility and a democratic basis for all institutions, the opposition to specialists and specialisation, and the struggle against any kind of hierarchy and bureaucracy, including that in educational institutions, schools, and universities.

47. In fact, referring to the 1968 events years later in his *Killing Time,* Feyerabend remembers:

> I didn't always accept the advice of the student leaders. For example, I didn't participate in the strikes they declared. On the contrary, I cut fewer lectures during the strike than either before or after. "Didn't you feel any solidarity?" Grazia asked when I told her. "With the students, yes; with the organisers of the strike, no. They presumed to speak for all students just as Johnson presumed to act for all Americans—the old authoritarianism again." Besides, I thought a student strike was rather silly. Industrial strikes cause a shortage of goods. Student strikes are a nuisance, nothing more. (I have changed my mind since then. Professors without students are as useless as screwdrivers without screws—and they feel it.) I would have stopped lecturing if my students had demanded it, but when I asked them, some said yes, some said no—and we spent the rest of the time debating the issue. Eventually, I moved off campus, first into students' quarters, then into a church. Now the administration got on my back: teachers were supposed to remain in the assigned lecture halls. Consulting the regulations, I found no such rule, and continued as before."

And this is the end of the story:

> For some of my colleagues, John Searle especially, this was the last straw; they wanted to have me fired. When they realised how much paperwork was involved, they gave up. Red tape does have its advantages. (Feyerabend, 1995, 126)

19 November 1968
University College London

Dear Imre,

After our recent discussions, and your lecture which again raised the spectre of rationalism, I think I now want to make use of your offer to write a page or two at the end of the Kuhn debate (in defence of irrationalism). I think I can summarise all my objections against you very briefly (though to argue against you is a much more sticky affair).[48] So please don't forget me when making your final editorial arrangements!

PNF

N stands for Novalis

And now I realise that Karl is a hidden Romantic, too, quoting Novalis at the very beginning of his book.[49]

48. See Feyerabend 1970c.

49. Cf. Popper [1934] 1959, 11: "Hypotheses are nets: only he who casts will catch (Novalis)."

18 December 1968

Dear Paul,

I showed your Appendix to Alan Musgrave and he said that we should try to include it in our volume. Thus, I sent it on to Kuhn, and I hope he will agree to the change. I shall write to you as soon as I get a reply from him.

London is a much quieter place since you left and it is much more peaceful, although, I have to admit, also more boring. Therefore, I hope you will be back in February.

I am now trying to close down my odds and ends; then have a fortnight's rest; and then start work again and try to minimise my participation in the great counter-revolution.

With warm greetings and best wishes for a happy New Year.

Yours ever,

Imre

31 December 1968

Dear Imre,

Here, at last, the *(Much Improved)* final version of the Appendix (I hope that "final" means *final*). There are a few changes. For example, I have introduced references to the *Open Society* and especially to the Appendix of the second volume (what Popper says there is the same as what I say in *Against Method*—this is now asserted in a new footnote of *Against Method*). I have also tried to improve the English. And I have added references in the footnotes. As regards these references—may I ask you a favour? I have noticed that in the old copy you occasionally inserted the word "my" in front of titles such as "Against Method," "Problems of Empiricism," and so on. I am sorry, but I absolutely despise the word "my" and I would be grateful if you could avoid inserting it into the present version. No doubt this is a silly quirk of mine (legitimate usage of "mine" according to "my" book) but I would be grateful if you would permit me to have my (again legitimate) way in this matter (there is no property in the third world).

Barbara yesterday sent me her final farewell, so we shall not see each other anymore (and you won't be able to see her either, when you come here).

Remember Newton's anger at Hooke who claimed to have discovered the law of gravitation although he had not spent days and nights calculating and recalculating, like Newton? Well, if it comes to making a choice I think I prefer being a Hooke to being a Newton (but I would prefer being a rich and well-married man to *both*).

20 minutes from now I shall have lunch with Liz. In the evening I am invited to a party but most likely I shall not go but watch TV instead.

(What I *actually* did was: after lunch we went to 2 movies in San Francisco, then dinner, then another movie, then something more to eat, then a party, then home. This is better than studying optics.) I have just seen too many people recently.

Be well, and my very best wishes for the New Year.

Love,

Paul

P.S. See *2001*—an *absolutely marvellous* movie. If you haven't seen it when I return then I shall see it again with you.

1969

10 February 1969
London School of Economics

Dear Paul,

I am sorry that according to your "beautiful chick" what you wrote is bull-shit and you have gone mad, but even if this is the case it will not alter our beautiful friendship.

Figure 3.2 The Heresy Goddess. Postcard sent by Paul to Imre.

Do not take it amiss that I have not written to you for weeks, but for the last few weeks I haven't written, I haven't read; I have only sat on committees (counter-revolutionary committees, of course). Our students have not yet burned down any buildings like yours, but they may do so soon, and the LSE is now certainly one of the centres of the vicious attack of the Cohn-Bendit/Feyerabend mob. We have been closed for three weeks now but may open next week, when we shall have further trouble and close for another four weeks, open again for two days and then close till October. All grants will be withdrawn, all Americans expelled from this country and the Special Branch of Scotland Yard has already sent a study group to America to study FBI methods on campuses. We are expelling a few lecturers along with the students so that the latter should not feel victimised. There are going to be big trials for conspiracy.[50]

In the last two weeks the LSE has occupied more of the front pages than Russia, space travel and Nixon jointly. However, University College is caving in nicely and in the Philosophy Department examinations have been abolished, and all the marking is done by students on lecturers and not vice versa. They are waiting for you because they need theoretical support, so come as soon as possible.

The Socialist Society has been banned and most of its leaders will soon be arrested. However the symposium which they meant to arrange between you and me will be held; we don't yet know exactly where, but probably at Westminster which will, by then, be occupied.

I am flying to Heidelberg tomorrow to give a talk to Tugendhat's[51] people on the morality of shitting in the Rector's office.

As you can see, there is urgent need for your presence here. Robin Blackburn will be deprived of his job on February 15th and possibly arrested on the 18th.[52] However, we still like each other—at least he says what he means, usually, which is becoming rare among our so-called leftish academics.

If you are fed up with your beautiful chick, wrap her up in a parcel and send her to me.

With warm greetings,

Imre

P.S. I hope you are not wasting your life on trivialities such as academic work. I'm not.

50. The darkest and possibly the most dramatic moment in the history of LSE began in the autumn of 1968. An anti-Vietnam War demonstration on a national scale had been organised for 27 October. Representatives from both the Student Union and the SocSoc (Socialist Society) had decided to participate. In his recent *A History of the London School of Economics and Political Science 1895–1995,* Ralf Dahrendorf writes:

> The School authorities were clearly and understandably worried: 1968 was a restless year; violence had been used in connection with Vietnam demonstrations. . . . Rumours had reached Governors that there would be trouble at the School and many felt that the Students' Union would not be able to control outsiders even if it wanted to do so. Thus the Director [Walter Adam] put up a statement on behalf of the Board of Governors on his notice board. The School buildings would have to remain closed even to academic staff on Saturday and Sunday; "without my express consent" no one would be allowed

> in. "The Governors have conferred on me at my sole discretion power to close the School." The Director had every intention of using this power without notice if forced to. (Dahrendorf 1995, 461)

Activist students, headed by Paul Hoch and Victor "Vic" Schoenbach (both American graduate students of theoretical physics), decided to take the opportunity to demonstrate against the foggy Rhodesian past of the LSE director at the scheduled Commonwealth Prime Ministers' Conference. The students accused the School of maintaining links with companies that did not respect international economic sanctions against Rhodesia or that were otherwise involved in investments in South Africa. In the summer of the preceding year unsightly and—as was later discovered—frail iron gates had been placed at certain strategic points in the university buildings to block access to the university should the need arise. On 17 January 1969 the Chairman of Governors, Lord Robbins, agreed to meet with the Students' Union and, following questioning about the gates, answered that they were intended to prevent non-authorised access during occupations. "The sharks smelled blood and drew nearer" (Dahrendorf 1995, 465). The students voted to have the gates pulled down within seven days, after which they would take care of the dismantling themselves. "Robin Blackburn, a lecturer in sociology . . . said that the gates were the material expression of class oppression" (Dahrendorf 1995, 465).

On 24 January, the students took action:

> When they found senior academics in position ready to defend the gates, the mood got angry. Suddenly, weapons appeared. . . . One of the defenders of the gates was Professor John Watkins, who clung to the third-floor gates. He testified later about a female student: "She had on her person the head of a pick-axe and the shaft was under her coat. . . . Did someone really shout: 'Don't hit him! He's a Professor'? Anyway, he was eventually prised away by five women." Hoch and Schoenbach recall: "We then went off to the St Clement's Building. As we ran through the refectory, brandishing pick-axe and sledgehammer, people barely looked up from their dinners. On the way over to the St Clement's building, we could see the members of the Chess Club, bent over and intent, locked in struggle. But then reports came that the cops had arrived. There was an almost immediate wave of panic: tools and implements were wiped clean of fingerprints, and dropped in all directions. Nobody wanted to get caught with a sledgehammer." (Dahrendorf 1995, 466)

Thirty students were taken into custody and spent the night in jail. The college remained closed for twenty-five days. Lectures were held in Bloomsbury and at Bedford College. LSE had captured the public's attention, and especially that of the media. Public authorities were also involved. "The Academic Board was exposed to all these pressures. . . . In any event, the measures taken were very strong. The contracts of two lecturers, Bateson and Blackburn, were terminated for misconduct, and their dismissal was upheld on appeal. A number of students had to appear before the magistrates' court and later before the High Court; four were found guilty. Three were suspended from the School for varying periods. . . . Two of the foreign students, including Paul Hoch, were deported" (Dahrendorf 1995, 468). The reopening of the college was announced on 12 February 1969. (For a detailed account of the same events, see also Harry Kidd, *The Trouble at L.S.E 1967–1968,* Oxford University Press, London, 1969.)

51. Ernst Tugendhat, born of Jewish family in Brno, Czechoslovakia, in 1930, emigrated to Venezuela, studied at Stanford, and received his Habilitation in Tübingen in 1966. He contributed to the re-establishment of analytic philosophy in Germany after the Nazi period. He has since held professorships in Heidelberg and Berlin, and is the author of *Ethik und Politik* (Frankfurt am Main: Suhrkamp, 1992).

52. Robin Blackburn, student and then lecturer in sociology at LSE, was expelled

from the School for having sided with the students during the protests of '68 and '69. He went on to become editor of the New Left Review and director of New Left Books (now Verso).

17 February 1969
Berkeley

Dear Imre,

My own Thermidor is close—it would be already a reality were it not for the fact that the status quo does not at all appeal to me and that such things as "academic liberty," which have some people pretty excited, are for me just academic affairs. Universities may once have had a progressive function and may have been in need of protection *then*—*today* they are schools for *specialists* and while there should be freedom for everyone, I have never heard any argument in favour of a *special* plumbers' freedom, or a gravediggers' freedom, or a pimps' freedom. But this is what "academic" freedom amounts to today, and so it means absolutely nothing anymore. But—the revolutionaries? What a sight! Still, in their strange way they are craving for more liberty and for a better education, hence I am on their side against that ancient dung-heap, the university. Yesterday the administration building was blown up in San Francisco State [University], today a new semester starts there. Hayakawa (the Japanese Lakatos) promised to keep the school open at all costs and "by whatever means necessary." The situation is getting very serious in San Francisco and it may have to cancel about 200 classes. The whole state college system may collapse.

Well, we shall see. Be well!

Paul

21 February 1969
London School of Economics

Dear Paul,

You are a very lucky man. We established that our agent, who was responsible for putting explosive charges in our last letter to you, sabotaged the act, and so now you are alive. In fact you are in no danger for a few weeks because we decided to execute the agent following the reference in your last letter to your forthcoming Thermidor.

As you may know, the LSE was closed for 25 days and now it is open again with the majority of students united against the majority of staff. This militant majority of students is supported by about 5,000 people in the country including four educational correspondents. On the other hand, we are supported by Lord Robbins;[53] Lord Robbins is supported by the government and the opposition, and they in turn are supported by 50 million people who are only waiting for the opportunity to send all intellectuals, whether teachers or students, to hell. I needn't say that I'm having a good time and enjoying the situation. I enclose a revolutionary leaflet mainly against John Watkins who is now our hero, and his statue will soon replace Nelson's in Trafalgar Square.

It was very kind of you to recommend me to Peter Caws,[54] but he has not yet written to me. I hope he will; in fact, we know each other and even like each other, I think. I am very anxious that he should invite me because I am determined to spend September and October in the United States to have a change from English women. I would even pay my expenses if it comes to that (at least to New York; I hope Caws pays from New York to Minneapolis). I shall then go and see Popkin[55] in La Jolla, you at Yale, and study the student revolution all over the country.

The great leader of the free world, Richard Nixon, is coming to London on Monday and I bet he will have a relatively friendly welcome. After all, it is under his nuclear shield that these student revolutionaries (and you) are playing their infantile games.

Your friends from New Left[56] have not yet contacted me about my preface to your opera omnia. I shall certainly consider such an offer favourably. I was pleased to hear that you are being made an honorary doctor by the Jesuits in Chicago. I am all for the Red-Black[57] dialogue. However, do not forget that you were for the cold war; and I hope still are.

We are all very much looking forward to your visit in the spring; as you know, Agassi is coming next week to give my lecture and to perform in Karl's seminar in two weeks' time.

With warm greetings (and many thanks for your intervention with Caws).

Yours ever,

Imre

53. Lakatos is referring to Lionel Robbins, economist and professor of economics at LSE from 1925 to 1961, and subsequently Chairman of Governors between 1968 and 1973.

54. Peter Caws, professor of philosophy at George Washington University, in 1969

organised a symposium on the relations between history and philosophy of science at the Center for History and Philosophy of Science at the University of Minnesota. The proceedings of this conference were published as volume 5 of the *Minnesota Studies in the Philosophy of Science,* edited by Roger Stewuer and with a foreword by Peter Caws. Lakatos was probably involved too late to be a presenter, and eventually decided the conference was not worth the trip.

55. Richard H. Popkin, lecturer at the University of California, San Diego, in the sixties and now emeritus professor at Washington University in St. Louis, is the author of the magistral *History of Scepticism from Erasmus to Spinoza* (1979)—dedicated "to the memory of Imre Lakatos." Lakatos and Popkin first met in the early sixties and then became friends. Since the 1950s Popkin's works, centred on theology and scientific revolution in the seventeenth century, had had an important influence on Lakatos (see, in particular, Lakatos 1963–1964b and this volume, appendix C). Thus it is not surprising that Popkin was invited by Lakatos to deliver the opening speech at the 1965 International Colloquium in the Philosophy of Science held at Bedford College in London (see Popkin 1968).

56. Lakatos is referring here to New Left Books, which published *Against Method* in 1975. The publishing house should not be confused with the New Left, which is the name of the leftist political faction.

57. Lakatos could be alluding to the attempts made to include the Afro-American freedom movements in the more or less "new" Left. This could also be taken as reference to the colours commonly used to symbolise the movements of revolution and counter-reaction in recent European history. More likely the passage suggests that professional proletarian revolutionaries (red) can generally be associated with the Jesuit society (black), as in the case of the terrorist Jesuit Naphta, the character in Thomas Mann's *The Magic Mountain* considered by some critics to be the young György Lucàks (see, for example, Lichtmein 1970, 87).

24 February 1969

Dear Paul,

[. . .] The LSE revolution is crumbling; since we announced that we intended to chuck out three staff and ten students, we have not had so much as a stormy meeting, let alone sit-ins or other terrible things. This shows the truth of the old Stalinist slogan: *if you meet with an irresistible force, oppose it with double the force and you will always win.* Of course, when the expulsions are announced, student solidarity will find it difficult to do nothing. In order to help them in solving this problem, on that day a parliamentary investigation committee will move to have its sitting at the LSE, and, I assure you, the place will be as quiet as a cemetery.

With warm greetings and many thanks,

Yours ever,

Imre

3 March 1969
Berkeley

Dear Imre,

I have carried out all your orders (letters to Yale, Minneapolis, Princeton). Have you carried out mine—i.e. has the latest version of my Kuhn-appendix (title: the role of reason in science) been incorporated into *Criticism and the Growth of Knowledge?*

[. . .] Here in Berkeley we now have tear-gas almost every day, the national guard has been moved here, windows on office doors (some right next to my office) have been smashed, a policeman has been beaten up, bombs explode during lectures, another fire has started, this time in Dwinelle Hall, and Reagan's popularity is at an all time high (78% now believe he does a good job mainly on the basis of what he *says,* on TV, about the students). In my classes I now only talk about the strike and you should hear my criticism. I think we both agree completely in our criticism of this *lumpenbourgeoisie,* however I think that the university is lousy, too, and must be reformed by giving the children a voice in the administration.

The Black problem (which is mixed up with all the mess here) gets worse every day. Four times now I have seen Black militants and members of the American Nazi Party agree on TV panel discussions. Here in Berkeley the Blacks have nothing but contempt for the white radicals and only make use of them where they absolutely need them. Crying girls with Mao buttons come to me: "They don't want us." Violence has long ago spread beyond campus. A school board meeting in SF dealing with the problem of busing school children turned into a war only a few days ago, for the Whites (trade unions etc.) now start fighting back. We shall see.

Has Joske already arrived? Are you civil to him? (Never mind how rude *he* is, just give a party in his honour and put him to shame.)

About six weeks, and I shall be in London again. Until then—all the best.

Paul

6 March 1969

Dear Paul,

Thank you very much for carrying out all my orders. On the other hand, you may be pleased to hear that now there exists a complete master-copy of your Kuhn paper with all your corrections. The only editorial correction which I made, and I hope you won't mind, is that I crossed out "Appendix" and instead listed "The Role of Reason in Science" simply as the last chapter of your paper which I thought was a splendid conclusion.[58]

[. . .] Both my students and I are very eagerly waiting for you; and I suppose we shall have a few symposia where we can jointly entertain an audience at other people's expense. I am still the favourite Fascist of the Socialist Society so their invitation to us still stands. I very much enjoyed reading the list of violences which you recorded in Berkeley; unfortunately, since we obtained the court injunctions here, everything has been so boringly quiet that I am playing with the idea of resigning. The only personal violence which occurred was that I kicked a rather nasty so-called Socialist, but unfortunately he did not dare kick me back, and the thing fizzled out. It is very difficult to introduce violence into British society. [. . .]

My promotion seems to be all right, although it will not be official before your arrival.[59]

Yours ever,

Imre

58. Reference is here to Feyerabend 1970c.

59. Lakatos was appointed "Professor in Logic with Special Reference to the Philosophy of Mathematics" at LSE in 1969.

7 March 1969
Berkeley

Dear Imre,

It seems I can get you $200 from Berkeley for Sep./Oct., so put that on your map, too. You may get even more if you are prepared to talk about *counter-revolutionary education* in addition to talking about counter-revolutionary philosophy of science. Maybe I can get you an invitation from Hayakawa, too. But, at any rate, put Berkeley on your map.

All the best,

Paul

12 March 1969

Dear Paul,

Thank you very much for your letter about the 200 dollars. I would certainly love to come to Berkeley to preach. In fact, if you by any chance come across any university on the way which would invite me, I am willing to do anything for money.

I am sorry that you missed my name in the list of professorial spies and academic assassins. The simple reason for this was that before the big police intervention I solemnly warned the Director and his colleagues present in the room that I was all against the use of the police because the British police are no good; but that if I had a machine gun I would do the job. They disagreed with me and therefore, at least temporarily, I decided not to cooperate in this despicable liberal ritual. On the other hand, in hindsight, I now tend to believe that I may have been wrong. The consideration of this problem takes up quite a lot of my time and energy.

The students are clamouring for you and everybody is looking forward enormously to your lectures and company. Especially after Agassi's visit you remain the only philosopher outside London for whom they have some respect.

With warm greetings and many thanks.

Yours ever,

Imre

14 April 1969
London School of Economics

My dear Paul,

Let me first congratulate you on "Against Method"[60] which I enjoyed *enormously,* without, of course, agreeing with it.

Also let me thank you for everything you have done lately to promote my Autumn Survey of American Girls.

I spent two months with Marsha which somehow in the end misfired. I have to confess that I have grave doubts about the wisdom of letting it misfire, for Marsha has more charm than any of her predecessors. I simply do not understand what went wrong. Anyway, the wounds are still sore on both sides so treat them with tact and care.

I miss you. Come soon.

Love,

Imre

P.S. I received in fact the invitation from Peter Caws and I am very grateful indeed to you for all you have done to further it. On the other hand, there is as yet no invitation for a definite talk, and as you know I, like you, do not go to conferences to listen but to be listened to. I sent Peter Caws an express letter requesting him to do something about this. But of course the most attractive possibility would be if we two could have a symposium: I with the title *For Method* and you with the title *Against Method.* If you really want to be a dear nice boy you will ring Peter right now and tell him that you are all for such a symposium.

60. Reference is here to Feyerabend 1970a.

21 April 1969
London School of Economics

Dear Paul,

I was very sorry to hear that you were unwell. I hope by now you are fine and bouncing.

Many thanks for your cheque. I shall give your talk on Friday.

I am depressed about Marsha but cannot make up my mind about making a revolutionary move. You are the only person with whom I could discuss it, but you are not here. *Hurry. But please come here before Heidelberg!*

The LSE has started to dismiss lecturers. Blackburn was sacked last Friday. I do not have the full account yet. Bateson was sacked too. AND: there is peace and quiet.

My father became ill in Sydney at the age of 77. He has not got a penny so I have to support him, which means that my professorial increment will all go to Sydney.

So my world is now not exactly jovial.

I shall start rereading your "Against Method" tomorrow.
Be good and come soon.

Love,

Imre

2 July 1969
Berkeley

Dear Imre,

I almost got a marvellous house up on the hills, with a large garden, and a separate room and a separate entrance for you. At the last minute I lost it, though I had quite a good lawyer looking after my interests (I was the highest bidder, but someone else had all the cash ready while I would have been forced to take a loan). So I am still looking and I might find another house before you come. [. . .] Also I must prepare my lectures in a slightly different way, there are Black Panthers[61] in them, they sit in the first row and grin at me, but after my first lecture one of them came up to me and said "beautiful" which, I think, is quite a compliment. Searle,[62] on the other hand (or, rather, his wife) thinks that I am evil and she means it seriously. Two days ago I was at Searle's house to look at it (he wants to sell it) and I started talking about my ideas of grading etc., and they started jumping about and then they told me in all seriousness and with great vehemence that I am the devil, that I am evil and so on. Well, I left in a hurry. But S [Searle] still likes you, especially after he read your famous letter (I have to get out of this house, all the neighbours have poodles and they bark all the time, it drives me nuts).

Enclosed find the contents of the books I am supposed to review for Cambridge University Press (remember?). It is depressing. Why, the book on confirmation and acceptance is almost all about the ravens. Are these people going nuts? Well, I sent them an appropriate evaluation.

With your evaluation of my quantum paper[63] I agree, you are right, *it is a lousy paper* (signed: Paul Feyerabend!) despite the fact that it has some nice ideas in it, but it is much too long, with much too much irrelevant stuff in it. When you come I shall explain to you what I think the nice ideas are and the rest should be forgotten as quickly as possible. Of course, a scholar writes the footnotes first, but you know how *I* proceed? I write the text first, and then I read lots of books and sprinkle footnotes all over it. But then, fortunately, I am not a scholar.

Incidentally, I am again getting dissatisfied with *Against Method* (the *book,* not the article) and may again rewrite it. It won't make any difference to you, for there will only be stylistic changes. I am also prepared to rewrite following *your* advice. The thing should not only *be* good, it should *sound* good.

So, you are in a bad mood? And you think this is a sign of old age? Have you never seen little children in a bad mood? Stamping their feet and shouting? During the last two weeks I had to take tablets against pain almost every day and to the people who come to me and say "smoke pot" (there are lots of them), I can only smile and reply: "Ducky, I have been on the stronger stuff since before you were born," which earns me lots of respect.

Well, best wishes, and I hope you soon find a nice girl to massage your back.

Paul

61. The Black Panther Party for Self-Defense (whose panther symbol was borrowed by the Lowndes County Freedom Organization of Alabama) was founded in Oakland, California, in 1966 by Huey P. Newton and Bobby Seale. It fought the exploitation and oppression of the Afro-American community. Its programme argued for peace, for the power to determine the fate of the black community, and for the end of capitalist theft from the black community. It claimed that because the racist U.S. government had stolen so much from Afro-Americans, they had a right to the 40 acres of land and 2 mules that were promised to them a hundred years ago in compensation for being slaves; however, they were willing to accept cash instead, which would be shared out among the community. The programme claimed the right of Afro-Americans to land, bread, housing, education, clothing, and justice. It demanded an immediate end to police brutality and murders of members of the black community, but claimed that this would be possible only by organising defence groups. The Second Amendment of the U.S. Constitution gives every citizen the right to keep and bear arms; thus, the Panthers urged that every black person have a weapon to use as defence. The books the founders used as their references were Frantz Fanon's *The Wretched of the Earth,* Mao's "Little Red Book," and *The Autobiography of Malcolm X.* Its militant members were in the hundreds, its sympathisers in the tens of thousands. Repression against the Panthers was brutal, with the police carrying out raids against party headquarters, and the leaders of the movement being persecuted or jailed. Huey Newton was murdered in 1989.

62. John Searle, philosopher of mind and language, is professor at the University of California at Berkeley.

63. Feyerabend 1961b.

25 July 1969

Dear Imre,

[. . .] Did I tell you that I was invited to Brandeis to a debate with you (to be added as a bonus to your talk)? I feel quite tempted to accept; on the other hand, I am becoming more and more irrational, so what's the use of a debate? Incidentally, there is a guy at Brandeis (an assistant professor) by the name of Burian who is going to write a thesis all about me, and has asked me for offprints etc. etc. So you had better get in touch with him and put him right. He didn't like the early PKF for being too rational, and he doesn't like the late PKF for being too irrational. He might just like the medium IL who is neither rational nor irrational.

It may please you to hear that I blasted the Radical Left in my lecture yesterday. The immediate reason was something that happened some time ago: an education professor, Jensen, published a paper to the effect that Blacks learn by rote while Whites learn by ideas and that the difference is innate.[64] They (the Left) made a lot of noise in order to get him off campus. My criticism: (1) are they so easily satisfied? If they don't see him, they are not worried about him any more? What if he gets a super position at another university? That doesn't seem to interest them for their principle seems to be "out of sight, out of mind." (2) Assume the guy was misled in his research by his ideology, assume that he experimented with racism in his mind. Wouldn't it be interesting to see, to study from close-up how allegedly "objective" science is influenced by very subjective ideological considerations? Is it not absolutely necessary to keep such people at hand close by in order to learn from a living example, that the "objectivity" of science is just moonshine? (3) Getting rid of somebody does not mean getting rid of his ideas. The proper way is to keep him here and, if one thinks he is using a faulty method, to criticise him *in class*. But for this the splendid *New Left* is too cowardly—they might get an F! (4) But assume he is right! Is that so bad? Is it so bad to live in a society where different people have different talents and learn in different ways, or is it not rather an advantage? Conversely, does the reaction of the lefts not show that they want to be like everyone else, that is, that they want an absolutely boring society? (All this in connection with a reading of Mill's *On Liberty*). There are black panthers in my class, and I think they agreed with me.[65] You would have loved this lecture.

Anyway, that is all for today.

Be well!
Greetings from all the little girls.
Greetings also from some BIG girls
to the great lover and mathematical philosopher,
and my greetings on top of all.

Paul

64. The reader should refer to lecture 6, note 1, this volume; see also Jensen 1969. In his autobiography, Feyerabend notes that in one of his crusades to promote "Reason, Law and Order," Lakatos visited Jensen in Berkeley "to recommend his methodology as a nostrum" (1995, 129).

65. Feyerabend recalls: "More black people turned up in my classes . . . and I was often at a loss. Should I continue feeding them the intellectual delicacies that were part of the white culture? I was a teacher, a professor even; I had accumulated information about all sorts of things and had strong opinions on a variety of topics and little respect for the leading lights of my profession. But now I felt ignorant and out of place. . . . I understood why the students welcomed Huey Newton, Bobby Seale, and their followers as if they were gods. Malcolm X came and gave a well-reasoned speech. He was dressed like a businessman, orderly, with a briefcase containing his papers. I read *The Autobiography of Malcolm X* (written by Alex Haley, who later wrote *Roots*). I felt sympathy, even love, and an urge to contribute, though minimally, to the life of this extraordinary human being" (1995, 123).

11 August 1969

Dear Paul,

I finally made up my mind to forget about Mary's tender soul and to dictate a nasty letter. I do not really know whether I should sound furious or simply deeply hurt; the most awful thing about getting old is that although one can sound very furious and very hurt, one is neither. So I shall be pragmatic.

You kindly engineered for me an invitation to talk in Berkeley and in Yale; I was delighted to have these two invitations (although I have not yet heard officially either from Yale or Berkeley) on the assumption that you will be there as you originally promised to be. Now it turns out that you will not be in Berkeley on October 1st and you will not be in Yale in mid-October. Since my main point in going to these places was not to preach the truth which I rather prefer doing in print, but to have a good time and enjoy your company, I feel betrayed.

So it is absolutely out of the question that I should come to Yale; in Berkeley the situation is different since I already seem to have fixed up for two talks in California and I may just as well go to Berkeley;

and also I know Berkeley and I have a couple of friends whom I would like to meet there, but at Yale I do not know a soul, I have never been there in my life, and I am too old to find my way in strange places where I have no friends. I will go to Toronto.

So this is the first blast to you, you bastard! Now comes the second.

You asked me to tell George Lichtheim[66] that you wrote that letter in order to be nice to me. Are you really that damned mad that you don't realise that with this gesture you may be nice to George Lichtheim but you are nasty to me; moreover, your niceness to Lichtheim will not help at all because he will despise you for your volatility. Do you really think that people who preach philosophical subjectivism have anything but contempt for practised subjectivism?

I think you will now realise why I have not written you a personal letter for some time, not that I want to break up our friendship because I am afraid you are still better than others, but I really think that your unreliability in matters professional should not also turn into unreliability in matters private. One cannot be consistently nice so you must make up your mind about a consistent sub-set.

I hope that by now you have cleared up the ambiguities in your references which, actually, might be at least partly my fault.

Will you come to Minneapolis? When on earth will I see you? In Pittsburgh?

With warm greetings.

Yours ever,

Imre

66. George Lichtheim, who was educated in Germany, had been living in Britain since 1945 and taught in various American universities. He contributed numerous articles to major British periodicals and was a distinguished scholar of Marxism.

15 August 1969
University of California, Berkeley

Dear Imre,

I have always been convinced that searchers for the Truth are deadly serious people who take everything they read literally, who are unfamiliar with the concept of an innuendo, or of a joke, who cannot read between the lines and to whom one must therefore talk as if one were talking to children. However, I thought that *you* were an exception. Your letter shows me now you are not. Hence, I must explain to you what I *really* meant when asking you to tell Lichtheim etc. etc. What I *really* meant was *not* that you should actually tell him. Nothing could have been further from my mind. What I meant was something like this: this guy (i.e. Lichtheim) opposes his subjectivism to my dogmatic objectivistic attitude. *If only he knew* that I am a better subjectivist than he, for I wrote my letter (and this is true), not in order to refute *him,* but in order to please *you,* the fact that *he* might get mad at my letter to the TLS being an added incentive. *This* was what I really meant, not an order to you to tell him. Why should I try to get in touch with him, even indirectly—he does not even interest me. Now, of course, you might have turned my joke into another joke yourself by actually telling him and thereby making him despise me (as you say he would). I would have had no objection to your proceeding in this way, for I couldn't care less whether L likes me, despises me, tells everyone that I am a contemptible fool, etc. etc. Altogether I care very little about the opinion people have about me as long as the money keeps coming in (there are a few people about whose opinion I do care, however, and you, you fool, are one of them).

Number two: your schedule. Originally we agreed that you would come to Berkeley *mid-September,* that you would stay here until the end of September and that I would be in Berkeley, too, at that time. Yes? Then I got a letter from you telling me that you would be in Yale *Sep. 22nd* and in Berkeley Oct. 1st. Well, I thought that you had changed your plans for some reason and that these were your new plans. Now I am not yet in Yale on Sep. 22nd, so I could not see you there. And I am no longer in Berkeley on Oct. 1st, so I could not see you in Berkeley either. I regretted that, but I hoped at least to see you in Pittsburgh on Oct. 6th ff and I firmly planned to come to Pitts. Now I hear that you will be in Yale mid-*October* and *not* mid-September. In mid-October I am of course in Yale. So there seems to have been some grandiose misunderstanding. Now when you changed your plans (i.e. not to come to Berkeley for two weeks in September) I thought there was some reason for it and not for a second did I think of calling you a bastard, for, not being a searcher for the truth, *I can afford* to be guided by benevolence rather than by careful analysis (bold hypothesis, refutation, another bold hypothesis etc.) of facts and this is one more reason why I am by now a most determined foe of critical rationalism (which, incidentally, has nothing to do with my relation with critical *rationalists*). Of you I shall always think fondly whatever you decide to think of me. Anyway—see you at least in Pittsburgh and I think *you owe me at least two dinners* (or, alternatively, 5 Feyerabend footnotes in your papers) for having so grievously mistaken a joke for reality and a misunderstanding for malevolence. I also *demand* (and this is *not* a joke) that Mary be given the opportunity to read this letter in order to improve her knowledge of you (her knowledge of me, I think, is not in need of being improved).

Be well, and good wishes to the fool from

the bastard Paul

Just returned from Hollywood where I resumed my singing yesterday.

P.S. Why did you wait so long, and why did you not write me a nasty letter at once? This would have cleared up the muddle much earlier.

20 August 1969
London

Dear Paul,

I have always been convinced that people with contempt for truth are deadly serious people who take everything they read literally, who are unfamiliar with the concept of an innuendo, or of a joke, who cannot read between the lines and to whom one must therefore talk as if one were talking to children. However, I thought that *you* were an exception. Your letter shows me now you are not. Hence, I must explain to you that:

(1) I understood exactly what you meant about Lichtheim and I thought it highly amusing that while you expect me to understand your jokes (which I understand) you expect me to be deadly serious, and refuse to understand *my* jokes.

(2) On the other hand, I was not exactly joking when I complained about your shiftiness. I should like to remind you that I complained to you that I shall lose some money on this journey and that therefore I unfortunately have to pinch the pennies. Therefore, I just could not afford to live in a hotel in Berkeley for two weeks for the $100 which I hope to get from John Searle. You promised that you would arrange something, but since you have not written anything (except that your original plan concerning your musical friend's house fell through), I had to change my plans and insert more lectures. Anyway, let us not go into all this; but let me point out that even at this stage my only lecture for which I have not yet got an official invitation or a time fixed is at

Berkeley. I even wrote a letter in my despair to John Searle to ask him about the date of my lecture and whether I should get some money but he has not yet replied either. So this is the story. My timetable still has a certain flexibility but even now I do not know when you leave Berkeley and I cannot really make any effort to try to be there to catch you. (I shall come to Berkeley from La Jolla where I am spending a few days.)

You wrote that "for having so grievously mistaken a joke for reality and a misunderstanding for malevolence" I must give you either two dinners or five Feyerabend footnotes. Since I did not mistake a joke for reality and since I never accused you of malevolence, *therefore* I inserted one new nasty footnote about you and Bohr in my paper. (If you don't understand how the conclusion follows from the premises—I am referring to "therefore"—do ask Tarski.) But let me end up by telling you that I arrive in Pittsburgh on Monday 6th and shall stay there until Wednesday 8th; and if you are not there I shall shoot you on the spot.

I meekly introduced all your corrections in the galleys, except for one. I refuse to put "it" instead of "Kuhn's defence" (on p. 9). This would render the thing meaningless and everybody would say that Feyerabend could never have written down such a silly sentence and would accuse Lakatos and Musgrave of being careless editors.[67] This is my last word on this particular matter and if you don't like my Decision do write a letter in 500 copies about my censorship. I am sure Bill Bartley will give you his address list.

Joske [Agassi] sent me a copy of his Schilpp paper about Karl which is a Byzantine piece of flattery.[68] However short it is there are two hundred footnotes which he did not enclose. I wrote to ask him for these footnotes and to tell him that I read his *Historiography* again, enjoyed reading it, but would make some critical noises about it in my next publication. I enclose his reply for your amusement.

Thinking further about your unusual unreliability (you have not even congratulated me on my letter to *Science* which, incidentally, they have not even acknowledged), I concluded that there must be a woman in the background. What will be my role in your fourth marriage? Shall I be witness at the wedding or at the divorce or both?

With warm greetings.

Yours ever

Imre

P.S. I am terribly interested in what you say about planetary orbits; but you had very exciting things to say about the history of optics. I hope that this subject will not meet the same fate as the other.

67. Reference is once again to the paper Feyerabend wrote for *Criticism and the Growth of Knowledge* (1970).
68. See Agassi 1974.

22 August 1969

My dear Paul,

[. . .] I am madly busy: Kuhn sent his final reply, which, incidentally, is both interesting and polite.[69] If you are a good boy I shall send you a copy. He stands up very nicely for standards and condemns your hedonism. It seems that, after all, you will be the only revolutionary left. However, I find your newly-found "dadaism" an acceptable position for a gentleman and a scientist and I am already writing a paper in partial support of it.

I am leaving London on September 3rd, so if you want to be nice, answer quickly.

One more point: Blackburn's dismissal was confirmed by the Appeal Court. The Appeal Court pronounced that non-condemnation of violence is incompatible with holding a university teaching job. With this principle we jump a generation ahead (or, as you put it, behind) of the United States.

With fond greetings,

Imre

69. See Kuhn 1970b.

18 September 1969
Berkeley

Dear Imre,

Of course, personal relations, emotions etc. can be just as phony as any search for the truth or as any work, but occasionally a change of phoniness is urgently required. I think I need one. I have now been in this business for about 15 years (teaching) and even longer going through the motions of thinking, so the alternative I envisage is: a farm, lots of dogs, cats, fleas, horses, children plus her, of course. Now, don't you worry that I shall run into disaster.

First, I am in disaster already, viz. I am in the philosophy of science. For years now I have been thinking that I am living at about 20% of my full capacity, not for work but for just being alive one way or another and it grieves me to have to go out of this world before I have ever been in it, one hundred percent. And being in the world one hundred percent does not mean filling every nook and cranny with work, it means developing as many as possible of the talents one has, and one of my talents is dealing with dogs and human beings below 3 years of age (this, of course, includes some of my colleagues, but I am not now thinking of them).

Secondly, don't worry, for she is much more reasonable than I am. Being connected with stars and witchcraft she is also much more solidly established in this world than I am, and is further helped in this by *I Ching*. So—what can happen to me? The worst is that I go mad and, considering what some sane people are doing, that might not be so bad after all. And as regards God, of whom you speak and whom I think I had kicked out more than 30 years ago—I am not so sure. Psychiatrists have the task to change the social surroundings and make them more humane. Why should they (or we) not change the physical surroundings also by the invention of cosmologies which are peopled by all sorts of spirits? The Christian God is *not* my friend, but Jupiter would be (especially when he turns into a bull)—at any rate, such a world would be more friendly than the world we are living in now, and all we have to do is *invent it* and then persuade everyone to participate in this particular research programme. The courage to enjoy a *meaningless* life can arrive only after there is a meaningless *life,* and to find one is just my intention, one way or another. So, you see, I have left Hegel, and have now arrived at Kierkegaard whom I greatly admire (I am now reading the *Concluding Unscientific Postscript*). Of course, Karl, and you, and Wittgenstein read all these things in their youth. I was never an irrationalist, so I was never a rationalist. It is a little late in life, but better late than never.

Incidentally, I have written my "farewell to Popper," and it will be published in German in the book I told you about. I have finished the paper, it is now Xeroxed and I shall let you have a copy as soon as I get one. It also has three pictures and replaces argument by ridicule and selfrighteousness.

Selfrighteousness—Searle is trying to get me either fired, or at least severely reprimanded by the Senate (whose academic freedom committee he chairs) for incompetence. Being the Chancellor's special assistant he may be able to do a lot—so we shall see. In the meantime, I do not intend to become competent (or honest, as he wishes me to be).

So, I do not know what is behind me, for the case of rationalism is finished for me once and for all. Nor do I know what is in front of me, except that she will be back in ten days and that I shall meet you in Pittsburgh (I am arriving in Pittsburgh Friday evening, October 3rd, and shall stay there until the Wednesday after—so we should have lots of time for talk. I have rewritten large parts of *Against Method* and shall bring the stuff with me. *Empiricism ii* won't be changed, it will just be incorporated into the new version of *Against Method*). Blackburn is in Berkeley, and I shall try to get in touch with him. I am not looking forward to Yale. It is a dump, intellectually and otherwise, but the money is good and I do want to buy a house soon. Once I have got *that* I shall drop everything except two quarters every academic year in Berkeley and partially retire.

So, be well, don't confuse too many people, I hope you are finding the girls satisfactory at the various places. Here I have made reservations at the Durant Hotel, I have also rung Olsen and shall come along with him when you arrive.

Finally: Kuhn has not arrived yet (his MS). You say he is most severe with John [Watkins]. Well, *we* agree with Kuhn on many points, but criticise him. John doesn't seem to see any point in Kuhn.

So, back to the most recent version of *Against Method*—and don't worry.

Paul

12 October 1969
Berkeley

Dear Imre,

Thousands of praises to the Vietnam Day Moratorium of October 15th! Here at Yale it is left to the "conscience" of the individual instructor to either hold his classes, or to talk about Vietnam, or to cancel his classes altogether. Well, my conscience told me to hop on a plane and return to California for a brief rest and here I am now, in the middle of sunshine, feeling like a prisoner who has ten days' vacation and has in this short time to make up for all the time he has lost in prison.

New Haven is a very nice town, plenty of trees, the surroundings full of woods, hills, but the sky is dark, even if the sun is in it and for somebody accustomed to the brightness of a California sky it feels like living at the bottom of a pit. There are lots of bright people here, some physicists are now after me with hidden variables, but to me sunshine means more than all the ideas in the world, past, present, future, taken together, and a sunny day is worth more than all the wisdom of philosophy. I have even a *double sun* here, and even though I may not see her all the time I am here, I still feel her presence.

Of course, I am a Romantic, also in my private life and this is another reason why I cannot be a friend of reason and rationality. Being a Romantic I may of course never succeed, but if I *do* succeed the results will be millions of times better (Romantic exaggeration) than if I had achieved them waddling along the straight path of rationality. This, incidentally, makes me very much regret that I could not debate with you, for, I think, I could have made life difficult for you, not *very* difficult, for you are much too skilled and ruthless a debater for that, but just a tiny little bit difficult, and that would have been sufficient. But I really had a cold, and by the time Brandeis arrived I was afraid to have it recur again, so I stayed at home.

Incidentally, it just occurred to me that Joske had no reason at all to be sore, I never told him I was coming, for I knew he takes these things so seriously (when one says "I shall come to Brandeis on . . ." he interprets this to mean "I solemnly promise before God and the world that I . . . "), I never was invited by him to his dinner, as a matter of fact my schedule did not include any dinner, even if I *had* come, so I did not owe him anything. I *did* owe Weisberg something, but that I had already cleared with his secretary, the very same afternoon. It seems that of all Popperians (and this by now already means "ex-Popperians" and in my case even "anti-Popperians") you are the only one with a sense of humour and perspective, a true critical rationalist. By the way—how *was* the dinner? How *was* your talk? Did you vilify me sufficiently? (You know, vilification by the right people just makes one's reputation rise.) And how is London today?

Please send:

(1) The exact date of the beginning of summer term at UCL [University College London]; (2) 50 tablets (or five phials, ten tablets each) of Redexon which is effervescent vitamin C, 1 gram per tablet (I need this, and they don't have it here for some reason). Let your communist

secretary send it to support a dying anarchist. Send it *airmail* to my *Yale* address and I'll pay you back all the expenses. I'll be back in Yale Oct. 21st, that is, I am leaving here on Oct. 21st early in the morning to return to my prison—so send a nice cheery letter to me in Berkeley.

All the best

Paul

P.S. How does the new version of *Against Method* sound?

Keep your fingers crossed that I see her (she does not like to be pressured, and I shall not. I may not even ring her once while I am here; we shall see. And don't reason with me on these matters. It is a waste of time).

P.P.S. Before I left I had a discussion with radicals in my class in Yale, and I found myself defending the American Constitution. But—what a narrowminded, bigmouthed, phrasemongering bunch of people they are! Wherever in the world you go, the same phrases (just as with Popperians). I am glad I am an anarchist. They also got very upset for I told them that while they were sitting around and talking for hours on end and complaining that everything was against them, Nixon, with two defeats behind him and everything against him started working and *won.* "He has much more guts than you" I said, and they didn't like it a bit. God help us if these people ever get into power, for what a drab and boring world *that* would be. And *both* of us would be in trouble.

15 October 1969

Dear Paul,

Many thanks for your letter of October 12th. The Brandeis meeting was my only unsuccessful meeting in the United States (at least, as far as I know; have you found out anything about the reactions to my talk in Berkeley?). I first explained to them your views to make up for your absence, but although I said that it is an extremely important position and only I can refute it, people simply didn't believe that you could be so wicked as to deny science's supreme authority. Then, when I defended authority in science, again they thought that I could not be that wicked and that I was obviously joking. Anyway, I have never seen an audience as stupid as the one at Brandeis, and it is quite clear that they are educated with empty slogans, both from the right and from the left. Being a Jewish university I am amazed by the pompous and vulgar illiteracy they have developed. (So you see that nowadays, when I give an unsuccessful lecture, at least I don't blame it on myself; incidentally, perhaps it would be better to blame it on your sudden absence rather than on the audience.) But, anyway, it went down very badly. [. . .]

Let me tell you finally how much I enjoyed your letter. I am afraid I am not a very good correspondent so I just hope that you turn up in London soon. However, I shall write to you in some detail about your *Against Method* very soon.

With warm greetings.

Yours ever,

Imre

17 October 1969
Berkeley

Dear Imre,

I see you are in the postcard stage. This I interpret to mean either that you are travelling or that you spend a great deal of your time in bed. Now the last postcard has Marsha on it too, so the latter interpretation seems to be the correct one—or am I wrong? If you still want to sink into the comfortable morass of married life, why not propose to her? Under the circumstances (assuming the circumstances are as I imagine them to be) this would seem to be the most rational action, so, why don't you act rationally for a change? Or do you want to wait until the next degenerating problem shift? And if that really is your attitude, should I then not point out to you that an action on your part now might eliminate degeneration forever? Anyway—let me know what is going on, in your brain, and in your lower parts.

You can tell John [Watkins] that I liked his inaugural *tremendously,* a simple but important problem presented in a simple and very pleasant manner, and an acceptable solution suggested for it. Of course, he could not help slipping, occasionally, such as when talking about "decency" and the like (which I interpret as sour grapes—a married man like him would just love to get involved in something indecent, but he don't have the guts, and so he tries to convince himself that decency is not only the only thing that is *actually* left for him, but also the only thing worthy of consideration by a rational being). Anyway, I loved his paper.

Heilbron's thesis is on atomic models and it is very good. Heilbron and Kuhn have written a paper on atomic models, and it came out in the newly founded *Journal of Historical Studies in the Physical Sciences,* University of Pennsylvania Press.[70] The title: "The Genesis of Bohr's Atom." In the same number Forman has an article with the title: "Why was it Schrödinger who developed de Broglie's ideas?" Again, very good. The Kuhnians are doing *very interesting* historical work.

AM [Against Method] is in the hands of the Minnesota Press and I have asked them to send you a Xerox copy (the only copy I have is already cut up into pieces for the book).[71] The book is proceeding swiftly and will, I think, be very good. I do hope I shall have finished the MS by *March,* for then I am going on a long journey, all by myself, to London, Vienna etc. etc. [. . .]

I have again two monster classes, one 700, the other 150, and this despite the fact that I am not giving any grades in the first. In my philosophy of science class I started by defending Max Rafferty who suggested that genesis and evolution should be taught as two alternative theories rather than the one a fairy tale and the other a "fact." There was a riot, and next time there will be a biologist professor (a bigshot) defending evolution, and a fundamentalist minister defending genesis. That should really lead into anarchy. By now students as well as professors say that I am immoral and that I should get out of university life, so, you see, I share the fate of every true anarchist. In the end it will be you who will be loved by everyone and I who will be hated by everyone, for no one likes me without principles (when students asked me why I taught at university, I said: "because of the money and the lots of free time I have there" which hurt them deeply. I think they would have preferred to hear: "because I think I can do valuable work in education," even if I gave Fs only. They like committed people and I am committed only to one thing: non interference). [. . .]

Paul

70. See Heilbron and Kuhn 1969.

71. Feyerabend is referring to his 1970a (the paper) and to his 1975a (the book). Feyerabend was in the habit of continually revising his work, proposing new versions and revised editions at every reprint. After having first appeared in *Minnesota Studies in the Philosophy of Science* (1970), "Against Method: Outline of an Anarchist Theory of Knowledge" underwent extensive rearrangements and was finally published in book form by New Left Books in 1975, and successively in revised editions in 1988 (2d ed.) and 1993 (3d ed.).

30 October 1969
London School of Economics

My dear Paul,

Thank you for the many letters that I have received from you recently. I am puzzled about when I shall get the really last version of your *Against Method,* which I am itching to demolish. Am also pleased with your hectic love life, because I hope that this will keep you off revolutionary activities. But now I want to pick four major quarrels with you.

Quarrel one: Mary bought you the Redoxon and sent it off to you airmail. Have you received it or haven't you? In both cases you should have said so.

Quarrel two: I asked you to explore how my talk went down in Berkeley. In spite of the fact that you spoke to Kolakowski,[72] you only wrote what he said about you, but not what he said about me. Also I should be most interested in what Bill Craig, Matson, Scriven[73] and possibly others thought about my talk. I quite understand that you are not interested in this problem, but I am, and I would have hoped that I have some place in your egocentric world. Or have you turned into a complete solipsist?

Quarrel three: not only socially but intellectually you are, I am afraid, becoming a solipsist. On my lecture tour I made for you probably the best propaganda you have ever got. To mention one characteristic instance: after my talk in La Jolla, Dick Popkin came up to me and said why didn't I tell him before that you were such a marvellous man and asked for your telephone number. I told him that you don't have one. I would not claim that I do this unselfishly: since the main point of my philosophy is the destruction of your position, I have to convince people that you are a very great philosopher whom it is worth demolishing. As soon as they discover that you are a nonentity, they will think that I am a nonentity. Therefore, it is in my interest to blow up your reputation, which I am doing all the time. On the other hand, I warn you that it is also in your interest to blow me up for exactly the same reasons. To put this point completely concretely: I want you to distribute my "Criticism and the Methodology of Scientific Research Programmes" paper among your better students and make them write essays about it. (This also saves your time and energy; moreover, they may tell you what is in my paper.)

In this respect I want to tell you that in the book version of my paper Kuhn will be relegated to the footnotes and you will get the first chapter and the last chapter in the main text. From this it also follows that you had better finish your *Against Method* finally to get my references straight.

Quarrel four: in your German paper there is a particularly nasty remark about me on page 37 at the bottom in the last few lines: "Ein Gegner wird nicht mehr zur kritischen Detrachtung einer *Idee* eingeladen, er wird mit *Schlagworten* eingesschüchtert (in solcher Einschüchterung ist übrigens Lakatos unübertroffener Meister)."[74] Do not misunderstand me—I like the sentence, but you should support it with footnotes or examples. I am asking this because bloody Carnapians accuse me of the same thing. (See the review of my "Inductive Logic" paper[75] in the last issue of the *BJPS* by Schick.)[76]

All for now—good luck for your anarchic, intellectual and love life,

Imre

P.S. 1) I am sending you under separate cover 12 copies of my Kuhn-Popper paper[77] for distribution, but if you do with these twelve copies the same you did with the three copies of my *Proofs and Refutations* which I sent you in 1964 and which you put in the bottom of a drawer in order to keep secret, I shall kill you.

P.S. 2) I shall be in the Boston area next year between October 20th and 30th (or 15th–25th). Couldn't you organise for us at last a good joint circus at Yale with some good money in it?

72. Leszek Kolakowski was professor of history of philosophy at Warsaw University until 1968, when he was exiled from his country following accusations of political corruption of his students. He now teaches at Berkeley, Yale, and Chicago. He is a scholar of positivism and has written a detailed criticism of Marxist research programmes.

73. These are all names of Feyerabend's colleagues at Berkeley.

74. "You do not invite an enemy to criticise your ideas, but you threaten him with strong phrases (in this kind of intimidatory act, Lakatos is an unrivalled *maestro*)."

75. Lakatos 1968a.

76. See Schick 1969.

77. Lakatos 1968b.

20 November 1969
Berkeley

Dear Imre,

I have already started collecting notes on our anti-Marcuse[78] which I shall start as soon as *Against Method* is finished. Title: *The Testa-*

ment of Dr Marcuse (do you know the story "The Testament of Dr Mabuse"?[79]—a madman in a lunatic asylum who conceives plans for crimes, smuggles them out of the asylum and his outside followers commit the crimes. I saw the movie, which has a marvellous scene in it; you see Dr Mabuse sitting with a vacant look on his face in a padded cell, writing furiously, while (intercut) people outside blow up bridges etc. etc. I want to get this particular picture and put it into the book. It's supposed to become a critical *satire,* laughing at the bloody Puritans). Does this sound OK? Important: it should not be *too academic* but contain lots of humourous abuse.

All. Be well!

Paul

78. Reference is to Herbert Marcuse, a pivotal figure of the New Left and the student revolution who preached nonviolence and the necessity to overthrow the "system." See, for example, his "Repressive Tolerance," in R. P. Wolff, B. Moore, and F. H. Marcuse, *A Critique of Pure Tolerance* (Boston: Beacon Press, 1965).

79. A famous Fritz Lang movie.

12 December 1969
Yale University, New Haven, Connecticut

Dear Imre,

I am typing this letter because I have a nasty tremor in my hand which makes my handwriting quite illegible. Don't think I mean it seriously when I write you a nasty sounding letter—I haven't written you a really nasty letter yet, and I never will. You are of course right about the effects of recommendations, but my attitude is that people such as F [Finocchiaro] are so much better than the top-Carnapians, or the top-Goodmanites, or many top-Popperians (I am on safe ground here because you are not a Popperian—you are a Pop-Hegelian)[80] that they *deserve* the best possible recommendations. Besides, philosophy of science is such a rotten business that taking standards seriously *in it* means having no standards at all, so I have no standards in it, because I have no standards. And if you point out to me that being an anarchist I cannot have any standards I reply that being an anarchist gives me precisely the freedom to use standards whenever they further my case rather than using them all the time.

Anyway—more about the Mach-Vienna Circle, Philosophy of Science in my forthcoming article: "Philosophy of Science, a Subject with a Great Past," which defends Mach, condemns the Vienna Circle, and ends in a eulogy of you. I entirely agree with your observations on the revolutionaries. My class is full of them, and I let them have it. Now I have included two lectures about the rise of witchcraft in my programme, from modest beginnings to the height of madness and this, I say, is a good example of how Marxism has developed. I defend Rafferty. I criticise their ideas, their behaviour, their morality. Their tactics. What I get back is the same aria, again and again, no progress, phrases and more phrases, and a primitive mentality where one talks about slaughter as if it were a picnic. Well, if ever they take over, I shall leave this country at once, for I do not like to be surrounded by barbarians who shout so loudly that one hears them everywhere. I must write this criticism of Marcuse, but from a "superliberal" standpoint (that's what I'm called here—a "superliberal"). [. . .]

And now, I have to go to bed. Be well, and don't be too hard on me. All the best,

Paul

80. The pun refers to Lakatos being a "bastard-child" born from a Popperian father and a Hegelian mother (see also letter, PF to IL, 10 March 1970, where, in the same way, Feyerabend invents a new philosopher named "*P*egel" or "*H*opper"). On Lakatos's philosophy as an attempt to stand between Hegel and Popper, see Motterlini 1999. On the importance of the Hegelian-Marxist influence on Lakatos's thought, see Kadvany 1995 and Larvor 1998.

[postcard] 29 December 1969

Dear Paul,

Haven't heard from you in ages. Or is it the Christmas mail chaos? Anyway, send me a *sober statement* of your position vis-à-vis London. Also, I am fed up with waiting for your *AGAINST METHOD*. If I do not get it instantly, I shall refute its Lakatosian reconstruction.

With warm greetings for a happy New Year.

Yours ever,

Imre

1970

Tuesday, about 1 January 1970
Berkeley

Dear Imre,

Rain and fog outside, ants, cockroaches, a smelly stove inside, depression still further inside, but all that does not count, now that I have your letter. I had to fish it out from the mailbox and then bake it in the stove (the mailbox has a hole and gets quite wet when it rains), but now the baked letter is in front of me and I hurry to write my reply. No, AM is certainly not expanding. The reason why it takes so long is that I want to make it shorter, and still shorter, and still shorter, and that is not always easy. Also the move from one house to another was a bigger interruption than I thought it would be; I have not yet managed to get back into the routine and am looking forward *with absolute terror* to the return of the kiddies in the nursery school at the corner. There has been quiet the last two weeks, because of Christmas, but the quiet will be over next Monday and then I shall hang myself unless the carpenter whom I have asked to put in double windows and soundproofing arrives earlier. But then *he* will be hammering around in the place. So, you see, AM has to contend with great difficulties but I am doing my best to make it succeed. I hope to have chap. 20 finished by today or tomorrow and then the only further difficulty ahead will be the chapter about you as well as some footnotes. I have written all the details of your chapter down on a single page: there are a lot of them and you will be surprised how good you look from where I stand. [. . .]

Somehow I would like to get a job in Italy. Right now nobody knows me there, but vol. one of my collected works in Italian will be out in a month or so (or is already out), and that may help. Of course, I would have to learn to speak Italian, but that does not matter. [. . .]

Things are now clearing up, the silhouette of San Francisco appears on the horizon, the view is really quite fantastic, the light is also better so that I can see Mara's picture on the wall to my right (and what a picture it is!), it is not such a bad place after all, even Grice[81] can be taken care of (I said a few rude things to him at a party the day before yesterday which left him speechless) but still, I have to return to Europe, one way or another. I also have an overdraft in my bank from all the presents and all the invitations to lunches, dinners, monster

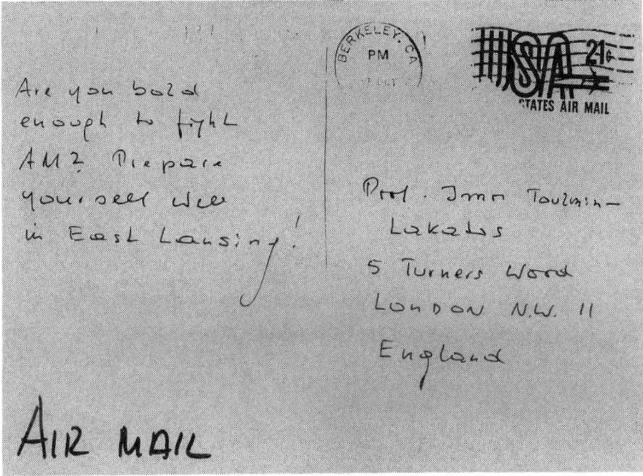

Are you bold
enough to fight
AM? Prepare
yourself well
in East Lansing!

BERKELEY, CA
PM

USA 21c
STATES AIR MAIL

Prof. Imre Toulmin-
Lakatos
5 Turners Wood
LONDON N.W. 11
England

AIR MAIL

Figure 3.3 "Are you bold enough to fight AM?" Postcard sent by Paul to Imre.

invitations where I sit surrounded by 10 people who are eating at my expense, so if you know any way of quickly making 100,000 dollars which you do not want to use on yourself, let me know. And all the best for the New Year, good wishes for an excellent Poppelian new year to you, Gillian and all the other planets, satellites, cosmic dust which have you as their centre of gravity.

Paul

81. Grice was Feyerabend's colleague at Berkeley.

28 January 1970

Dear Imre,

[. . .] There are now two alternatives open to you. (i) you apologise properly as befits a decent, moral, selfrighteous person who has realised that he has done wrong to others; (ii) you *don't* apologise which will make it clear to me that you are just as big a bastard as *you* think *I* am. Of course, I prefer the second alternative to the first, for the more bastards, the fewer selfrighteous people, the better the quality of life is going to be (Erste Paulinische Grundregel [Paul's First Fundamental Rule]). So, now you have your choice.

Cheers

Paul

31 January 1970
London School of Economics

Dear Paul,

(1) I do not think you are a big bastard. I think you are an occasionally charming, occasionally dangerous monkey.

(2) I do not perform, and do not demand verbal rituals like apologies. If I want to kick somebody, I do it myself and do not ask him to kick himself.

(3) You are muttering about *choice* (my choice: what to do now). I have *no* choice: I prefer monkeys like you to the average donkeys available in the market. But I have now bought a barge pole.

(4) *If* I get a copy of the final version of your *Against Method,* I suggest we call it a day and start quarrelling on a different subject.

Love

Imre

5 February 1970
London School of Economics

Dear Paul,

[. . .] Karl is again working on his autobiography.[82] You will see that he had the idea of continuity, irrelevance of refutations already in 1902 (inborn idea).[83] I got a letter from Robin Blackburn. He is now in Japan, may return through Berkeley. He is desperate about your book. I am already working on my reply; I shall fill in the questions later. (You may sue me for libel, since this statement implies that you are predictable and, therefore, dead.)

Popkin is here. He delivered an absolutely superb talk on 17th century Bible criticism. Said John: "This is uncriticisable. It is a work of art."

(Incidentally, John and KRP had the following exchange in the seminar:

JW: "Karl, you are dishonest. You hate criticism."

KRP: "*You* are dishonest. Your statement refers to my state of mind; it is irrefutable. Only dishonest people raise irrefutable criticisms." *(Conversation adapted to the stage by the well-known dramaturge IL.)*

Be good.

Love

Imre

82. See Popper 1974a.

83. Lakatos is here being ironical about Popper's accusation that he plagiarized the very idea of research programmes from the second volume of Popper's *Postscript to the Logic of Scientific Discovery,* which Lakatos had read in galleys. The whole controversy is documented by a series of nasty letters from 19 January 1971 to 5 December 1972 (Archive, section 13). The following are just two relevant passages:

From Popper to Lakatos, 4 December 1972:

> [. . .] I must protest against your manner of treating me: this is my third letter about the galley proofs of my *Postscript,* which I have asked you [. . .] to send to me: you have raised a different issue, and you just avoid, no doubt on purpose, saying one single word about the proofs. [. . .] In your letter of November 24th, 1972, you write about your letter of June 7th: "My letter contained evidence (in your handwriting) that I indeed had your permission to quote the passage in question." But the evidence in my handwriting supports another story; that is, the following. You asked me for permission to quote the Postscript, and I said that I wanted to see what you were saying, and in which way you were quoting me. You then showed me the passages, i.e. two pages from your MS. I was not satisfied with the context of some of your quotations, and changed some of it. *It is only due to this fact* that my handwriting occurs on your pages. The implication is, clearly, that I said that you can quote me in the way indicated by my handwriting. Obviously, I never gave you carte blanche to quote me in a different way, and in a different context. [. . .] thus you had no permission to print what you printed, or to print it in the manner in which you actually printed it. I am sorry this angers you; but this is all I can make of the evidence submitted. [. . .] I do not expect that we shall reach any agreement [. . .]. I do not intend to continue the discussion [. . .]. I ask you now for the third time to hand over the proofs of the *Postscript* without delay.

Lakatos's reply to Popper, 5 February 1973 (which was apparently never sent):

> According to copyright experts the two pages, the Xerox copies of which I sent you on June 7th, 1972, constitute perfect evidence for your permission to quote your *Postscript* in the way I did. According to legal advice, the copy of your Postscript in my Library constitutes my property to which you can claim no right whatsoever. [. . .] I regard this particular piece of our correspondence as closed. [. . .] Incidentally, I am delighted that (on my recommendation) you were elected to an Honorary Fellowship of the LSE. It was well deserved and I wish to congratulate you on the occasion.

See Lakatos 1974a for his criticism of Popper's approach, and Popper 1974b, 999–1013 for the reply. On the Popper-Lakatos affair, see Watkins 1997. On the novelty of Lakatos's concepts of research programme and positive heuristic with respect to Popper's falsificationist account, see Worrall 1995, 101.

6 February 1970

Dear Imre,

AM *(the paper),* postdiluvian version number one, is finished and on the way to Frognal Gardens.[84] There are many new footnotes, the Preface and chapter one have been completely rewritten, and there is new material in the chapter on the usefulness of *ad hoc* hypotheses. The material is from Galileo's early writings and it seems to confirm my guess that Galileo changed his mechanics in order to accommodate Copernicus. Even a cautious student of history like Dr Schmitt from Leeds (whose paper on Zabarella and Galileo's *De Motu* is a masterpiece, I think, though written in a slightly constipated fashion), agrees that there may be something in my hypothesis. Can you refute it, can you further confirm it, or is that again a "Popperian triviality"? There is also additional material on the relation of the general theory of relativity to classical mechanics.

AM *the book* is *in statu nascendi* and it won't be finished before March, I think. It will contain AM the article unchanged (though broken up), *Empiricism i* and *ii* as well as parts from *Explanation, Reduction and Empiricism*[85] viz. the chapter on incommensurability, the example of the impetus theory. I have just bought a house on the hills which is big enough to accommodate you and whatever biped you are going to bring with you on your next visit, and it is on the hills, with a fantastic view. But I am now broke. (I also have jaundice and am very tired.)

[. . .] Last week members of the so-called "Gay Liberation Front" (homosexuals) gave a talk in my lecture, in the end I gave a rousing speech in favour of sexual anarchism, goats, incest, and so on, which upset even them. So, you see, I am still on my way. This week somebody is going to talk about yoga and in the meantime I am encouraging people to elaborate Genesis in order to make it a useful alternative to Evolution. Thinking about it, our chairman broke a leg and is laid up with it. In addition I am digging in for our next quarrel.

Be well,

Paul

84. Imre's home in London at the time.
85. Respectively, Feyerabend 1965a, 1969c, and 1962a.

20 February 1970

Dear Imre,

[. . .] A few days ago a little revolution was going on in Berkeley with about $200,000 worth of windowpanes smashed, bombs thrown into police stations, one policeman killed, many wounded; I was on the street most of the time when it happened, out of curiosity, and almost got hit by a stone; next day, in both of my lectures, I preached (a) against integration ("it means that you are all supposed to go through the same sausage machine and come out exactly the same old stinking sausage, all American type") and (b) against "the revolution" (which, at any rate is no revolution at all, it is ridiculous). Slogan (introduced here by me, but stolen originally from the KPD in Germany, 1929): *The enemy is on the left.* My lectures ceased to be systematic lectures long ago; I have all sorts of guests coming, I have also invited some guys from the SDS,[86] but they do not want to speak. Even in the philosophy of science class we are now talking politics. Well, we shall see. And now I must run. (Incidentally, do you know that a crucial experiment concerning hidden variables in the quantum theory is going on right here in Berkeley? Some people will report on it in my philosophy of science class in about a week. I shall also have a debate between a biologist and a reverend about the usefulness of *Genesis* as an alternative to evolution etc.)

Be well.

Paul

86. SDS, Students for a Democratic Society, was the main leftist (white) student group in the years of the protest against the Vietnam War. It took radical stands and was often directly opposed to the Black Panthers. In the summer of 1969, a group split from SDS to form the Progressive Labor Party, and subsequent discord between that group and the faction known as the Weathermen (named after a verse in one of Bob Dylan's songs: "You don't need a weatherman to know which way the wind blows") was obvious, since the latter was progressively given to "exemplary actions" in their protests, meaning the use of terrorism.

28 February 1970

Dear Imre,

Got Redoxon plus two letters today. [. . .] St Thomas, in my book, is still a better philosopher than anyone who exists today, and Plato much much better still (compared with him, LSD[87] is a *biblis pauperum*). I sent you the last version of AM (the one that will appear in the *Minnesota Studies*) about five days ago, so you should have seen it. BUT, it is already behind the times again, for I am again rewriting the whole stuff for the book. *Style* becomes more and more important for me, and I am never satisfied. I have never written as elegantly as you have (in *Proofs and Refutations*). And my only consolation and chance of survival is that your stay with the Popperians has ruined your style and has made you constipated. So I shall not be completely wiped out by the light emerging from you. You want me to participate in an anti-Popperian, anti-Carnapian etc. etc. conference? My dear friend—I am not an *anti*-Popperian, (*anti*-Carnapian etc. etc.), but an *a*-Popperian (*a*-Carnapian etc. etc.), i.e. I don't even find it worthwhile any longer to throw my darts in this direction. Incidentally, I shall defend Bacon in the new AM as a truly critical rationalist (and also Mach).

Question: why should one *argue* when one can use *ridicule?*

Question: what is this thing "reason" to which everyone appeals?

Question: why should one refute when everyone is already on the verge of forgetting? And why should one refute when one can make people forget even more quickly?

And so on.

If I were not so damn lazy I could have finished off all you damn rationalists with a nice play, exhibiting to ridicule the strange games you play. But it would be hard work (and, besides, I don't have any talent for really good *satire*).

I met Polanyi in Chicago—he seems a very nice man. I also met

Schilpp—as a matter of fact he attended my lecture and almost embraced me afterwards. Well, we had one sentence in common in which we both believe: Wittgenstein ruined Schlick. Did you ever read the *Allgemeine Erkenntnislehre,* second edition? A *marvellous* book.

Future events in my philosophy of science class:
Tomorrow: discussion on the *New Left.*
Next meeting: hidden variables.
Then: Genesis and Evolution.
Then: St Thomas's angelology as a contribution to the theory of light.

Searle is again trying to get me fired. At the last department meeting (which I did not attend, like I do not attend any meeting), he brought up the scandal of my teaching (some responsible graduate students who want to be indoctrinated and are frustrated by the fact that I don't give them any doctrine to indoctrinate themselves with, have complained, saying that it is a scandal that there are such lectures etc. etc.).

Well, one of these days I may ask you to take me on as an assistant ("under" Helena, perhaps).

Be well. I am getting more and more depressed as I get older, and want to get out of the philosophy business. The only bright things are the little chicks here, and the freshmen altogether. They are very naive, but nice and undogmatic. But give them two years of an education and they become pillars of salt. Do you know that Yaffe writes like you talk? How many more Lakatos-zombies are there going to be?

Anyway, be well, long live Popper$_{-\infty}$,[88] and happy fornication (NOT with him, God beware!)

Paul

87. LSD stands for *The Logic of Scientific Discovery,* but the word play is self-evident.
88. Feyerabend is making fun of Lakatos's habit of using subscripts next to a name in order to indicate a person's various points of view. See also lecture 2, note 1.

10 March 1970
Berkeley

Dear Imre,

A postcard from you always cheers me up—but now, get well, take a few weeks off from politics and from the crusade for rationalism (which is a lost cause anyway as is shown in AM) and be a rationalist yourself for a change: find yourself a nice broad who is at home in the kitchen, in bed, and, if must be, even in the Library of the British Museum, and marry her. She should be younger than you, she should not acquiesce in all your follies so that you don't sink into the quagmire of married life and end up like Philipp Frank. Perhaps you should even think of producing some (I am sorry to say, but this is what they will be, poor things) little Popperians, for there is nothing better to keep a guy young than the necessity to keep up with the young. And by "the young" I do not mean those ageing pseudo-intellectuals, the students, but people who are really young, such as babies or, if you cannot stand crawling and belching monsters, four and five year olds. And don't you want to have the foolishness of your *Pop-Hegelism* propagated in the flesh as well? Here is a unique opportunity for indoctrination—how can you resist it? Anyway, whatever you are going to do on that head—get a new house, get married and continue writing your case studies.

You told me that you have material on *political* revolutions such as the French revolution, or the Russian revolution. This is a most important preparation for a more reasonable attitude in the sciences. For once we see what goes on in events of this kind, we shall discover, with our vision thus sharpened, that scientific revolutions are not very different and we shall be better prepared to understand them and to deal with them. This is one of the reasons why I always return to witchcraft. My question is: how did witchcraft finally *disappear?* Trevor Roper raises the problem which is an important epistemological *and* political problem, and he says that no one has a satisfactory answer. It was *not* stopped by the intellectuals, by Descartes and others of the same kind. George Lincoln Burr who studied witchcraft in Cornell (*he* was in Cornell, not the witchcraft he studied) criticised Lecky and H. D. White for holding this thesis (do you know Lecky's *Rise of Rationalism*[89] which, roughly speaking, holds the thesis that the new rationalism of Galileo, Descartes and others swept away that old superstition?). And Burr sums up his historical investigations with a very nice saying: "The pedants" (i.e. the new rationalists) "were just as cruel as the bigots." *His* thesis is that witchcraft stopped because people became *kinder* and *not* because they became more reasonable. This leaves the problem still open but indicates paths of possible solutions. However that may be: realising the difficulties in giving a naturalistic account of the disappearance of the *witch*mania will perhaps make one more prepared to realise the difficulties in giving a naturalistic account of the disappearance of the *Ptole*mania, after Galileo. So, on with your studies, for it is you and you only who can carry them out in the right way. *I* am much too sloppy to gather all the material and to arrange it in a manner that will satisfy the pedants. But I promise to be an excellent *populariser* of whatever you will discover once you have left the barren field of abstract epistemology.

[. . .] Today I paid $9,300 in taxes for the last year, I also just bought a house, am having it fixed and am about to buy a second house: the smog here is getting worse and worse, so I am now looking for a house away from Berkeley, about two hours away. If that is of no use, i.e. if the smog spreads even that far, then I shall look for a place outside California, a small college in the countryside, perhaps even a small university in Scotland, somewhere in the mountains, reachable only on a mule.

The fact that Lester Maddox does not allow me to come has set me back by $9,500, I have the fall and winter free and shall not be paid (*unless* I get the NSF grant I asked for, which is not likely; everyone knows by now that my intention is not to understand and praise science but to destroy it). So, as a result, I may have to cancel all my plans to go to Vienna, Heidelberg and shall just sit in London putting the finishing touches to AM (which, I hope, will make us both rich).

[. . .] Today there was a fire at the university library here, I already hoped that all the books would be destroyed and that we would be allowed to sink back into barbarism, but no such luck, so we are stuck with the unhappy barbarism of the intellectuals.

And now I have to do some further work on my taxes.

Be well.

Feigl, whom I rang yesterday, spoke very nicely about you, he admired you a great deal and thinks you have better ideas about science than has KRP (and *certainly* better ideas than I).[90]

[. . .] So, six more weeks, and I shall be able to refute you *in person.*

the grand-master of the neo-Bakuhnian church

Paul

P.S. How shall we call the father figure of your church: Pegel, or Hopper?

89. See Lecky 1910.

90. Lakatos and Herbert Feigl exchanged a few letters in 1970 which contain interesting remarks about the members of the Vienna Circle.

11 March 1970
Berkeley

Dear Imre,

Thank you for your book on the Bible but, alas!—*I've already got it,* for I am the favourite pupil of the local chapel of Jehova's witnesses. Every Sunday they send a guy around to my place to tell me a truth or two and they have arranged matters so that I shall know all the truths worth knowing in about 10 years. They are very encouraged by my attitude, and the older teachers bring apprentices along to show them their masterpiece, me, for I listen, make intelligent remarks, and generally seem to make tremendous progress. Well, one thing I must say for them, they sure try, and they are nice people, and they are patient, and they never give up and maybe one day I shall join them.

In my class on the philosophy of science I had a debate on Genesis vs evolution, the guy from evolution was mean, the guy from Genesis was cowardly and it was generally *a great bore.* Of course, the guy from Genesis said, Genesis was not meant to be taken literally, it is a *moral* document etc. etc. [. . .]. Well, first of all, he is not correct, and secondly, science often made progress by taking parts of silly doctrines seriously and using them for an attack against well-established, scientific, precise etc. views. Thus Copernicus turned to the Pythagoreans, those strange people who had a secret society, thought that women were equal to men, ate no flesh, revered the number ten, revered fire and therefore made the earth move as it obviously was not fire and, as the most dignified element could not be supposed to run around, Copernicus turned to *this* strange philosophy, picked up one element of it, the motion of the earth, and knocked Ptolemy to the ground with it.

Anyway—thanks for correctly guessing where I am headed at the moment (incidentally, I am going to found a counterchurch to yours, namely the *Neo-Bakuhnian Church of Intellectual Freedom and Self-Expression* and I shall be secretary, only member, Pope, with Hans Arp the Dadaist as the Patron saint). Before doing that, however, I have to think about my whereabouts next year, for I want to survive and do not want to be killed by the smog. Every day now my eyes burn, my chest hurts, the smog is really bad here and moving up to the hills will not really be much help. I have various alternative plans to be tried out in the next two years. One is to have another house away from Berkeley, but that will not be much use for the Bay Area is going to grow, grow, grow, and I would have to move five hours away in order to be safe. A second alternative is to look for a job at a small provincial university which is guaranteed to stay provincial for the next 15 or 20 years. This second alternative I am taking more and more seriously and if they offer me a job in, say, *St Andrew's,* I would now take it (or Aberystwyth, or some other provincial place in England). So, if by accident you hear about jobs in such places, let me know. I would even be prepared to take a position as a *reader,* or as a *lecturer,* for I have always been a clear air fanatic and I prefer living on a low salary in an intellectually uninspiring place to not living at all. (Besides, for stimulation I shall always have your letters and MAM.)[91] I shall look around myself when I come to England, for this is now the one great project for me in the next 5 years. If that does not work, then I have a third alternative, viz. to *quit* here (in about two years), take all the money I have (about $20,000, excluding the value of the house I have bought) and move to a little village in Scotland and live there until something turns up. For $20,000 I could live about four or five years, I think, and I might try to write a bestseller. Anyway, these are my serious plans for the next 2–5 years. I'll talk with you about them when I arrive on May 3rd or 4th. I am really depressed about the way the Bay Area degenerates, no view anymore etc. etc.

Just now I am working on the third chapter of AM, called "Empiricism" and I shall give a rather long account of *Aristotle,* the only reasonable empiricism that ever existed, then of Bacon who was an anti-empiricist plus an anti-rationalist ("rationalism" in the sense of Descartes) and in this respect the father of science (of course, good old *Galileo* did all this without advice from Bacon, but Bacon was one of the few writers to make this new counter-empirical and counter-rational philosophy explicit—only, no one understood him!) So it will later on be Bacon vs Popper, and I shall support *Bacon.* People say I am getting crazier every day.

[. . .] Be well.

Paul

91. MAM refers to Lakatos's reply to Feyerabend's AM; it probably stands for "Methodology *Against Method*." In many of the following letters, Feyerabend plays with these initials, referring to his reply to his rival's reply as AMAM, that is, "Against MAM."

31 March 1970

Dear Imre,

The enclosed lady (who should be treated with care—see below)[92] is a better teacher than either of us (which makes her the best teacher in the world, naturally) and I think all the little Popperians, crypto-Popperians, paraPopperians, counterPopperians, popperoPopperians, lakatopoPopperians should be given an opportunity to profit from her presence, so let her give a lecture at LSE. Moneywise this will not be a problem, for she is prepared to spread wisdom for free, *provided* you are nice to her, and take her out. Beware, however, for she is a witch (truly). She is a good witch, so do not be afraid, but still, she is stronger than you, so do not try to overpower her with your presence. Good luck!

[. . .] All the best. I am quite depressed about this stinking smog, but AM is proceeding nicely.

Paul

92. Feyerabend is referring to his friend Joan McKenna: "a bigmouth with a heart of gold and a certified witch" (1995, 124).

5 April 1970
Berkeley

Dear Imre,

You have powerful allies in this country: Ayn Rand (know her?) is after me. She wrote a long article against a short (three and a half pages) article of mine, and sent copies to 4,000 philosophers in order to prevent American philosophy from deteriorating further. Sample: the author (i.e. me) "heralds the retrogression of philosophy to the primordial, pre-philosophical rationalism of the jungle [. . .] But what is innocent and explicable in an infant or a savage becomes senile corruption when the snake oil, totem poles and magic potions are replaced by a computer . . . ".[93]

You must admit, she writes much better invective than you. I'll send you a copy as soon as I have made one. And, mind you, the whole commotion is about my article "Science without Experience"[94] which is only 3 and 1/2 pages long. But, like every good ideologist she smells the devil even from a small footprint. At any rate, how about marrying *her?*

Much love,

Paul

93. See Rand 1970, 88.
94. See Feyerabend 1969b.

27 April 1970[95]

Dear Imre,

I annihilated all your arguments, I annihilated empiricism, rationalism, critical rationalism, but when I left I saw everyone yearning to return to safe and sensible Popper-Lakatos commonsense. So you will win in the end NOT because your *argument* is better, but because your position is more comfortable; so as a matter of fact your rationalism will win for an irrationalist reason, and the *abstract* victory is mine anyway. Everyone loves you in Minneapolis and Herbert Feigl[96] finds you the most reasonable philosopher. Tomorrow I shall talk about Cohn-Bendit in Atlanta (they don't know it yet). I have a slight laryngitis: I *may* have to ask you to give my lecture on Friday though I shall be there. Grover[97] is a convinced and stable anarchist, the only one.

Be well,

Paul

95. This is a handwritten letter mailed at the Sheraton Hotel in Atlanta, Georgia, where Feyerabend was staying during the conference he refers to in the letter.

96. In those years Herbert Feigl was director of the Minnesota Center for the Philosophy of Science in Minneapolis. Feyerabend acknowledged him as an open-minded positivist: he applied empiricism to psychology and psychoanalysis, and also dealt with the mind-body problem (see, for example, Feigl 1950, 1970–1971; Feigl and Meehl 1974).

> [Herbert Feigl] was tall, distinguished-looking, with a roving eye and a knack for short and pungent phrases. Some people, de Santillana among them, have criticised him for being repetitious and behind the times. "Er ist in seiner Entwicklung stecken geblieben" ("He's got stuck in his development"), said Popper. The charge of repetitiousness may be correct—it applies to every salesman of ideas. (It certainly applies to Popper.) The charge of backwardness, however, is pure nonsense. Feigl made important contributions to philosophy in the traditional, nonpositivistic sense. He was one of the philosophers of science who addressed the specific problems of psychologists and psychoanalysts, and

he changed the philosophy of science by combining it with a rather large dose of realism. (Feyerabend 1995, 116)

97. Grover Maxwell was one of Feigl's assisting colleagues at the Minnesota Center: "Grover came from Tennessee. He spoke slowly and haltingly with a grim look on his face. I often thought he was about to hit me—but that was his friendly expression" (Feyerabend 1995, 117).

13 June 1970
an hour after you left, in my office
London School of Economics

My dear Paul,

I already miss you, but I hope that our LINKING OF CONFLICTING BANDWAGONS will increase our fun in life.

[. . .] I am trying to organise a little conference on AGAINST METHOD next June (when the book is out), possibly in Jerusalem if there is no war. I do not know when I'll succeed, but let us cross our fingers.

Dick [Popkin] is here for lunch, so I have to go down to the porter's lodge to let him in.

Gill sends her love to you. She likes you, she is a lefty-liberal: I took an enemy in my house.

Till soon.

Fondly,

Imre

30 June 1970

Dear Imre,

Every morning from six to nine I sit in my new study and retype AM. My typewriter stands on an ancient desk, to my right is an ancient bookcase, to my left an ancient Japanese lamp, there are purple curtains in front of me and the whole thing looks like a whorehouse. After nine, I start lecturing, eating, socializing and: *looking for a new house,* for the house I have here is a lemon.

It is too noisy, a street right in front of it, too old, it smells, it has no character and on top of all that it has its own house *rat.* I am having it cleaned (the house, not the rat), furnished, the garden will be improved, but as soon as all this is done, I am going to leave it and move elsewhere, provided, of course, I find an elsewhere soon. Otherwise I shall go crazy and turn AM into a Fascist pamphlet. A friend of mine who is studying the history of Marxism in Germany in the last century suggested that I call the rat Kautsky.[98] Well, Kautsky wakes me up every night; last night he sat at the foot of my bed and now we have started a battle of wits. Traps seem to be of no use for Kautsky is an old and wise rat and won't go near them. I have closed some holes, Kautsky soon finds others. The method that works at the moment is that I leave the lights on all night, this worked for two nights, but Kautsky soon found out that lights need not be harmful and is back again. I don't want to put out rat poison, for then Kautsky will creep into some hole, die there and start stinking, and it will take days before one finds his resting place.

So you see, I have an interesting life and unfortunately it interferes with the progress of AM. I now have 30 wonderful pages, brief, to the point, hardly any footnotes, but progress is slow and will be even slower when I start moving. My intellectual level has experienced a great drop, in London I felt intelligent, now I feel rather dumb as I always do after a few days here in Berkeley. Your presence makes a decisive difference. It is not so much *what* we are talking about but the *attitude* behind it that makes me wake up and turn into a true intellectual. *Here* I seem to turn into a *boor.* Today I am going to look at three houses with Mara, the girlfriend of the guy who is applying for Post's post,[99] who also helps me with the interior decoration. [. . .] I am going out with her again tomorrow, and if I am not careful this enterprise will soon turn into something less businesslike. But no time for that. You have no idea how I miss London, England, even with Heath[100] in it and all the conservatives.

Please give all my best regards to John [Watkins], Alan [Musgrave] (I have made reservations for him at a hotel), *Gillian* (tell her to be patient, soon you will be too old to be unfaithful and will want nothing but a good meal, a warm room, a pat on the head, and at least a footnote a day in some article or other) and take all best wishes from us, that is from me *and Kautsky.*

Paul

98. Cf. Feyerabend 1995, 113: "Remembering that Lenin had called Kautsky (the Austrian socialist) a rat, Robin called my rat Kautsky." Karl Kautsky, who devised the Erfurt 1891 programme, which was supposed to trace the movements of the German Social Democrats, had campaigned in the name of Marxist orthodoxy against the revisionist

wing headed by Eduard Bernstein. Persuaded that the end of capitalism was inevitable and that it would dissolve in socialism, Kautsky was also convinced that revolutionary violence had no standing, and severely criticised the Bolshevik *putsch* in Russia of October 1917 and the "revolutionary" methods of the government led by Lenin and Trotsky.

99. Heinz Post was at the time head of the Philosophy of Science Department at Chelsea College in London.

100. Edward Heath, conservative leader and British Prime Minister (1970–1974). He supported Britain's entry into the EEC.

30 June 1970
Berkeley

Dear Imre,

AM progresses slowly, will be much shorter and much more compact. No appendices, almost no footnotes, greatly reduced text. Am now labouring on some transition from chapter three to chapter four; and when that is out of the way I shall be able to leap at once to the chapter on incommensurability, leaving out all the mess about the mind-body problem. So, although you get the MS later than expected, it will be much shorter, and much easier to read and to refute. The monster MS I intend to rework for Cambridge University Press into a *history* of empiricism from Neanderthal to Lakatos, with a *long* chapter on Aristotle worked into it, and an interesting thesis about the origin of 16th century empiricism (Aristotle plus magic plus confusion).

Joan is working on my curtains, in three days I am going to Los Angeles for some singing and to celebrate the birthday of a chick there, this might make me forget my little house which was a big mistake. Kautsky the rat is still about, last night I spent two hours from 2 to 4 looking for him, I heard him, did not see him, the rest of the night I dreamt of living in the stomach of Kautsky like Jonah in the stomach of the whale. It's pretty difficult to remain sane under such circumstances. But I shall concentrate until I have AM out of the way, that is, until I have sent the new and greatly improved version to you, and may then go to an insane asylum instantly. Christopher, the big lady with the big dog that frightened you so much, is back, she sends her regards and wants you to know that she might accept an offer of marriage from you *provided* you get her a nice big house in England with a doghouse in the back. Joan seems to be in good spirits and I, well, I am homesick for England. Pray to the gods that Brighton turns out OK.

All the best,

Paul

30 June 1970

Dear Professor Feyerabend,

This is only to tell you that unless you send me your *Against Method* by August 1 (that means that it should *reach me* by August 1), unfortunately I cannot consider the possibility of demolishing it. A further condition is that you let me have, practically immediately, a good Xerox copy of the page proofs of the version of *Against Method* which is being printed for the Fourth Volume of the *Minnesota Studies.*

Yours sincerely,

Imre Lakatos

P.S. I sent a message to Kautsky to torture you until you fulfill the conditions laid out above.

15 July 1970
Berkeley

Dear Imre,

Kautsky has gone, but his stench pervades the whole house. It is so bad that I shall have to leave and live in a hotel for a while. Yesterday I called in an EXPERT, a ratman. He arrived in a big Cadillac, smelled the air, said: "Oh yes, a dead rat, most certainly," and said that he was prepared to tear down the walls and look for the rat for a small fee of $2,000. Otherwise I would have to wait two to three weeks for the stench to go—this is how long it takes a revisionist rat to decay (human stench, of course, being propagated by books, lasts for centuries, so I am still lucky). So I may go to a hotel. It also seems that the rats have dug up the sewers so I am getting the contents of the sewers as well as the rat bodies (did I tell you that I found four little dead rats? children of Kautsky's, no doubt). An interesting life. To make it even more interesting, I have started to study cuneiform: having invaded, and ruined, the philosophy of science and physics, I now want to invade one of the more recondite subjects, such as Babylonian astronomy. I am also reading your research-programme paper, for I am determined to finish AM here in this house, even if the stench kills me (if the final sections dealing with you are overly nasty, just ascribe it to the spiritual presence of Kautsky). And I must say: your article is a *masterpiece,* almost as good as PaRs (i.e. "Proofs and Refutations"). But your new rule is still too strict (it is certainly not *empty*). So I am now rewriting the final two sections in order to give you the proper build-up. Incidentally, can you do me a favour? I brought you the book with the article on Lorentz in it (you asked me to buy it in the US as it is not available in England). Now the book is sold out here, I cannot get it anywhere: can you send me a Xerox copy of that article on Lorentz?

Chris has gone off to British Columbia with her new lover, but she is prepared to sleep with you if you buy her a return ticket to London. Joan will send a letter to you. Do not omit Mara from your considerations: she is an excellent cook, very clean around the house, very attractive (she was once in the centrefold of *Playboy*), but she loves *parrots* and takes one with her wherever she goes. But of course, you can always defuse the parrot by teaching it some principles of sophisticated methodological falsificationism. That is all for today. You will get AM unless the fumes kill me.

Love,

Paul

P.S. This morning at 9 am the body of Kautsky was found. Now I can breathe freely again, I hope (unless there is another rat somewhere). I had promised a prize of $100 to anyone who found him, so Mara started looking, but it was not she who found Kautsky, but an exterminator, and for $12.50.

P.P.S. Just got your long letter indicating you got the proofs of AM. Now, I must warn you: I am going to *completely rewrite* what I say about you, I shall give you a tremendous build-up, and after that only a slight and polite criticism, for your point of view is indeed quite something. I am also rewriting the passages about incommensurability so, in the end, we shall be much closer than it has seemed so far. You have got some marvellous *ideas* and I only get put off by some of your *words* ('honesty', 'integrity' and so on). [. . .]

I have already written half of my *praise* of you; *next week,* I hope, I shall finish the rest. So you should have AM in about 10 days ± 4 days. Here in my lecture I am defending Velikovsky and other nuts. Get well soon, and above all—WRITE SOON.

18 July 1970

Dear Imre,

With peace returned, I am now reading Giordano Bruno and his views on Copernicus in the evening and am working on AM during the day.

There is now a very long chapter 2 which gives my opinion on the history of empiricism from Aristotle to Popper, all the stuff on mind and body has been eliminated and the final chapters on rationality and incommensurability will be rewritten to give a better account of your views and a better criticism of them. I think you will like what I say about you just as I like what you say about Michelson, Planck, Prout and so on.[101] When AM is finished I shall start studying cuneiform; I also bought myself an old Egyptian grammar to study hieroglyphics, so science is more and more receding into the background. My house is slowly becoming more livable. So far I am still living in the basement while the girls are working on the rooms upstairs. My upstairs bedroom has now scarlet curtains, a purple bedspread, deep blue walls and a green carpet, so it looks like a room in a whorehouse (which it may soon be). Mara cleaned my kitchen and washed all my dishes, so it is now absolutely spotless; today she is making curtains for the kitchen and painting the bathroom (with stars on the ceiling); Joan is still putting up curtains in the other rooms, there are a few Persian rugs and I have also bought an old, very attractive, but completely torn couch, so Mara is re-upholstering it in deep saturated velvet green (she is a woman of many talents).

Well, for the next year I seem to be stuck here in Berkeley, and I must say I regret that I made such a mess of my London stay and of my appointment in Yale. *This* university is slowly going down the drain, also salary-wise; the President has now suggested that the content of courses be determined not by the *Faculty* but by the Regents, which to me is not at all objectionable (it does not make much difference to me whether I have an expert philosopher telling me what the theory of knowledge is really about, or whether that is done by some administrator, and experts should at any rate be supervised by others), but, of course, that will cause an *outcry* in the whole faculty and, maybe, there will be further student riots as well. More and more people are leaving and only the bores remain and teach "relevant" subjects which to me are the most boring subjects of all. I am badgering Joan to learn a little more, and on Monday she will go and become a graduate student, and study languages, Chinese among them. So, you see, I am continuing your good work on her. She is preparing a long letter to you; she has already written 6 pages, typed, single spaced, and has told only a quarter of her story in jail. Your fame is spreading, not because of P&Rs [*Proofs and Refutations*], but because of the stories about your charm and your sexual valour. "Who is Imre Lakatos?," ask children, little girls, big girls, matrons, grandmothers, and, when they have obtained *that* information: "When will he come here?" You owe it to your admirers to come here, as soon as possible. Even my admiration for you has further increased, as the result of a more careful reading of your research programme paper,[102] and I had thought that PaRs had been your intellectual peak. [. . .] And now, back to work on AM.

Love,

Paul

Give my love and best wishes to Gillian, Sabra and the chicks from King's Road.

101. Cf. especially Lakatos 1970. According to Lakatos, William Prout's programme is the typical example of a research programme which develops in an ocean of anomalies. Another of Lakatos's favourite examples is Bohr's programme, which grows on "incoherent foundations." Michelson is often quoted in reference to the retrospective character of crucial experiments, while Planck is noted for the ad hoc-ness of the formula he put forward (Planck 1900a,b).

102. Lakatos 1968b.

27 July 1970
Berkeley

Dear Imre,

AM is being Xeroxed, a copy will be ready tomorrow and will be sent to you at once. I am not at all satisfied with the new version. The older parts are much improved, the section on incommensurability is much better, and there is a fairly good section on the history of empiricism, containing much praise of Aristotle. But I am not at all satisfied with what I have said about you. I drastically cut what I wrote originally and I am not sure that I have selected the material well. I started reading your paper about three weeks ago and it was quite impossible for me to comprehend all the things that are going on there. There was a time, about a year ago, when I thought your paper rather thin, but what was thin was *my reading of it.* You are such a subtle bastard—a real city slicker—compared to the bumbling peasant Popper. So, take a really critical look at the sections concerning you and give me your advice on how to change them if you think the change is necessary and that I have simply missed the point. The trouble is, I just cannot work under pressure and it is quite impossible for me without help to get good results in a short time. But such is life, the number of disturbances, rats, girls etc. much larger and it is difficult to fit the ones into the other. So, I repeat, I ask for your indulgence, and I am awaiting your suggestions for improvement. Also, I think the essay got rather long again, so make suggestions also for cutting it down. Anyway—what I want is *a good book,* and if what is necessary to get a good book is that you write my part under my name, then so be it. I am curious what you think.

[. . .] I also have a plan: I shall write a short story of *empiricism* from Aristotle (including his *empirical* theory of the continuum) to Mao on the one side and Democritus to Lakatos on the other and I think it can be *very good,* for there does not yet exist *any* good history of the empiricist ideology. There is much material on Berkeley, Hume, etc. but that is just a side issue. I also got invited to spend one semester in Göteborg to lecture in 1971/'72 and I shall certainly accept. '71/'72 I am completely free, no obligations anywhere, so I shall drift around in Europe, giving a few lectures here and there, but most of the time taking a vacation. *You* go to Cyprus etc. I haven't had a single holiday since 1965. It is true that I am constantly disturbed by painting, cleaning, baking, scrubbing women, but I am always close to some deadline, so what could be pleasure is only interruption . . . and now, there is another interruption at the door, so, farewell, until the next letter.

Paul

30 July 1970

Dear Imre,

I am now slowly starting to *rewrite* AM [. . .]. My criticism of you is twofold. First, that your standards do not exclude anything. Secondly, that if this problem could be fixed, they would be too restrictive. The standards contain a reference to content and increase of content is a virtue. But (and this is a historical assertion) theories which succeed each other in some domain are hardly ever in the relation

but much more in the relation

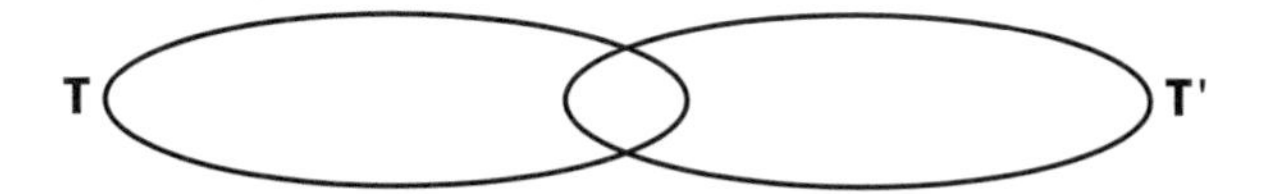

and the second case occurs especially when the successor is the result of an exciting revolution (example: Galileo, Einstein, as described in my part of AM).[103] And (this is an assertion concerning standards) if we want our theories to be interesting, coherent, simple etc. etc., we shall have to give up on increase of content. Secondly, there is the problem of incommensurability. This is all I have to say about you. I also say that you are the last stronghold of mechanical rationalism, so, if you fall, mechanical rationalism falls too, to be replaced by dialectical rationalism, or anarchism (this bit about mechanical rationalism vs dialectical rationalism will be included in the new version).* I am sorry to be constantly running away from you in this way, but our book should be really good.

In two weeks another girl, *Janet,* is going to turn up at your door with a letter from me in her hand. She is entirely different from Joan, takes Quine seriously etc. etc.

Be well,

Paul

*As a matter of fact, I am constantly *debating with you,* as Bohr (= I) did with Einstein (= you). And Bohr also constantly reformulated his views.

103. See Feyerabend 1975a, chap. 15.

7 August 1970

Dear Paul,

First of all, let me thank you for your most recent letter. I am very much looking forward to the last version of your *chef d'oeuvre.* I am sure you will understand that I somehow feel inhibited from getting down to serious concrete work before I have the final manuscript. Your criticisms seem to be reasonable except for your claim that I am the last stronghold of mechanical rationalism which, of course, is utter nonsense. There is absolutely nothing mechanical about my "rules," as you had pointed out yourself earlier (and, I am afraid, rather well, too). Your other point seems to be unfortunately good, but I hope to be able to deal with it. I am looking forward to the volume which I hope will turn out to be a very constructive and even serious exercise, and this very fact will refute your flippant anarchism. On the other hand, if you go with this *mechanical vs dialectical rationalism,* pinning me down with that mechanical nonsense I shall cut off your head. I quite understand your difficulty because you do not believe your own scepticism and therefore you keep veering over to my position. On the other hand, if you do so then the controversy falls. Anyway, I let you work this out by yourself.

I am constantly pestered by the New Left people about the book and keep encouraging them that (1) it will be an absolute masterpiece and a classic, and (2) that the final version will be there within a few days. So don't let me down. If you really have made up your mind about your position and finish the book, we could get together to agree on the final shape of the whole edition with my appendix and your final kickback.

What will Janet's letter contain? If she really turns up in mid-August I shall be in the middle of a nervous breakdown when I shall not want to see anybody (having just finished the bloody paper and having just bought a bloody house).

Sorry for being so abrupt. I am struggling with the end of this paper which I am now doing and which, incidentally, will also be the *machine de guerre* which will kill all your arguments in one go, but I am exhausted and wish I was already on holiday. Also, it is absolutely idiotic to have a non-English girl in one's place when one writes a paper in English.

With warm greetings.

Yours ever,

Imre

11 August 1970

Dear Imre,

Just got your most recent letter. Please regard the MS with the corrections I sent a few days ago (and in which there is additional praise for you) as *the final thing.* There are many things I don't like about it: especially my treatment of you I would like to have improved, but any *basic* improvement would take too long to fit into our time schedule. Hence, there won't be anything about "mechanical rationalism" vs "dialectical rationalism" either, and there would never have been, for this was only a jocular remark in my letter. So go right ahead annihilating me. However, if you have some suggestions for major (or minor) alterations in *my* text, to make the battle between us even more a battle of giants rather than of midgets, please write them down and tell them to me when I come to London.

I think that some sections of my part are very good, I think that your essay in *Criticism and Growth*[104] is marvellous, and if we can get some sparks out of the tension between our two attitudes, well, it should soon overshadow Kuhn, Polanyi, and perhaps even the old Kronos[105] himself [Popper]. So, in order to make the thing excellent, if you make a suggestion, I may be prepared to say some things in a way I do not quite believe. To sum up: this is it, and there will not be any further basic changes, there will only be stylistic changes and those basic changes which you suggest.

Incidentally—a flippant anarchist, a *really* flippant anarchist is of course *also* prepared to engage in a rational debate and to defend Spiro Agnew.[106] It is only the mechanical anarchist who is either too shortsighted, or too cowardly to do such things. Thus, in my lecture, I defended Nixon and made the New Left the laughing stock. It all started with some silly questions from the audience, such as: "If your Teaching Assistant gets shot, what will you do?" My reply: "Take care of his girlfriend." "If there is gas on campus, what will you do?" Answer: "Buy a gas mask," which reduced the puritanical leftists to quiet fury. And, of course, I am against reconstitution. My *argument:* in times of crisis, one should learn more, not less, and one should try to increase freedom rather than reduce it. Increase of freedom means increase of alternatives, not turning everything into a dull political soup where people, who have perhaps no idea even about where Vietnam is situated geographically, make speeches about it. And so on.

Incidentally, I recommend that you read Mao's "Oppose Stereotypes Party Warning" which is full of useful rules, a really marvellous piece. For example: "Hurling insults and threats is certainly not fighting." Or: "What is scientific never fears criticism," or: "The main question is: what is the problem?" And it has marvellous things to say about *style.* Of course, I read the article in class.

Who will retype the MS so that it is in a shape fit for the printers? Mara is prepared to do my part, and this seems reasonable, for she is right here, and she says she could do it in a *month,* when the final form is available. Are the New Left [Books people] prepared to pay her for this?

Do you know the play *Die Hose*[107] from which I have chosen my motto? Well, if you don't, let me tell you: the wife of a Prussian civil servant loses her panties in the middle of the town, and the episode leads to all sorts of complications and severely endangers law and order in Prussia. The motto itself is from a long monologue from her husband complaining about the disappearance of law and order in his own life. Don't you think this is an appropriate motto for the stink-bomb? (In talks to friends I call our common enterprise "the stink-bomb," and I think the name is right, for many people will regard it as rather infuriating.)

I am getting more and more restless and I may end up in Alpbach, after all, for the last five days: September 1st to September 6th. Hans Albert[108] is running a seminar there with Spinner[109] (I just got the programme), and after that I may go to Vienna with Hans, and then, on the 15th, to London. Right now I am working in the garden, removing bricks, levelling the ground, shovelling sand with a redhaired cutie always stumbling across my path. *Where is your paper for the Popper volume?* I certainly want to add a footnote or two concerning it (and Kraft).[110]

So, you have just bought a bloody house? No trouble. One can also *sell* bloody houses. My own house is quite nice, especially after Kautsky has gone, but too noisy, and so I am just improving it a little in the inside, and also the garden, so that I can sell it at a higher price when I move away. I haven't seen Janet for a long time, for I am now in my isolationistic period, telephone cut off etc. etc. Just before I did this, I got a call from Chris in Oregon with the question whether she should allow herself to be impregnated by a guy she loves, so I said: yes, go ahead, if things go wrong Imre will take care of you and the baby; so, look out, one of these days a bedraggled Chris may turn up at your doorstep, about to burst with child.

All the best

Paul

104. Reference is to Lakatos 1970.

105. Hesiod *(Theogonia)* gives a detailed account of the origins of Kronos: son of Uranus (the sky) and Gea (the earth), he cripples his father, who, afraid of losing his throne, keeps his sons, the Titans, prisoners. He marries Rea, and, scared himself of losing the throne to his offspring, gobbles them all up. Rea escapes to Crete, where, after giving birth to Zeus she hides him in a cave, and presents Kronos with a stone wrapped in swaddling clothes instead of his newborn.

106. Spiro Agnew was governor of Maryland between 1967 and 1968. Nixon chose him to run as Vice President, and he was elected in 1968 and reelected in 1972. In 1973 he resigned following accusations of tax evasion, and was found guilty by a Baltimore court. His resignation enfeebled the Nixon presidency, and Nixon himself resigned the following year following the Watergate scandal. At the end of the summer of 1969, many demonstrations against the war in Vietnam were held, and on one of these occasions, Spiro Agnew inflamed passions by claiming that "the [protestors] were Communist dupes

or worse, and, in a barrage of sophomoric adjectives, later called them 'an effete corps of impudent snobs who characterise themselves as intellectuals'" (S. Karnow, *Vietnam: A History,* New York: Penguin, 1984, 599).

107. A comedy by the German playwright Carl Sternheim.

108. Hans Albert, professor at the University of Mannheim. A keen Popperian (see his *Treatise on Critical Reason*), he is retired now but still runs the Alpach Seminar.

109. Helmut Spinner was at the Hans Albert Department in Mannheim. He visited LSE in 1971. He is now professor at the Institute of Philosophy, University of Karlshruhe.

110. Feyerabend is referring to Lakatos 1974a and Kraft 1974, respectively.

20 August 1970

Dear Imre,

[. . .] Have you got McMullin's "refutation" of you and me?[111] *Well, I shall add a footnote to AM (of which you will approve) refuting him.* History-wise, he is much too naive, and in philosophy he just stumbles around the categories he has set up himself in the first place. But this is the stuff people pay attention to nowadays (the "careful scholar" who is the guy who says: "so and so has gone too far"; but how shall we know whether so and so has gone too far until he has *as a matter of fact* gone too far? And, in our case, even I don't believe that we have gone too far. Quite the contrary: it is McMullin who has gone too far, into the *mud* of conservatism).

Quotation from Brecht: "Concerning the freedom to teach and the freedom to think. If I were asked which I prefer: the terror of those who refuse to know something new or the terror of those who assert that they know it better, then I would prefer the former always, and under all circumstances. I now want to express myself quite precisely, so that no one can accuse me of being afraid of being caught: I admit that the willingness of conservatives to accept responsibility occasionally manifests itself in tough ways, but at least it testifies to real experience concerning the interests of the large masses. And I also believe it is the period of enlightenment that is to be held responsible for this terrible fear of responsibility . . ." To be found in the *Schriften zur Literatür und Kunst* (which I bought you some years ago), Vol. 1.[112]

Be well

Paul

111. See McMullin 1971.

112. For more on the relationship between Feyerabend and Brecht, see Feyerabend 1978a, 114–20.

20 August 1970
London School of Economics

My dear Paul,

I have been so madly busy in the last couple of weeks refuting you that I did not have time to write you a proper letter. Since your vicious attack on reason upset me very much I could not sleep and I had to take double sleeping pills from which I have a headache each morning. So do not be surprised if I am temporarily a bad correspondent.

First of all let me thank you for your very kind report about that quantum paper which I now hope Mellor[113] will reject. I joined in with you and I think you put the case very clearly. It was terribly nice of you to respond that quickly.

McMullin sent me his stuff[114] and I already included a short reference to his criticism of me in MAM. In fact we arrived independently at the same joke: I referred to him as "the very learned historian." Of course McMullin is a very nice man and if he was brighter everything would be all right. So anyway, do defend yourself against his silly comments in your AM and just instruct me where to insert it.

I am looking forward with increasing delight to our joint venture and I hope we shall have a lot of fun. Unfortunately, the New Left [Books] doesn't seem to have money for your crossing the Atlantic: although I mentioned that you might bring a pretty typist with you. This they seemed to take rather as something of an insult because they think that they have all the pretty girls in the world. Frankly, I do not think that it is necessarily a bad thing if we do not meet for a long time in September. If you started stimulating me before I finish my contribution, it will be immensely longer and the whole thing will get out of control. I am already overstimulated and I want to avoid personal meetings with you (except for pictures, gossip etc.) until I have actually finished my version.

Not that I am very enthusiastic about this pragmatism. In fact my heart breaks that I cannot join you and Spinner and of course Hans Albert in Alpbach where we could have a wonderful time. However, what I need is to have a rough first version which will be there in about a week's time; then I want to have 2 weeks of complete peace and quiet on a Greek island (I take Gillian with me on the condition that she won't talk to me) and then return to London to finish the thing off. We should, I think, stick to the original idea that we get together to get the whole thing into absolutely final shape in Pittsburgh. I arrive there on Sunday, October 25th and would stay there working most of the time with you until you have to leave for London, Ontario. Please do let me know express when you have to leave because when you leave Pittsburgh I will return immediately to London. As a matter of fact, I would not mind at all staying for a whole week but if I remember rightly you have to leave earlier for London. When is that bloody conference going to take place? (Don't try to have me invited once more by Butts; I just cannot take any more conferences for 1970.) On the other hand, as you know, I accepted the invitation to Bloomington. This will be from November 19th to 22nd. I wonder if we could meet then again to prepare the very final corrections for the printers. I promised them a fully edited copy by December 1st and *I want to do the whole editing work.* This last statement may surprise you but after my bad experiences with publishing houses I learned to insist on this because what they usually do is give the typescript to a little fool called a "sub-editor" who corrects one's English in such a way that you do not recognise your own text anymore. I insist on editing the whole book and that the typescript that I provide go immediately to the printers without the publishing house interfering. This means that I shall take care somehow of the final re-typing; I shall see to it that the New Left Review pays my work as editor and also the typist's work who undertakes the final re-typing of your manuscript under my supervision.

As you may remember, the New Left people promised that if we get them an *unedited* manuscript by December 1st it would be published in May. I have the impression, after having talked to a chap there, that even with the best of intentions (which I hope Robin Blackburn will provide), even if I give them a completely edited typescript which can go directly to the printers, the volume will not be out before, say, the end of July. In my experience the main thing is that it should be out before the beginning of the American academic year, which I think we shall succeed in achieving.

I have been so busy lately that I could not write to Spinner for a long time. Please do persuade him to translate this book in Alpbach.

A mortgage company will decide this afternoon whether they will advance money on an absolute monstrosity of a house which an architect friend of mine describes as a classical example of "romantic" Edwardian architecture. By "romantic" he means that the windows are thick, there is some elaborate ironwork in each window to make it romantically dark, and from any one room to another one needs to climb at least 10 stairs which, for instance, means that one can only eat in the kitchen because one could not carry food to any other room. This is all the more romantic since the kitchen is 4′ by 3′. The most romantic touch of all to my mind is that the bathroom, which is on the third floor, was built in 1900 and one never knows when water comes out from the tap and when not. On the other hand, the garden is half an acre of woodland and although the house is in the middle of London (22 minutes' walk from Golders Green tube station through the Heath) one cannot see any neighbours from the garden or from the house. The whole thing is surrounded by a wood full of squirrels, rats, and scorpions.

If the mortgage company says no, then the whole thing is of course off. If they say yes, then I may buy the house. The only trouble is that the lease is only for 967 years and I have no idea where I shall live afterwards.

With warm greetings,

Imre

113. Hugh Mellor, professor at Cambridge, was editor of *The British Journal for the Philosophy of Science* between 1968 and 1971.

114. See McMullin 1971.

16 September 1970
Berkeley

Dear Imre,

My visit to England seems now definitely off. I have bought one house, am selling the other, must move my furniture, and I am a nervous wreck.

As regards AM I have had the following idea.

I know you are not too happy with the New Left as publishers. Maybe the situation could be slightly improved by finding a good American publisher for the American distribution of the book. Usually Penguins are sold worldwide, but *some* Penguins are restricted to the UK and some American publisher is then given the paperback rights for the United States. Maybe this should be done in our case also. Then, at least, we would have a good American publisher. You could negotiate this in England with the New Left, i.e. first make the suggestion to them and then work out the details, and while in Boston you might even talk with some US publishers. I am thinking of Harper Torchbooks, for example. At any rate, think this over and work on it, if you think the idea is good. I give you power of attorney as far as my part is concerned, for I know you are good in these matters, much better than I at any rate.

[. . .] Also, send me Alan's [Musgrave] address in New Zealand.

I want to go there on a lecture trip in the summer and take over the business when Alan leaves for his next job in England. I am finished with civilisation (do you know that the last line of my *Naturphilosophie* will be: "Back to the Stone Age"?)

Carnap's death made me very sad. He was a nice and gentle man. Did I tell you that he told me that he was afraid of meeting me in person for he got the impression from my writings that I must be a very nasty person? And that he was happy to find out that I am gentle etc. etc.? Kraft's wife also died. Well, it will soon be our turn. Until then.

All the best

Paul

24 September 1970

Dear Imre,

You like *classifications,* so here is one for you. Philosophers of science can be classified in two different ways: viz. (1) by paying attention to their attitude towards justifications, or (2) by paying attention to their attitude towards history. In (1) we distinguish justificationists and those who admit that unjustified hypotheses are fully fledged members of the domain of knowledge. In (2) we distinguish abstract epistemologists who pay attention to the (historical, cosmological, physiological etc.) conditions which must be satisfied if a certain epistemological ideal (such as "naive falsificationism") is to be realised in the form of a science. Now the distribution of characters is as follows:

	justificationist	**conjecturist**
abstract	Descartes Plato *(Republic)*	Mill *(On Liberty)* Popper
historical	Hegel Plato *(Laws)*	Vico Lenin You I

Greetings!

Paul

26 September 1970
Berkeley

Dear Imre,

Knife in one hand, pencil in the other, I am starting to cut down AM so that you will be able to develop your point of view fully, and will not be hindered in any way by considerations of space. As I told you I intend to eliminate chapters 3, 4, 5, 6, 7, 8, 10, cut down 25 and 26 into a few lines, and conflate 1 and 2 into a shorter introduction. I may even completely eliminate 25 and 26 which are too technical by far. The rest remains as written, except there will be stylistic changes in order to have smooth transitions. What do you think about this plan? Please let me know all at once for I want to send you the new, shortened version *before* we meet in Pittsburgh. [. . .]

All the best,

Paul

PS I had an offer from Oxford University Press, about 7 years ago, to publish an anthology of mine, but I put them off, and now I think I'd rather appear in the same Press as you—if they will have me.

6 October 1970
Berkeley

My dear Paul,

I am leaving tomorrow for Germany with an awful cold. If I were you I would certainly cancel the journey. Alas, I am a petit bourgeois who sticks to his agreements. I envy you for your anarchistic habits!

I arrive in Boston on October 21st and I am booked into some Sheraton Hotel. [. . .] I very much wanted to see Feigl and the Minneapolis people, but for reasons of health I would like to cut down the journey and make it as short as possible, and since I am sure I shall see at least Feigl and Maxwell in Boston, this would make things psychologically easier.

I am very much looking forward to our volume, which I am sure will be an extremely good one. On the other hand, I have very mixed feelings about your silly agreement with the New Left. I think Cambridge University Press or something similar would have been much better for the commercial part but, again, not being an anarchist but a petit bourgeois, I don't want to break up your contract, even with the anarchists. What we have to insist on will be a good American publisher (Basic Books would be good, or Prentice Hall) and we have to see to it that it appears in paperback in Penguin.

[. . .] Please do let me know whether I have to bring my copy of your book to Pittsburgh, where obviously we shall have most of our discussions. I would be terribly grateful if you could send a copy now to Laudan, which I could then take home to London if needed. I am afraid my concern is the size of my luggage, which can become very uncomfortable when I travel for more than two weeks in the States. Please do let me know immediately.

With warm greetings to you and to all your charming household.

Yours ever,

Imre

10 October 1970

Dear Imre,

Pity you cannot come to Minnesota and Berkeley. At some time it seemed you had two lazy weeks available between London and Bloomington, and so I rang Feigl, Hayakawa and Matson, and got you an invitation from Minnesota and a conditional invitation from San Francisco. And in addition, I was prepared to pay you an assistant's fee for coming to Berkeley, reading AM and entertaining the members of "my household." Now all that comes to nothing, and I am quite sad. Anyway—I shall come to Boston (unless something serious intervenes) and Pittsburgh, and London: I think we can ring Butts from Pittsburgh and ask him to send a car to Toronto. I shall also bring my MS so you do not need to fill your suitcase needlessly (though it would be nice if each one of us had a copy). I am looking forward to your papers—the ones you have so far hidden from me: but why? A word from you would have sufficed, and I would have disregarded their contents for AM and used them only for *Schlusswort.* At any rate, my only changes on AM from now on will be stylistic, and, perhaps, further cutting down. I am glad you like the new version, I think it is much better, too: not so longwinded and repetitive (I think I shall be able to cut it down still further). In this, of course, I followed Chairman Mao who says ("Oppose Stereotypes Party Warning" in Selected Works, volume III, p. 66): "Rule 4: After writing something, read it over twice at least, and do your utmost to strike out non-essential words, sentences and paragraphs, without the slightest compunction. Rather condense the material for a novel into a sketch, never spin out the material for a sketch into a novel."

I still want to change some of the stuff in chapter 24 where I deal with you, I am not quite satisfied with it, and I hope you will help me to bring it into good shape. I agree that it was silly of me to sign the agreement with the New Left. Now I would not have any compunction in breaking it, were it not for the fact that I signed it with a person I know and like, namely that leftist philosopher (what is his name?). I am not loyal to *institutions,* and I am prepared to cheat them left and right, but I have some loyalty to *individuals:* this is how anarchist ethics work. But it does not matter: as long as the stuff comes out in Penguin, and after that either in Prentice Hall or in Basic Books (I think Harper Torchbooks would be still better), at least *part* of the disaster is undone. And I have the feeling that the New Left will be quite liberal as regards arrangements for translation etc. etc. (The German version *must* come out in Suhrkamp!)

Incidentally, Searle has written a marvellous book on the crisis of the universities which, among other things, is heavily critical of all of sociology. I am reading the MS right now, and I may send you a Xerox copy. All the people you know [. . .] are turning conservative, not of the Agnew type—God forbid—but conservative nevertheless, and I am thinking of inserting a footnote right into the preface that I am for anarchism in *thinking,* in one's *private life,* BUT NOT in *public life.* Which reminds me: I am what one calls "the placement officer" of the department which means that I have to find jobs for graduate students, male and female. Let me know if you know of any opening.

And now I must go to bed. I have swallowed tons of barbiturates because of pain,[115] and have a pill hangover. Good luck for your house and all your enterprises.

Paul

115. Feyerabend suffered intense pain in his spine throughout his life, caused by the wound he incurred in the Second World War.

20 November 1970

Dear Paul,

Thank you very much for your missives. I am delighted that you got rid of your illness and acquired a girl instead. God knows which is better.

Yesterday I was at a New Left [Books] party. They are a very nice gang. I saw Jacob, Blackburn and Branka Hoare and told them roughly that you have changed your position from *Against Method* and thus you have taken the carpet from under the book whose underlying idea was a spectacular clash between a Hegelian-type anarchism and strict orthodoxy in the philosophy of science. I have invited them to lunch on Tuesday and shall try to come to a realistic agreement with them then.

As far as I can see, the basic fact is that the book was genuinely based on the assumption that you are a wild anarchist and I a wild reactionary, and the whole thing set against the dramatic background of the student revolt and the New Left uprising. All this background has certainly gone in England (much to my disappointment); and, more importantly of course, the duel has become a mock duel and our lines are more parallel than dramatic because of your change of heart. Thus you, as my favourite intellectual enemy, vanished into thin air and instead I got an ally. I shall try my very best to explain this situation to them on Tuesday. On the other hand, frankly I am in despair that with the decline of the New Left, intellectual apathy and dreary boredom will set in, and I see my audience already shrinking, since students this year seem to want solid syllabuses and not intellectual experience. This situation of course establishes a new emotional link between us on the one hand and Robin Blackburn and company on the other, since I see us growing old and looking back with nostalgia on the heroic revolutionary and counter-revolutionary days of '68/'69. (I am sure they disagree with this evaluation.) I would therefore still be interested in doing something for them.

First I thought that they might republish your *Against Method* as it appeared in the *Minnesota Studies* Volume IV, and that I could still fulminate against Feyerabend$_2$ while mentioning in passing that he is now dead and replaced by Feyerabend$_3$. But somehow I think that this exercise would be so contrived that there would be no point in it. So I am afraid that what I may propose to Robin and his friends on Tuesday will be to try and get a new, real anarchist or marxist in philosophy of science and then let the conservative speak and retain the last word for the revolutionary. The only trouble is that for the moment I do not see who would be the person on the Left.

Especially after the party yesterday where I met quite a few nice men and pretty girls, I am rather sorry for you that your career as the hero of the New Left will soon be nothing more than a faint and quaint memory.

Anyway, I shall write to you again—and probably Robin or even Branka will write to you too—after the Tuesday lunch.

Love,

Imre

24 November 1970

Dear Imre,

The talk with you returned me to sanity. You have a marvellous sense of reality, inside of philosophy and out, while I sometimes get completely out of touch with reality. Now, at last, I can present my problems in a clear fashion. There are two different problems.

Content of my essay. I think you misunderstood me when believing that I was going to change the tone of the essay. My intention is the same as in Minneapolis: to streamline AM so as to increase its effectiveness, to cut down on footnotes and omit unnecessary epicycles.

I also intend to devote a whole chapter to you and to treat you much more favourably than in all previous versions. I have realised what one might call "the rich complexity of your philosophy." As far as I am concerned, it is the only philosophy worth considering at this stage of the debate. But while I think it very important, I am still opposed to it, only I realise now that the argument against you must be strengthened considerably in order to gain some plausibility. I think I can strengthen it and I think I can argue with you on the basis of those same "basic statements" (such as: "Copernicus is better than Ptolemy") you use. In addition I now feel the *temptation* of your ideas, not only their intellectual force, for I certainly have not much sympathy for the Mellors of the world. Also I would cut down science rather than permit all sorts of weeds to obstruct our view; I am, after all, "disrespectful Feyerabend," as you called me completely correctly. Hence, my task is now much more difficult, but it still exists, and I think I can still carry it out in a way that leads to a real intellectual debate. More concretely, I shall carry out the following changes: the Preface and Sections 1 and 2 remain unchanged (except for a few stylistic changes, and an explanation of "progress" in accordance with your suggestion in Minneapolis). Just read Section 2 and you will see how apparently insane the whole business still is. Sections 3, 4 and 5 will be shortened; all I shall do is discuss the case of the Brownian particle, draw the consequences of Section 6 which comments on Voodoo, and show that a study of Voodoo can help us to revise the sciences. Sections 7–21 will remain almost unchanged, only I shall make it clearer why I insert them and I shall emphasize their weak points and their highly hypothetical character [. . .]. Section 22 remains, Section 23 will be streamlined, Section 24 will remain intact up to the point where I start talking about you, then comes the new Section 25, viz. Lakatos's *rationalistic cliffhanger* (I shall not use these words, of course). I am very doubtful whether I shall insert *incommensurability,* for although I think I am right in a very trivial manner (proof: your "Proofs and Refutations"), I present the case in a way that can only lead to increased intellectual pollution and is actually leading to it. The conclusion, finally, will be rewritten to make the issue between us as clear as possible.

Now I think that the best way to publish the whole business would be in two separate, but identical looking, volumes. In Minneapolis you said you had no time for such an extensive enterprise (money problems etc.); this is why I dropped the idea. But it would be absolutely the best idea. We would then have two volumes, each volume of about 150 pages, each one with an appendix containing a reply to the main text of the other volume. The title of both could be quite neutral, such as: "Problems of Science" (not "of Empiricism"), Volumes I and II, or Part I and II, and inside my volume my subtitle "Against Method," and inside yours your subtitle, exhibiting the difference in philosophy *as well as in style.* This would be ideal. Next best is a single volume, title to be found, with one essay by me, one essay by you, and a *final dialogue* which we might write together. Or one book with two essays commenting on everything we have written, the essays themselves excepted. Or any other form.

Now, as regards the *publisher,* this is problem (2). If the chance to get fundamental debate going is greater when we publish with the NL [New Left Books] than elsewhere, then I would prefer the NL. I also prefer them because they seem to be nice people, at least they seem to be nicer than the people at CUP [Cambridge University Press] from all I have heard (from you alone, for I have no other sources in that matter). And, besides, I am bound to them by contract. Of course, I can always substitute something else, for originally they only wanted to publish "Empiricism I" from me to make it more accessible to students (this is what Robin and Branka told me when we first met). But AM is *not* changed in its fundamental structure, so that argument cannot be used against them. Still, on (2) I am doubtful and await your sentiments, especially considering (1).

I am moving to my new house at the end of this week (if everything goes well, that is). My new address: 1168 Miller, Berkeley, Calif. 94708. I have been in bed the last three days in lots of pain, Mara is ill in bed, her parrot (which I bought four days ago for 98 dollars) is ill too and close to dying, Christopher is sneezing, only Searle is healthy. Is there a moral in this?

All the best

Paul

25 November 1970
London

My dear Paul,

It was very nice of you to ring me from Berkeley. I think that the decision should rest with you. There are two apparent possibilities.

(1) You publish it with CUP [Cambridge University Press] or Oxford University Press. If it is with CUP we could make it twin volumes—I expect to finish mine some time next year but I do not know exactly when. The title would be *Problems of Empiricism.*

(2) The second possibility is to publish it with the New Left. Incidentally, Robin [Blackburn] tells me that your contract is for a volume called *Problems of Empiricism* and not *Against Method.* If you choose this alternative, I may write a "Postscript" and if you wish you could have the final word although I think that our disagreements are so minimal by now that I am bound to be frustratingly polite to you and you may not want to have a final word. In fact, I may have to praise you since you praise me; and in order to make your praise more authoritative I have to turn you into the authority (or I may opt out). In this case the volume would be published in hard cover by the New Left; simultaneously by a first class American publisher and all that they would request is that paperbacks outside the United States should be held up for a year. If the volume is successful Penguin would publish it in paperback, says Robin.

I suppose that your decision will depend on the character of the volume and of course on your personal predilections. I feel neutral.

[. . .] Incidentally, I do not mind if you take my "Rational Reconstructions" and Schilpp paper [116] into account in your volume, whether it goes to the New Left or to CUP. I now have new arguments against what I anticipate to be your new arguments. So I can waste some ammunition.

Coming back to Minneapolis, I just wrote a conciliatory note to Paul Meehl. The letter was written on the spur of the moment, after having read a new paper of his which I liked.[117]

No news otherwise. Alan Musgrave is feeling homesick in New Zealand and wishes he had never left London. Julia is very happy because they are rich and comfortable. All for now and again, let me tell you how I much I enjoyed hearing you on the phone.

In a rush.

Yours ever,

Imre

116. Lakatos 1971a and 1974a, respectively.

117. This piece of the correspondence between Lakatos and Paul E. Meehl (Regents' Professor of Psychology at the University of Minnesota) neatly brings to light some aspects of Lakatos's personality. In his letter, Meehl expresses his high opinion of Lakatos's intellectual achievements, but also his deep concern about Lakatos's rudeness. In fact, Meehl claims that Lakatos owes him an apology for the way he was treated during Lakatos's last visit to Minneapolis. Here follows Lakatos's reply (Archive, 12.1):

Dear Paul Meehl,

Your letter gave me immense pleasure and relief. You are one of the few people I know whom I respect and admire and also of whom, instinctively, I am extremely fond; and our clash in 1969 distressed and indeed depressed me.

Let me also tell you that your letter provided me with something of a shock—one of those rare shocks which may even change me (or at least my behaviour) a bit.

I do not remember *at all* that I told you that if I was running the US I would have you shot. But I believe you, since my friend Arpad Szabo claims that I told him in 1951 that if I was running Hungary, I would have him shot (for some anti-Soviet remarks). He still smarts from my violent attack which I forgot immediately as I uttered it.

Alas, for at least two decades of my intellectual life I had thought of history as an apocalyptic struggle between angels and devils, even if my criterion of demarcation between good and evil has changed immensely. This, I suppose, led to many dramatic, or rather violent, scenes with my friends when I suddenly "realised" that they were slipping into "evil." I confess that even until recently (and perhaps even now—I have not yet sorted *this* out) I have regarded Anglo-Saxon military power, even if partially evil, as the main (and, in fact, *only*) safeguard of the world from some new Middle Ages. You must remember that most of my family and childhood friends died in gas chambers, and too many of my later friends were hanged, tortured and/or ruined by the Communists.

To my horror, I equally deleted from my memory that I had "accused" you of being a "charmer." I do however remember my naughty reference to your Regents' Professorship: I was trying to kick you since I felt badly hurt.

Let me indulge, for a moment, in some sort of self-analysis. I lived, for various reasons, with a constant guilty feeling throughout my childhood and youth; my last superego being the Party. In 1956, I decided that I was *not guilty* and was amazingly successful in getting rid of my superego. Perhaps as an overreaction to my youth I altogether ceased to think seriously, or even to pay any attention to other people's reaction to my behaviour. It now seems to me that this may have been a major source of our 1969 friction.

Anyway, let me thank you for your very kind letter; for your very interesting paper, and for your book which I shall have read by the time I arrive in Minneapolis.

With warm greetings and many thanks.

Yours gratefully,

Imre

26 November 1970
Berkeley

Dear Imre,

Before I fall into bed again with that bloody liver of mine, let me give you an outline of my last will and testament, regarding AM. First thing: AM is still AM as regards content—just read section 2 of the latest version you have and you will see that in the sciences I am as mad as ever and, I believe, with good reason. This being the case, AM should still be *called* AM, and not receive some harmless title such as "Problems of Empiricism." Now, if AM retains its title, "AM," *then I do not see any way in which I could get out of my contract.* I could of course break it and damn the consequences, but I would not do this, not because I am such a moral being, but because I like Robin and Branka etc. etc. Moreover, the New Left, whatever else the disadvantages may be, are a *nice* group and it does not seem to me that one can say the same about Cambridge University Press, or at least about some people in there. Besides, one of the reasons why I did not send anything to Oxford University Press about 8 years ago when they wanted to publish part of my then Collected Works is because I did not like the idea of having anything published with an *academic* publisher—and this was at a time when I was still relatively sane, and gave some Ds on [examination] papers, and not all As. My aversion to academic publishers has been mostly increased as the result of my having to write a recommendation for Princeton University Press, re. translations of Hans Albert's *Traktat.*[118] They sent me a questionnaire with the most inane questions, and the word "scholarly" *which I despise* occurred there at least 8 times, five times in italics. No, Imre, my wish for respectability is not so great and my concern about the fate of my "first book" (you on the telephone) not so overwhelming that I would swallow all that crap. Besides, if you can really make sure that Basic Books will be the US publisher, then things will be OK anyway. [. . .] So, my decision, as you call it, is: *stay with the New Left* and have the thing out as soon as possible and with some tight guarantees concerning US publication, translating into German (Suhrkamp is bound to take it) etc. etc. And I do hope that you will stay with me and will annihilate me, as you promised. I shall send the final version, with smaller footnotes, larger Lakatos chapter etc. etc. in about a month, after I have moved and gotten rid of my liver. So, unless you have some new arguments, tell them everything is OK and production proceeds according to plan. And now I shall see Christopher, Mara, Shirley and Mickey for a Thanksgiving dinner—if I can stay on my feet, that is.

Be well, and think kindly of your confused anarchist friend and *still intellectual opponent* (what on earth gave you the idea I was *no longer* your intellectual opponent?)

Paul

118. Reference is here to Albert's *Traktat über Kritische Vernunft* (1969). The book was eventually translated into English and published by Princeton University Press in 1984.

2 December 1970
London

Dear Paul,

I was terribly amused to have on the same day two letters from you dated 24th and 26th November, from which it transpires that you must have changed your mind about 6 times on about 6 major issues within two days. This is all wonderful, since I do not think that there are essential differences between your six different views and the main thing is to publish your paper, so do finish it. I am sure by this time it will be very good indeed. If you like, do publish it with Robin. It won't do you any harm if you have a good American publisher and I trust Robin will be able to arrange that.

Frankly, I am relieved that you are still my intellectual opponent. However, I have to remind you that all philosophers differ from each other if they are worth anything, but I still think some are more allies and some more opponents. All is a matter of degree. You and I have to agree that we are more allies in the given set-up than opponents. Anyway, all this is irrelevant, just go on and work.

I am very much looking forward to receiving your ms.

The expression "Lakatos's *rationalistic cliffhanger*" is a very good one and of course, as you bloody well know, I find it a very good description of my position, so your promise "not to use these words, of course," is a disappointment for me. Of course you must use it. It is a wonderful and faithful description of the situation.

I hope your health serves you well and that your ms. will be ready for the New Year.

I was amused by what you said about Hans Albert and Princeton University Press. Do not think that the New Left Review is very different (you keep forgetting that for false pretence they match any academic orthodoxy). Anyway, you have my blessing with Blackburn (incidentally, after I had him over for lunch the other day I was attacked by several idiotic colleagues of mine who thought that I should not bring *that* man here. I needn't say that I was rather quick with the answer, namely that I enjoy much more having lunch with a man whom I succeeded in kicking out than with people with whom I have not yet achieved this success).

In a rush.

Yours ever,

Imre

30 November 1970
Berkeley

Dear Imre,

I can't tell you how nice it is to know a person like you who cuts one down to size when one gets too silly and self confident, and who has good advice when one gets confused. I am now working very hard at AM despite the fact that I am tired with that bloody liver of mine—every hour I have to go to bed for about a quarter of an hour—and despite my moving to another house which means that I have to pack all my books, clothes, dishes, and there is quite a lot of them. Then, after the move, I have to unpack them, clean the house etc. etc. Still, I have already arrived at section 17 and have, I think, improved things quite a lot.

The message is the same as before: anarchism of knowledge or, as you say, extreme scepticism. It seems to me that it is worth arguing this position which now seems to get additional fuel from case studies. A good debate of the position will make lots of people interested, they will really eat it up. So, all that is needed is a good debate. In order to achieve this, the following conditions must be satisfied. (1) I must give this extreme scepticism some *plausibility,* so that it looks like a real possibility, and not just like a quaint aberration. This, I think, I can do. (2) The debate must not be sidetracked by superficial misunderstandings. Now, in the past, I have not given a very fair account of your position, and I have not understood it too well. I think that I shall now be able to say what you are really up to and what it means for the philosophy of science. (3) There must be some objections left on both sides. Well, I think I still have one objection or two, and you can't possibly be without any objection against my final version (which will also deal with your Popper paper and your Boston paper).[119] How these objections are arranged is a tactical problem that is best solved after you have read my final version, and I hope to have it finished before the year is over (I want to have it finished by mid-December, for I want to spend Christmas in Palermo, in the sun, with a visit to you on the way—but if I don't succeed I shall stay in rainy and smoggy Berkeley until the last letter has been set right).

Finally, when I consider my own intuitions, I think I am temperamentally a sceptic and have little sympathy for rationality, quite apart from the fact that I have never understood things like "the truth" etc. etc. So I should really get rid of these intuitions in the form of some good arguments and counterarguments or, rather, not *get rid* of them, but *test* them. So, objective and subjective conditions demand resumption of the argument between "dogmatism" and "scepticism." [. . .]

Lots of things to do and for relaxation I am going to the wrestling on Saturday in order to see the woman world champion, the fabulous Moolah, as she is called. (I shall meet her tomorrow, for I know a journalist who makes propaganda for her). With me will be Robin,[120] Mara, Shirley. They and all the other broads send their regards.

All the best.

Paul

119. Lakatos 1974a and 1971a.

120. Feyerabend's assistant and "lifelong friend" at Berkeley (see Feyerabend 1995, 121).

7 December 1970

Dear Imre,

This is my first letter written from my new abode, right above Grice's house so that I can spit on him. I moved on a day when it rained cats and dogs, there were 6 assistants who helped me, Mara and Christopher among them, all the books got wet but they survived except the *Logic of Scientific Discovery* which dissolved completely and was washed away. When I arrived, the house was full of ants who, fleeing the flood, had sought refuge. So first I had to fight them. Next problem are the various noises one hears when one enters a new house. Now every house has its own specific noises, it breathes, it sighs, it creaks, it shakes itself and these noises must be distinguished very carefully from the outside noises which do not belong to the house proper. It is only *after* I have identified all the noises that I can forget about them—but to identify them takes me a week or two. Last night I moved around in the garden for about 40 minutes with a flashlight to look for a strange dripping noise until I discovered that my roof leaked. Then there are raccoons in the neighbourhood, they have a riotous sexlife right on the top of my house. So all this has to be taken into account. It drives me nuts.

In order to recover, I went to a wrestling [match] on Saturday with Robin and Mara and Shirley and we hoped to see, and afterwards dine with, the fabulous Moolah, the world's woman wrestling champion, but, alas, she had been beaten up in an earlier wrestling match and was resting in a hospital.[121] Today, Monday, we, i.e. Robin, Mara, Shirley, Christopher, Abigail and a few more are going to see Muhammed Ali in his second boxing fight, and we are all looking forward to it, for he is so graceful and when he knocks an opponent to the ground it looks as if he had just stroked his cheek with affection. Wednesday Mara and I are going to *The Merchant of Venice* and *all by ourselves* and, I am sorry to say, I am becoming more and more fond of this wicked woman.

So, you will say, slamming your fist on the table, what about AM? Patience! I am restarting work today and hope to have the final version finished before the year is over. It will be much better and really paradoxical, for what you will get will be a sane presentation of and a sane plea for an insane view. [. . .] Incidentally, you call me "darling of the New Left," I just heard that some students here on campus try to induce others not to go to my lectures because I am a Fascist, so they say. Now, this puzzled me no end and I tried to find out what made them say this. Well, the reason seems to be that they know I was an officer in the German army and regard everything I do now as an attempt to cover up my past. So, they say, I am not "authentic" which, considering what people now mean by "authentic" is just as well.

Liz looks more and more luscious although she has grown to tremendous proportions. Mara is still the most beautiful thing around especially when she is in a good mood, and very intelligent. But I must finish AM first. The Deutsche Buchgesellschaft asked my advice on various matters, for example about an introductory series in *Mengenlehre,*

Metamathematik, Didaktic der Mathematik, history of ancient mathematics, and I gave them your name adding that you would either do some of these things yourself, or give them good advice. Why not extend your power basis to Germany? The Deutsche Buchgesellschaft is an excellent *Verein.* I also recommend Szabó[122] for their history (by the way, I lost Szabó's address when moving from one place to another, so could you send me his address again?). And this, I think, is all I can tell you *today.* Be well, be economical with your sexual powers and be kind to your enemies.

Paul

P.S. Got a really charming letter from Hans [Albert]. With you, him and John [Watkins] around it is no longer true that Popperians are a sinister Puritanical crowd.

121. Here is how Feyerabend remembers the show: "Even from the distance you could hear the roaring of the crowd. Little old ladies threw their knitting into the ring: 'Kill him! Kill him!' Married couples drove up in their Buicks or their Fords: they would look straight ahead when they arrived, straight ahead during the first few minutes; then gradually they melted, became attentive, raised their fists in unison, smiled at each other, and may have had their first real contact in weeks. Like everybody else I had my heroes; I trembled when they got into trouble and was relieved when they won. In Hawaii, where Robin had a job and where I stopped on my way to New Zealand, I met some of them in person. But then Robin convinced me, by word and demonstration, that it was all a fake: the outcome was predetermined, the blood artificial. I still have not forgiven him" (Feyerabend 1995, 121).

122. Arpad Szabó, historian of Greek mathematical thought, had taught Lakatos at Debrecen University. Lakatos (1976a) expressly recalls his teachings (Szabó 1958) when he considers "thought experiments" as the pattern of mathematical proofs in the pre-Euclidean Greek mathematics. From Szabó again (1960) Lakatos takes the idea that in Euclid's time, "postulates" and "axioms" meant propositions in the critical (dialectical) dialogue put forward to be tested for consequences without being admitted as true by the discussants. Szabó showed his appreciation of his old disciple by dedicating the English version of his *The Beginning of Greek Mathematics* (1978) "To the memory of my friend Imre Lakatos."

8 December 1970
London School of Economics

Dear Paul,

I am disgusted with you. You haven't written for at least 4 days from which I conclude that you must be happy and content. This is not the thing I expect from you.

Mary sent you the vitamin tablets. Are you already living at your new address? If yes, send a picture.

What are your plans for 1971? I think I shall go to Penn State for a week in September.

With warm greetings, in a rush.

Yours ever,

Imre

18 December 1970

Dear Imre,

Cheer up! In 2300 when Popper will be known as the Kant, you as the Hegel and I as the Engels of the 20th century, one will have to go to the darkest corners of libraries to find out about Cohen, Hesse, etc. etc. I hope you got my flood of letters by now just as I got the Redoxon.

Many thanks! You saved my voice! And my energy! And AM!

Paul

21 December 1970
London School of Economics

Dear Paul,

[. . .] I cannot stand megalomaniacs, apart from you and me. Spinner is writing a three volume book on philosophy of science and he doesn't have the faintest idea of what he is talking about. He has already finished the first volume. 600 pages. At the same time, there are people like Howson, Worrall[123] and even Zahar, who are at least 250 times cleverer than he is, and know about 700 times more than he does, and still don't dare to write anything down, let alone three volumes, most of which is filled with quotations from you and me, and then some irrelevant and twisted criticism in order to get on top of us. No thank you. He will either leave London next summer with an improved mind, or with a destroyed body.

[. . .] I have already discovered that nothing is worse for work than taking criticism into account. I have always tended to ignore criticism and, as you know, all that I have been writing recently is a theory of how to avoid it, but now in my reply to Noretta Koertge,[124] which you will soon have, I develop my position further. Poor girl, she criticises me because, for her, research programmes are too autonomous for her progressive, critical, liberal mind. She would like to see them indented (as she puts it) by criticism. I destroyed her position very gently, as you will see.

I am worried about AM. Since you have not finished it I suspect that you are again destroying it by lengthy footnotes and appendices and other Popperian paraphernalia. Don't.

Yours ever,

Imre

123. John Worrall, Lakatos's assistant in the late sixties, joined the LSE staff as lecturer in 1971, and is now professor of philosophy of science there.

124. See Koertge 1971 for the criticism and Lakatos 1971b for the reply.

26 December 1970

Dear Imre,

I just rejected New Zealand—tell me I did the right thing or else I hang myself. It was a choice between one sabbatical year visiting former girlfriends and ex-Popperians, and immediate travel to New Zealand, and I chose the former (I didn't want to get into another sticky situation). Now, when in Europe, I shall try to find a job there, either in Germany or, what would please me even more, in Italy. Of course, England is first on the list but I wonder whether I shall find anything. A Fellowship in Oxford, or a post in Brighton would be marvellous—maybe I will talk with Ayer about the former when I come. Anyway, I repeat, tell me at once that I did the right thing.

Two weeks Christmas vacation, no screams from the nursery school, no ants on the floor, no teaching, but AM does not proceed very fast, I got stuck at the bloody 20th chapter and it always becomes too long. After that there is another difficult mountain to be overcome—*the chapter on you* which should be brief and to the point, and tell more about you than you do yourself in your giant dinosaurian monster essays. (I think I can do it.) By the time I arrive at that chapter, the kiddies will be back and will fill my ears with noise and my heart with murder. Well, we shall see.

Very depressed at the present moment. More than once in the last few months I was on the verge of selling all my books, furniture etc.

etc. and of simply *disappearing* into an altogether different profession. I do not find being a full-time intellectual very congenial. It wears me out and does not make me happy either, especially as I am not a theoretical type. And yet I sit here day in day out trying to think! What a fucking way to live! Why did I not join Brecht and become a producer? (Any fatherly comments on that?)

Of course, it may all be connected with the fact that I do not live in the proper surroundings. A little theatre, a little flirtation, no discussions at all, but mostly reading and writing, well, most of the time I feel I am living at about 3% of my natural capacity and am wasting away. And my recent excursion into anarchism really is not anything to please me very much. [. . .] But keep your fingers crossed that I get a chair in Palermo and a position in the Mafia.

A bleak day, fog, I can hardly see the Bay Bridge. Be well, give my best regards to Gillian; yes, an anteater would not be such a bad idea, but a lion to eat all the little kiddies on the corner would be even better.

Paul

1971

5 January 1971

Dear Paul,

Thank you very much for your letter and for your card. I am very much looking forward to your finishing the book. If I am tempted to write an Appendix (and I probably shall be) it would take me a little time. So I am looking forward to it very much. In a couple of days' time I shall send you my reply to Noretta Koertge and to Richard Hall.[125]

It would be very nice if you came to London for '71/'72. You are very cagey about your plans. What are they?

[. . .] The weather here is unusually marvellous. It has been frosty and cold since Christmas Eve and I only look back with horror to last year when it was mild and I was ill. Now it is cold and I am completely healthy except for my nerves which are ruined by the house. (Incidentally, I too wanted double windows and sound-proofing but they are quite out of reach of my pocket.)

I took over the *BJPS*.[126] I inherited some rubbish but from August

Figure 3.4 Feyerabend's favourite Bela Lugosi poster. "Dear Imre, do you recognise your anarchistic countryman Bela Lugosi whom I greatly admire?"

on it will be primarily my responsibility. Don't dream of escaping my demands on you in this connection.

With warm greetings and best wishes.

In a rush,

Imre

125. See Lakatos 1971b.

126. Lakatos became the editor of the *British Journal for the Philosophy of Science* in 1971 and held the appointment until his death in 1974.

January 1971
Berkeley

Dear Imre,

Just emerged from two catastrophic weekends. First I was in pain almost every day, then there were lots of parties, then Arne arrived, and now I am back to normal, though very tired. This is all the news I have. [. . .] Karl Popper is in town for the Tarski festival which is going on here and I may ring him up. But then again, I may not ring him up. I shall tell you what I shall do, I am in a pretty isolationistic mood these days.

Be well!

Paul

January 1971

Dear Paul,

I smell a Sussex Chair in the air for you. After all, Roy Edgley may do some good. JWNW [Watkins] and I are telling everybody that you are the Greatest even though Roy thinks so too. Now: are you really interested?

I shall be in Los Angeles on Nov. 1st. Will Berkeley invite me to preach against Pope Paul?

I am so busy winding up my administrative affairs before leaving for Boston that I cannot prepare any arguments. But, unless my back gets very bad, you will get Lakatosian rhetoric in Topless Form. I wrote a song to Reason and I now need a girl to sing it. (A rhyme is still missing in the 7th line: what rhymes with "Paul Feyerabend"?)

John Watkins is ill. Latsis,[127] the Rising Star, gives his lectures. Watch that man. Have you read his *Situational Determinism* (*BJPS,* Aug. 1972)? [. . .]

Love. Till soon,

Imre

I broke up with Gillian, and we are selling the house.

127. Spiro Latsis studied Economics and Logic and Scientific Method at LSE in the early seventies. He was awarded the Ph.D. for a thesis in the methodology of economics. Lakatos and Watkins were his supervisors and Terence Hutchison his external examiner. He is a leading figure in an immensely successful family business created by his father, John S. Latsis. Together with Lakatos, he planned and organised the Nauplion Conference in 1974; see letter, PF to IL, 20 April 1973.

14 January 1971
London School of Economics

Dear Paul,

[. . .] I completely stopped being a human being, let alone an intellectual. I am completely the slave of this bloody house[128] because one has to make decisions, for instance where lighting should be in rooms which do not even exist yet, and where central heating radiators should be when I do not know what sort of furniture there will be if my life is long enough ever to get into a position when I can afford to *buy* furniture. Also, Gillian hates every colour which I like and vice versa. We have now agreed that each room will be painted up to two yards in her favourite colour and above that in mine. Where we cannot agree, the room will be painted black.

On my doctor's advice I am going to Jerusalem tomorrow for a week to live in a tent because houses whether built or unbuilt drive me mad. I shall write to you again on my return.

With much love

Imre

P.S.$_1$ Karl gave a wonderful radio interview lasting a full hour which appeared in the *Listener.* It turns out that he invented fallibilism, criticism, and also language. He really must be a great man. About Mercury's perihelion he said that it did not refute Newton's theory because it was an irrelevant falsification. The great scientists know what is relevant falsification and what is irrelevant falsification.

P.S.$_2$ Peter Achinstein gave a big talk yesterday and we were so fascinated by his stone-age views that John Watkins invited him to come again next week because some students didn't believe their ears. Fortunately when he talks next time, I shall be sleeping in the middle east sun. [. . .]

128. Feyerabend recalls: "Imre had bought the house for representational purposes. It contained a kitchen, bathrooms, a large sitting room, and Imre's library on the upper floor. Visitors were first shown the garden, then fed, and finally led upstairs for serious talk. I was often invited as an extra guest. I enjoyed the garden and the dinner but, anticipating the drift of the conversation [. . .] I remained in the kitchen and helped Gillian with the dishes. Some of the guests didn't know what to make of this. Men, especially scholars, were supposed to engage in debate, while dishes were the domain of women. 'Don't worry,' said Imre; 'Paul's an anarchist' " (1995, 129–30).

24 January 1971

Dear Imre,

Back from Los Angeles, not a tone left in my throat,[129] for I sang too much and strained my voice. But my hands are strong and capable, and I continue with chapter 25, that is, with you. When *that* chapter is finished, the main difficulty will be over, then there is only the chapter on incommensurability and that should not be too difficult for, although my ideas on that matter are pitiful, the objections are even more pitiful. Did I tell you that they are prepared to wait for me in New Zealand until 1973, so that I might end up there after all? Well, in the meantime I am going to see you, so you will have every occasion to convince me to do something better. Incidentally, I may be invited to go to New Zealand this summer to look it over, so they really seem to be eager to get me. Which means that, if all comes off well, I shall spend about one month in Australia and New Zealand, and come back to Europe the other way around. I have now at last with me my "reply to criticism,"[130] and there, on page 224, I have three arguments for proliferation only the third of which is psychological, while the first two are logical. But it is so already in "How to Be a Good Empiricist," which came out in 1962. For there I argue that the Brownian particle does not form part of the class of potential falsifiers of the phenomenological theory of thermodynamics *unless we use the kinetic theory to put it into this class.* I am telling you this not in order to make you revise your *writings* but in order to make you revise your *opinion of me:* I am not so much of a psychologist as you make me out to be (Of course, being a Quinean, I am a psychologist in a wider sense.)

And now, lunch

Paul

P.S. I also say, on page 224, line 13 from the bottom: "The principle of proliferation not only recommends the invention of new alternatives, it also prevents the elimination of older theories which have been refuted. *The reason is that such theories contribute to the content of their victorious rivals . . .*" (*me,* 1965!) Cf. with this also what you say on me and Reichenbach on page 136, footnote, of "Criticism."[131] I think I say everything you say a wise man should say and you say I don't say. On page 225 of my "reply to criticism" I even say that theories "possess a utopian component in the sense that they provide lasting, and steadily improving, measuring sticks of adequacy for the ideas which happen to be at the centre of attention . . . ," which means that I am here speaking of research programmes in your sense (cf. also my explanation of my term "theory" in footnote 5). So you see, it is simply not true that in my reconstruction of science a falsified theory "can be removed from the rational reconstruction" (your Boston paper, page 16, footnote 1).[132] I explicitly say the opposite (see the passage in italics just above). Nor is it true, as you say on page 11, line 4, that I developed a *psychological* thesis of Popper's. If Popper is psychological (this I do not know), then I changed him and insisted on a logical function of proliferation. Also (your footnote 3 on that page) I think that alternatives must not be removed, but are *necessary parts of the falsifying process,* see again the passage in italics in the above quotation from my "reply to criticism." So, I am much closer to you than I am to Popper (assuming he really is as psychological and externalist as you say he is, a matter about which I have no opinion). Of course, later, in *BJPS,* I say that "negative instances suffice in science," but then I take it for granted that negative instances can hardly ever be obtained without the help of alternatives.

Added February 21st: on page 145 of your "Falsification,"[133] footnote 2, you say that in earlier papers on the quantum theory I "was more Popperian than Popper" on the issue Bohr vs. Einstein. Now, it just occurred to me that I *never* published "Popperian" papers on quantum theory. Already in 1959 (*Current Issues* volume—this is the report of a monster conference in Chicago) I supported Bohr and criticised Karl, and I did so even more in "Problems of Microphysics."[134] Even I was surprised by this discovery—but this is how matters seem to be!

129. At this time, Feyerabend used to fly to Los Angeles once a week to sing with Ina Souez, one of Fritz Busch's "Glyndebourne stars" (see Feyerabend 1995, 120).
130. Feyerabend 1965b.
131. See Lakatos 1968b.
132. See Lakatos 1971a.
133. See Lakatos 1968b.
134. Feyerabend 1962b.

21 February 1971
Berkeley

Dear Imre,

For all my scholarly attempts in the past few decades I never got a cent. Now I have just received $1,200 for my article "Experts in a Free Society"[135] (of which I sent you a copy) which is irresponsible, unscholarly, loosely written and so on. But: I am getting the message. So, farewell Paul the responsible scholar (if there ever was such a guy), and welcome Paul the flamboyant and irresponsible journalist.

This brings me to another point which, as a matter of fact, will be the topic of my first lecture in London (if I ever get there, that is).

I guess it is true to say that a philosopher has to choose two things for his philosophical life. He needs an *ideology* (Marxism, critical rationalism, etc. etc.); and he needs a *father-figure* (Marx, Thales, or what have you). A party secretary like you needs a father-figure even more urgently than a normal party member or even a stray visitor such as myself. Now, having chosen one's ideology, the question is: who will be the father-figure?

Now, a critical rationalist has a large crowd of father-figures to select from. We can start from Thales, that first capitalist and world traveller, to Democritus, to Voltaire, Mill, Boltzmann, Russell, Einstein, up to our *lapis irae,* Popper. Now, what I shall say in my first talk is that a critical rationalist should choose *Mill* as a father-figure (provided he needs one), and should avoid Karl like the plague. Reasons: everything that is in Karl is also in Mill. And what is in Karl over and above Mill is not worth special mention (the Carolinian additions, technicalities such as corroboration, are due to the Vienna Circle background and are of interest only to technicians).

Just listen to Mill (I am always talking of *On Liberty,* for Mill was imaginative enough to also have been an inductivist): "The beliefs we have most warrant for have no safeguard to rest on, but a standing invitation to the whole world to prove them unfounded." "There is the greatest difference between presuming an opinion to be true, because, with every opportunity for contesting it, it has not been refuted, and assuming its truth for the purpose of not permitting its refutation." Discussions cannot be decided by experience alone: "There must be discussion to show how experience is to be interpreted . . . very few facts are able to tell their own story, without comments to bring out their meaning." "Even in natural philosophy, there is always some other explanation possible of the same facts" (the only exception which Mill makes and where he errs, just as Karl erred here before you came, is in mathematics: *"the peculiarity of the evidence of mathematical truths is that all the argument is on one side. There are no objections, and no answers to objections"*). He invites to "free and daring speculation on the highest subjects." Even Newton is acceptable only because one has so often tried to refute him and has failed, and so on.

So far Mill.

What are his advantages?

I notice several.

First, his theory of *science* is more liberal than Karl's (it tends more in the direction of Neurath), and less hampered by technicalities. *Second,* his theory of science is part of *a theory of man* that aims to find conditions for the full and free development of individuality. In Karl the historical path seems to be: Hume's problem → new theory of knowledge → generalization → open society. In Mill the path is: development of individual → open society → critical rationalism → critically rational science. The second type of path seems better to me, for the fate of the individual is more important than the fate of science. *Third,* the *atmosphere* is so entirely different. In the case of Karl: self-righteousness, rigid rationalism and constipation; in the case of John Stuart: openness, charm, elegance; he had an affair (Harriet Taylor) which led him to *On Liberty* (compare that with Henny), no constant self reflection, no constant reflection on the position of the shadow of one's own behind (footnotes on footnotes on footnotes), generosity etc. etc. These are much more important properties than expertise in the handling of some type of rationalism.

Just found a letter from Einstein to Born. Born had pointed out to Einstein that Whittaker denied him any merit etc. Einstein's reply: "Don't give your friend's book any thought (Whittaker was a friend of Born's). Everyone behaves as seems right to him, or, expressed in deterministic language, as he has to. If he convinces others, that's *their* problem (my opinion entirely). At any rate, I found satisfaction in my efforts, and I don't think it is sensible to defend my few results as 'property,' *like an old miser* who has laboriously gathered a few coins for himself. I don't think ill of him (Whittaker) . . . And I don't have to read the thing."[136] *How marvellous is this man Einstein! And how small Popper is, compared to him. And this smallness is not an accidental property; it will pervade his philosophy or, at least, the brains of his students who have started quarrelling long ago (Bartley, Agassi etc.). I* can say that, and I can utter this criticism for I have never pretended I have invented anything *myself.* I defend others: now Mill, now Bakunin, now Lenin.

I am telling you all this because the topic will be a special section in AM with the title: "Mill or Popper?," and I shall say exactly what I said above, only in about twenty pages. This is advance warning so that you can annihilate me. To repeat, I say: every movement (if there has to be a movement) has to have an ideology and a father-figure. I am prepared to accept an anarchistically loosened form of critical rationalism as one finds it in Mill, but I say that everything of importance in that connection, even for science, is already in Mill (and as regards science, we just take along Einstein and Boltzmann, for the details), and much that is undesirable is in Karl (compare his autobiography with that of Mill, or of Einstein, or with Boltzmann's sketch of an autobiography); so Mill will be my father-figure and should be everyone's father-figure. Of course, when dropping Karl we bypass the Vienna Circle, but that is no grievous loss (of course, all your own recent efforts on the various Poppers, and on confirmation, will be in ruin, but after *Proofs and Refutations* you can afford wasting yourself forever).

Best greetings,

Paul

135. Feyerabend 1970d.

136. Letter from Einstein to Max Born dated 12 October 1953 (Einstein et al. 1969). The comments in parentheses and the italics are Feyerabend's.

21 February 1971

Dear Imre,

[. . .] I am now well into the incommensurability chapter and the end of AM is *imminent.* I shall make no further changes after that, except in style and such *substantial* changes as you suggest. I hope NL [New Left Books] agrees to have it brought out in the U.S. and England *simultaneously,* otherwise it would be a loss, considering how fast things move nowadays (only ten years ago Popper was tops, and now he is antediluvian, thanks to you). Also annihilate all earlier versions of AM in the PKF library at LSE for the latest word from *The Prophet* which supersedes all earlier words is going to arrive any day now. [. . .]

I have not done much in the last month and am a little disgusted with myself. The fact is, I got disgusted with the whole world and started sleeping 16 hours a day for a few weeks. Now that my dreams are becoming as bad as the real world, I am reversing the trend. I am also taking long walks in Tilden Park, and yesterday almost got killed when I slipped on a steep hill and rolled down 80 feet. But I need some exercise. The university here is absolutely peaceful, even after Laos.[137] A few people marched around campus, but they look very dispirited. Somebody tried to break up my seminar: I invited him to come in and let him have it for the stupid tactics of him and the likes of him. I myself am one of the "get out of Vietnam" freaks, but with these tactics they will never achieve it. I then went to the class of the resident Marxist Lichtman who, incidentally, will get fired next fall, and his attitude was that there is nothing anyone can do right now except to work out some long-term plans such as influencing people etc., pretty much in the democratic tradition. So Nixon turned out to have more brains than all these campus intellectuals. I also just taped a talk for Radio München, for a series called *Beyond Right and Left,* and in it I made one long propaganda speech for Mill's *On Liberty,* also recommending more tolerance towards Fascism—they will love it. And now back to footnote 248 of AM.

Be well,

Paul

137. Feyerabend is referring to the American bombings of civilian and military targets in Cambodia. "Nixon unveiled the Cambodian 'incursion' on the evening of April 30, 1970, in a televised address that was, as Henry Kissinger derisively put it later, 'vintage Nixon'. . . . An allied force of twenty thousand men, supported by American aircraft, were attacking the two main North Vietnamese and Vietcong bases in Cambodia as Nixon spoke. . . . A large proportion of the American people, traditionally loyal to the president in crucial moments, supported the Cambodian incursions. Once again, however, the opinion leaders set the pace. Press commentators lashed out at Nixon . . . . Universities and colleges across the country were then seething over one issue or another, but Cambodia suddenly crystallized the unrest, and disaster struck at Kent State University in Ohio. There, as elsewhere, antiwar students had attacked the reserve officers' training building. Echoing Nixon's inflammatory rhetoric, Governor James Rhodes assailed the rioters as 'worse than the brownshirts' and vowed to 'eradicate them'. He ordered national guardsmen onto the campus to impose order. On May 4, 1970, nettled by the demonstrators, they shot a volley of rifle fire into the crowd, killing four youths. . . . The Kent State killings sparked protests across the country. More than four hundred universities and colleges shut down as students and professors staged strikes, and nearly a hundred thousand demonstrators marched on Washington, encircling the White House and other government buildings." S. Karnow, *Vietnam: A History* (New York: Penguin, 1984), 609–12.

24 February 1971
London

My dear Paul,

[. . .] My visit to Boston now seems to be fixed. I would not be particularly enthusiastic about a lecture in Los Angeles since I cannot stand the town but if Avrum Stroll and Dick Popkin can put up some money for me then I am very willing to refute any of them (including Marcuse) except for Stanley Moore with whom I finally agree. On my way to California, if you know some fool who would pay me for stopping over and giving a circus, I should of course be very pleased. [. . .]

[. . .] Gillian is coming to America simultaneously with my return and will stay there for a month including a visit to California. I cannot stay longer to meet her because I don't have money for the difference between the 45 days' ticket and the longer stay. I am very much looking forward to seeing you and your girls.

I have some funny stories for you but I reserve them for our meeting. It was terribly nice of you to ring.

With warm greetings.

Yours ever,

Imre

P.S. Ian Hacking has two quite talented students who write silly theses about you and who want to have the final version of *Against Method.* I promised to give them a copy on my return from the States. One of them thinks that you are right and I am wrong, the other thinks that it is the other way around, but both agree that they do not quite know what the difference is.

4 March 1971

Dear Imre,

Rain outside, jaundice inside, unfinished papers in front of me—this is my situation at the present time. I don't lecture anymore—fortunately—for all my lectures have by now been taken over either by students or by faculty from other departments, but there are two things which still take up my time: AM and the book in German on the philosophy of nature. Also, as I told you, I have written a long application for the NSF [National Science Foundation] which goes from Aristotle, *via* Thomas and Bacon, up to the present time, and whose topic is the development of two views about knowledge, one Aristotelian, where knowledge arises on an [empirical] basis and where the basis itself is not a result of research, and one Baconian, where both our knowledge *and* its basis (if there is any basis) is the *result* of research.

Of course, behind all this, there is my perennial question: *Why be an empiricist?* There is no answer anywhere in Karl to this question. Carnap *(Testability and Meaning)*[138] realises that there is a problem (in this respect he is far ahead of Karl, at least the Karl of LSD), but he decides not to discuss it. He formulates a *decision* (*Decision 1* in *Testability* etc. on page 9 of the second part), according to which every primitive descriptive predicate of the language he wants to introduce should be observable in the wide and generous sense he describes (*Explanation 1* on page 454 of the first part). Having formulated the decision, he leaves it to *science* to decide what particular observable predicates should be chosen ("Only after constructing a system of physics can we determine what bases are sufficient with respect to that system," page 471 of the first part). But this means that the scientist whose work is a presupposition for making a selection of this kind *does not yet have a definite empirical basis* (class of basic statements for test) at his disposal, he works without such a definite class as only the philosopher finds a definite class *after* the scientist has done his work. The work of the scientist is *empirical* (see Carnap's fundamental decision), but without a fixed class of basic statements. Bacon goes a step further. The research of science *need not even be empirical,* and if it seems to be empirical then this is the feature of a particular *stage* of science, not the essence of science: science occasionally becomes empirical, but that is a transitory historical stage.

Incidentally, one little item in Carnap's paper which everyone takes for granted is that in testing, everyone refers to his own observations and to the observations of others only insofar as he has become acquainted with them through his own senses. "No scientist," says Carnap on page 424, "denies this trivial fact." Well, I have refuted it in my short paper on "Science without Experience."[139] Now, a science without experience has already been described by you in *Proofs and Refutations.* The arguments which you describe there can be empirical arguments and they can be "purely intellectual arguments"—the interpretation is left entirely open, so the whole debate is independent from the mathematical vs. empirical dichotomy, though it is of course possible to *illustrate* large parts of it by "experience."

Anyway, to come back to my NSF proposal. There are these two modes of research, the one *involving* an empirical basis, the other *constructing* it for particular occasions and dropping it on other occasions (I call them the Aristotelian and the Baconian mode respectively, and I classify Karl with the former, on account of his remarks against Neurath who is a Baconian in this sense), and the contemporary debate about the foundation of the quantum theory (hidden variables or not) is not only a debate about two different *theories,* but even more about two different *modes of research,* and this should be made clear. So this is one of my intentions.

Another one is to find out whether the moon looked different to different people at different times. There existed a theory that the moon is but a strong glare—but anyone who holds such a theory must see the moon differently from us, for to us it looks like a big stone, as

Aristotle and Plutarch after him had already said.[140] *This* kind of research I shall (I hope) carry out with Gombrich.

Finally—but now I return to my pet gripes—Karl and Mill. My thesis (which is being fortified by research done by our good friend Larry [Laudan]) is: Mill has a general philosophy that entirely suffices as a philosophical background for a critical philosophy of science that is based on falsification. For this, just consult my Mill chapter in AM. There occurs even the word falsification, experience is excluded as a final judge, for "there must be debate to interpret it," even Newton's theory can be criticised and can be trusted only because it has been criticised to such a large extent in the past. Now these principles of criticism are not introduced in order to solve an *epistemological* problem such as Hume's, they are the natural consequence of a philosophy *of man,* which suits me much better. (Incidentally, it should suit you too, for open debate is recommended as a method of solving problems only for people "in the maturity of their judgement": small children, students included, are still supposed to be led with a strong hand until they are mature enough to participate in the discussion. All these details are missing in Karl.)

Now, if we combine this with Boltzmann and Einstein, notably with the latter, then we have everything we want. Popper does not follow these authors, except incidentally; he starts with the Vienna Circle so that we have a forward motion (as regards sophistication of doctrine), viz. Mill plus Einstein, a large backward motion, the Vienna Circle, and then again a small forward motion, Popper, which leaves us worse off than before, although better than if we had stayed within the Vienna Circle.

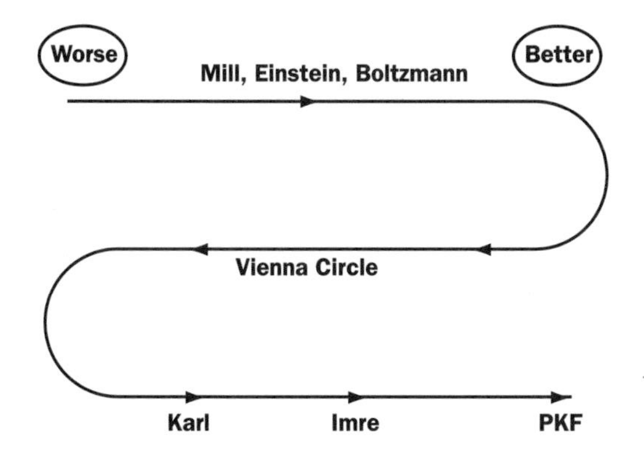

There is only one element that is missing in Mill, and that is the idea of *content*—but this idea is implicit in Einstein; it can be easily supplied, the technicalities are superfluous and besides, content does not develop as Karl thinks anyway. Larry sees the matter somewhat differently. For him the starting point is Whewell, the backward movement the early Vienna Circle, and Karl again a small forward movement, only that according to Larry, Karl misrepresented the Vienna Circle by insinuating that they wanted to *exclude* ideas which were not verifiable. No, he says, they did not; they wanted to retain them under the name of *hypotheses,* just like Newton who permitted hypotheses to enter the *body* of science, though reluctantly.

I am giving my last lecture today, on Brecht. On Friday the members of *Venceremos,* the Spanish SDS, will speak in class, and a new experiment to test hidden variables in QM [quantum mechanics] will be discussed in my philosophy of science class. And then all is over, and I shall be able to devote myself to AM fully. Incidentally, let me tell you a typical Berkeley story: a lady I know who is married and has a child, decided that she might be a lesbian, and that she should find out by some practical tests. So she talked it over with her husband, got permission, and set out to find a companion. Soon, another lady all clad in leather caught her attention. She followed her around, spied on her, discovered her address and one fine day knocked on her door for some discussion and, perhaps, a little lesbian sex. Well, she was cordially welcomed but, as it turned out, the leather lady was a man, and she is now living happily together with the leather man. Her original problem, however, remains unsolved. Is she a lesbian? The answer might be yes, for she followed what *looked like* a girl. No, for she followed what *was* a man. Yes, for she followed a man who wanted to be a girl. No, because she followed a man who wanted to be a girl but who still was very good in bed as a man. What is *your* answer? (Or perhaps I should introduce you to the lady when you come here?) [. . .]

Paul

138. See Carnap 1936–1937.

139. See Feyerabend 1969b.

140. In his autobiography, Feyerabend recalls that he "was puzzled by Anaximander's idea that the sun and the moon were holes in dark structures containing fire. Did Anaximander *see* the moon as a hole or was he just speculating? According to Plutarch's *Face in the Moon,* some people interpreted the face as a visual disturbance while for others the moon was a glare issuing from a single luminous point. Did they perceive the moon in that manner? And is it possible for simple perceptions to vary so drastically? Often when wandering around in the countryside I stared at the silver disk, trying hard to make it appear as a hole, or a glare; I didn't succeed. Galileo's telescopic observations now became much more adventurous than Galileo had thought them to be. They not only increased knowledge; they changed its structure" (1995, 140–41).

15 March 1971
Berkeley

Dear Imre,

Welcome to the US (if you have arrived, that is). Thank you also for your most interesting letter to Alan [Musgrave] re. Karl and Yoske. I am so glad I am not a Popperian. Your letter, incidentally, gave me the idea for an additional footnote to AM: Karl, Yoske and you talk about the role of *metaphysics* in science (which you call "research programmes"). The difference is this. For Karl metaphysics is a reservoir of ideas, for Yoske it is a reservoir of ideas and a theoretical background that makes us concentrate on one theory rather than on another. For you however, metaphysics is the *essence* of science, for without metaphysics there would be no science (all *theories* are refuted).

I have also made other minor changes, mainly of a stylistic nature but now I want to forget about AM for a while and prepare my *Naturphilosophie.* I shall return to it as soon as I have your final suggestions and, remember, have no compunction about making the *most drastic* suggestions if you think it will improve the mess. I guess somebody has to go over the English. I am thinking either of the New Left (maybe they will do it themselves) or a guy at the LSE whom I would pay out of my own pocket. It may also be necessary to retype *large* portions to make it easier for the printer to read. And, if you write a *large* criticism I would still want to add a few final words. But we can talk all this over when you are in Berkeley. UC Berkeley already got $200 and I hope to get another 200 from Stanford and perhaps a little from San Francisco State. I shall keep you informed. [. . .]

Here it is raining and raining and raining but this, at least, keeps the air clean. I now know a little more about Beirut: they want a philosopher as the director for 1971/'72, but they need State Department Funds for that purpose, and the funds have not yet been granted. If they are, then it all seems to be plain sailing, for I am the only volunteer. They want somebody in American philosophy, but US philosophy is at any rate an adjunct of the Vienna Circle, so I qualify on account of being from Vienna. If all goes well I shall go to Beirut in the fall of 1971 and have my Sabbatical in 1972/'73, one year later. I am already studying the routes of the prophets, for I want to examine them all and perhaps pick up a bone or two for miracles. Origen, St Augustin, St Athanasius are all really so much better than Popper as I shall try to show in my *Naturphilosophie.* Keep your fingers crossed. I have also started reading the Bible in order to find out when the devil first occurred in the Jewish tradition. [. . .]

And that is all for today—be well, and don't turn too many Bostonians into anarchists, type Lakatos.

Paul

16 April
The Busoni Society,
250 West Fifty-Seventh Street, New York,
New York, Circle 6-2270 -
Executive Vice President: Paul K. Feyerabend

Dear Imre,

Am I an unrepeatable phenomenon? Certainly! Am I therefore unscientific? Decidedly! For that is my whole point: down with science, up with Novalis, Schlegel, Pindar, Amenophis IV, Pherekydes, Ogotemmeli, Dionysos and the Neanderthal man, though I do not trust the latter on account of his predilection for law and order. He may well be an ancestor of Sir Karl's, not mine, and I hope I have directly descended from the joyful baboon, omitting all these (in my view) degenerating branches of evolution. Using your terms, however, I would gladly admit that my research programme is degenerating—I would not like it any other way. Opportunists, slaves in mind and body, cowards and creeps: these are the people who assemble around successful people and their advancing research programmes. I, however, shall have *interesting* people on my side, people with character, or rather, I *would* have interesting people on my side if having people on my side in a capacity different to that of friendship were my cup of tea. So, I shall keep my research programme degenerating as long as possible, and shall join you in case I become advancing and you degenerating. Of course, there is a little difficulty here, for if I am advancing, so are you, as I have shown in chapter 25 (or is it chapter 26) of my masterpiece. Right now, you are as degenerating as I, only people haven't realised it yet, and so, I am afraid, you will get more than your fair share of creeps—but that is your problem (this part of my letter you may read to your class on April 26th as my contribution to their education—it's very doubtful that I will be there in person, even if I should be in Boston. I just don't like the idea of lecturing, or intellectual debate, these days).

[. . .] Criterion for becoming a lefty: you lose your sense of humour and become a self-righteous bastard (or bitch, as the case may be). In this, of course, you do homage to an age old American tradition: Puritanism.

Now, I am un-American to the extent that I despise Puritanism, whether it comes from the right, left or centre. "But what about the truth?," people ask. *The truth,* whatever it is, *be damned.* What we need is *laughter.* You have got the gift for laughter, even where your own position is concerned, so, as far as I am concerned, you are a good guy (and you are "good" even theoretically, for your theory is equivalent to mine as I have said above and in chapter 25 of my magnificent AM). Bunge, since you ask about him, is another humourless bastard and a paleo-Popperian as well. But his wife was once very charming. Give my regards to the Eastern Establishment and be well yourself.

Paulus Empiricus[141]

141. Feyerabend is obviously hinting at the ancient sceptic Sextus Empiricus.

4 May 1971

Dear Imre,

Think a little more about Murdoch's[142] complaint (medieval science, according to you, is not science), and you will realise that it can be transformed into a most interesting account of the difference between medieval science and the science of Galileo, Newton, and others. There was a time when they were both thought to be entirely different (medieval science is not science at all, Galileo's was the first science there ever was). Then Duhem reversed the trend: Galilean science was well prepared, so well prepared indeed, that to talk about a "revolution" only reveals the historical ignorance of the talker. Now *you* come along and point out that *there is a difference after all and the difference is:* there are various possibilities:

(a) that medieval "science" was based on naive falsificationism while Galileo, disregarding lots of refuting instances, turned Copernicanism from a theory into a research programme, and introduced sophisticated falsificationism. This has much plausibility, considering that *Aristotle was a tough naive falsificationist;*

(b) that around the time of Copernicus and Galileo, the so-far degenerating research programme of Copernicanism turned into an advancing research programme (explain why); etc. etc.

So why don't you get Murdoch to work on this business? *Again, it proves the fruitfulness of your method, and you didn't even notice it!*

Incidentally, they are interested in me in Brighton, and have invited me to come over for a week in fall to look me over (and for me to look them over). Maybe there is a chance for me in Brighton, after all, and I would take it *only* if *you* are prepared to build a London-Brighton axis with me.

Christopher is still ill, but happy because in love and too weak to worry. She has given up her dog.

Be well,

Paul

142. John E. Murdoch, professor of history of science at Harvard, is a scholar of medieval history and philosophy. On Lakatos's attitude toward medieval science, see lecture 1, this volume.

25 May 1971

Dear Paul,

(1) According to the *Philosophical Journal of the Soviet Academy of Science:*

You and I are the major figures in contemporary philosophy of science but since my *Proofs and Refutations* is translated into Russian I am more major than you. Between us we have 7 pages in *Voprosi filosofii,* 1971.

(2) The LSE is being reformed and changed out of all recognition. The change is so big that I have nothing else but to participate in it and therefore teaching, life and other minutiae have stopped. One of the changes is the Mathematics/Philosophy degree course, a copy of which I enclose and to which you can now send all your girlfriends with sufficient qualifications. Also Logic and Scientific Method will become a major field of study at the London School of Economics and most other departments will be closed down to give us space. John Watkins mentioned that the demand may be so big that we may have to climb over our usual standards for new staff and hours, meaning that even you may have a chance for an assistant lectureship.

(3) Alas no such job will be forthcoming for you because you had an engagement with Gillian last Thursday at 8 o'clock and you didn't turn up. Incidentally, I made a £10 bet that you wouldn't and I won it.

(4) Spinner re-translated my paper "Popper on Demarcation and Induction" and now it is better than the original. Therefore I had to make peace with him and try to turn him into a real scholar. In order to do so I shall persuade his wife to leave him because marriage is not conducive to scholarship.

(5) Hans Albert and Szabó are coming next week. The Alberts will move into my flat, Szabó will move into Gillian's flat and the only question is where I shall live.

(6) The building costs which were originally estimated at £6,000 are now at £10,000. I told the builders that you will cover it. So you'd better quickly organise something for me in America.

(7) I also enclose the draft of a letter from me to Popper. I say it is a draft but may send it off as it stands. I only want to get John Watkins's support and if he agrees to give me moral support it will be sent off. (Sorry the copy is missing, you would have too much fun with it and I do not think you deserve it in view of (3) above.)

(8) A student of mine proved that Robin Blackburn does not exist. The proof seems to be correct and therefore I cannot get hold of your manuscript. Spinner cannot get hold of it either because of the same theorem.

Yours ever,

Imre

31 May 1971

Dear Imre,

It is pleasant to hear that we are being praised side by side in Russian. I shall get the issue of *V* [*Voprosi filosofii*] and let Liz Aschenbrenner translate the eulogy (incidentally, Liz is going to return to LSE the coming fall). It is also pleasant to hear that, and to what extent, you are involved in the rebuilding of the department. Exactly the opposite is happening here. Grice has become chairman, examinations, lectures, hiring etc. *everything* is now becoming Oxford and I am being paid for *not* giving lectures (too dangerous for the young). It has already started this quarter. As a result, I have a salary, but absolutely nothing to do. And, strange to say, I miss the stage (i.e. the lecture room), so I may ask you to reconsider and to hire me as an assistant lecturer.

The Gill-case does not really speak against me, as a matter of fact I was *very* good in this instance; I turned up at her hotel door *twice* within two hours but she was gallivanting about in San Francisco, next day I got a telephone message from her, I responded *immediately,* a thing I never do. I also saw her, for about five minutes, but she had to run off with a few old men whom she preferred to me (or perhaps there was a young man whom I did not see and whom she hid from me), then, in the evening, I admit, I had to go to bed, so I left a message for her that I could not come and I am sure that twenty men of all ages were overjoyed at the news. So, I beg you to reconsider your verdict.

Had to participate in an exam last week with Grice where one of the main things was a discussion of the sentence "there was a rhinoceros in my refrigerator," which sentence has been made famous by a guy called Donald Davidson who seems to study words in Princeton. This almost made me a rationalist again, but only almost.

I knew that sooner or later you would discover Spinner so, now that you have discovered him, don't lose him again. Give my love to Hans [Albert] and Szabó, even more so to Gretl and Szabó's wife, I envy you, being in London with famous people in an expensive house while I sit up here on the hill looking at the view with nobody to talk to. Also tell Szabó how much I admire his book. I hope this is not his (her) only visit to London, so that I can see them again when I come (*if* I come: I did not get my NSF travel grant, for I applied too late).

Robin Blackburn may not exist, but he still has consequences (the nothing nothings—remember!) at any rate—I shall send you a copy of AM and please make another copy for Spinner, at my expense, if need be. My German paper drags on and on, then comes Giedymin and by then I may be on my way. Robin got a job in Hawaii, I may go to Auckland, and we may introduce the philosophical triangle Honolulu-Auckland-Peking.

Be well,

Paul

1 June 1971

Dear Paul,

I gave a talk against you last week in Cambridge. Unfortunately it transpired that we agree at least on some issues. Braithwaite got up again and delivered his London-Ontario tirade about Hiroshima and the conservation of mass. I smiled and said: "Richard, do not make a fool of yourself." Funnily enough he shut up and that was the end of the discussion.

I hope that this story convinces you of my intellectual superiority to you.

Otherwise, the audience consisted of a pack of illiterate and aggressive little fools and the level of the discussion was about this:

Bob Young: I have to tell you that in my historical work I can make no use of your theory.

Myself: I knew that my theory is of no use to a hopelessly bad historian.

Hacking giggled.

So this is how it went.

At the end of the meeting Gerd Buchdahl asked me: "You know, Imre, the trouble is that you never quote anyone from Cambridge. Why don't you quote Braithwaite, Hesse or myself? You always quote Feyerabend."

Yours ever,

Imre

June 1971
The Electric Symphony Orchestra,
Suite 529, 250 West Fifty-Seventh Street, New York,
Paul K. Feyerabend: *Chairman Educational Advisory Committee.*

Dear Imre,

[. . .] I have never denied that you are intellectually superior to me, but the Braithwaite case is certainly not evidence. You had plenty of forewarning from him (he told you his BIG ARGUMENT in advance) and from me (I told you that he shut me up). It is however evident that you are intellectually superior to me *on account of being a better irrationalist.* For your method of arguing, though speedier than mine and certainly more elegantly applied, is identical with mine in *spirit:* you do not argue, you *annihilate* with a slap of your tongue. This, incidentally, will lead to a situation where the more you get ahead of me, the more you are behind: you may overpower people, one after another, you may even overpower me, but in doing so you prove the weakness of *argument* (e.g. mine) and the force of *personality* (yours). So while *reason* may be on my side, the *cunning of reason* is on yours and thereby also on mine. So in the end I am really the better rationalist than you for I *admit* the force that makes people agree, while you *use* it with great effect without acknowledging its irrational character.

Anyway, you will find I said all this in my German "How to be [a Good Empiricist],"[143] a copy of which is enclosed. As regards AM: I am prepared to cut out another twenty pages to make room for your 50 and my final 20, for I shall have the last word even if it should only be to thank you for the marvellous burial you have given me (however, there is one hitch: you don't believe in ghosts, while I do; so watch out after my demise). If you really want to come to Boston, let me know, for then I shall apply for your job at LSE and promise to be a good rationalist for the rest of my life. [. . .]

Christopher has completely recovered, may get married within 3 weeks to a great-great-great-grandson of the great John Quincy Adams, and her first son's name will be Agamemnon (his idea). She is as happy as I have ever seen her and sends her regards. Raveneugh is here as well, for a few days, but always tired and exhausted ("these strong young bodies"—he says; after all, he is now 37 years of age, has just celebrated his birthday and his most recent girlfriend is the athletic type; both of them visit me every day, she glows, he yawns and falls asleep on my couch). I have just passed through two very painful days, yesterday Mara planted lots of flowers in my garden. The department here gets more and more Grice-ized, and they treat everyone else like a heap of dirt. Depressing. I wish somebody like you were here.

With love,

Paul

143. Feyerabend 1963a.

5 July 1971

Dear Imre,

Right now I am in great pain, legs (usual) as well as heart (quite unusual, for I have been quite blasé in this respect for quite some time). As a matter of fact, it is not such a far flung possibility that I may do something really foolish in the not too distant future, nobody will suffer from it, AM is finished, and Giedymin[144] will even profit from it, for his life-span may be extended by as much as a year. However, we shall see. Popperian Reason may take over in a day or two, and I may return from her feet to my papers. We shall see.

Popperian reason—I saw Karl last Wednesday (June 30th) for breakfast and was shocked at how much he has aged. He was very kind, and very unhappy, he complained that nobody ever told him anything anymore (he heard about your appointment in Boston from Tarski and was a little disappointed that he had not heard it from you), he complained that I did not send him my papers, he told me that van der

Wärden and von Weiszäcker had praised his quantum paper (in the Bunge volume),[145] van der Wärden agreeing completely, von Weiszäcker almost completely, which latter thing means quite a lot, for von Weiszäcker was very close to Bohr for some time and regards himself as one of Bohr's foremost interpreters. He told me that he gets only £30 a month now and is afraid he will be financially insolvent soon, he praised your *Proofs and Refutations* as a masterpiece but said that your criticism of him was not very good (no stronger words were used). He seems to be very lonely so, Henny or no, I shall visit him, or try to visit him when I am in London just to cheer him up. There are many things about him I do not like, but I shall try to make him happier with a few visits and by writing to him now and then. Seeing him made me feel much more kindly towards him and I remembered the impression he made on all of us when I first met him.[146] Anyway, I also asked him about Hedonism. He told me he had thought about it for a long time and had not come to any conclusion, then I showed him the two monkeys you sent me with the inscription: "Feyerabendian philosophy leads to happiness," to which he said in surprise: "But Imre is a Hedonist himself," against which vicious slur I defended you manfully. He thought that Bela Lugosi's[147] portrait I have on the wall (a much bigger one than I sent you) was the portrait of a great scientist (I drove him up to my place and showed him the view), and then we parted, I promised to go to his lecture but could not as I was lying at the feet of see above (Popperian reason). Do you know that Yoske will visit him on the 4th, he urged Karl to give him an opportunity to speak to him. It was very strange to see Karl again at breakfast, and I felt very, very sad. Ring him up sometimes, tell him a little gossip, do not mind if he criticises you, just try to make him happy.

Pains are getting worse, for the last three weeks I have had pains almost every day, I am on pills almost continuously and may end up a dope addict in addition to being an addict to a new lady—see above. And now I am too dizzy from the pills to continue writing.

Be well,

Paul

144. Among Feyerabend's opponents, Jerzy Giedymin (1970, 1971, 1973, 1978) came to the forefront for having pointed out that the roots of the incommensurability thesis lay in the so-called "radical" conventionalism alive at the turn of the century in France, and that the same trend of thought was present in the work of the Polish philosopher Kazimierz Ajdukizwicz. In answer to Giedymin's objection, Feyerabend progressively strengthened his interest in the anthropological characteristics of the incommensurability

thesis over its logical aspects. For further developments see his *Farewell to Reason* (1987), chap. 10, in which he examines the possibility of "translating" different cultures considered "incommensurable," in possible analogy with Poincaré's solution (1905) against Le Roy's nominalism. Feyerabend claims that *"every culture is potentially all cultures* and that special cultural features are changeable manifestations of *a single human nature"* (1995, 152). And Poincaré himself (1905) had already mentioned the need for a "minimum of humanity" in order to achieve mutual understanding.

145. See Popper 1967.

146. Feyerabend describes his first meeting with Popper at Alpbach in the summer of 1948 in the following words: "I had skimmed through his *Logik der Forschung* and had formed a mental image: he would be tall, thin, serious, slow and deliberate in his speaking. He was the very opposite. Walking up and down in front of the participants, he said, 'If by a philosopher you mean one of those gentlemen who occupy chairs of philosophy in Germany, then I certainly am not a philosopher.' The German professors—and there were many of them—were not amused. We students, however, found his speech rather refreshing" (1995, 71).

147. Bela Lugosi was a Hungarian actor trained at the Budapest Academy of Theatrical Arts. He played lead parts on the Hungarian stage and in films. In 1918, during the collapse of the Hungarian monarchy and the establishment of the Communist regime, he was active in politics and organised an actors' union. When the Leftists were defeated, he emigrated to the United States. His most notable success was in the title role in the stage presentation of *Dracula,* which he played for a year (1927) on Broadway and two years on the road. When he repeated the role in Tod Browning's 1931 screen version, introducing himself to film audiences with a heavy, deliberate, inimitable Hungarian accent, "I-am-Drac-ula . . . ," it was clear that the American screen had found itself a worthy aristocrat of evil. On the screen Lugosi portrayed mad scientists and demented megalomaniacs. He allowed the vampire image to become a part of his life. He began giving interviews while lying in a coffin, and when he died, in 1956, he was buried with his Dracula cape.

15 July 1971

Dear Imre,

What a fantastic advertisement for Reason. I never had such a magnificent poster. But then, of course, Reason, fighting a losing battle, needs the best propaganda it can get. For anarchy has its own rewards.

So you admit that your view is propaganda dressed up as a theory. Magnificent. Now we completely agree about the nature of your theory. But there are still some things we disagree about. For example, you love science, and I think I now almost despise it. Also, you plan your travels months in advance and cannot stand indeterminacy while I don't plan at all and cannot stand definiteness.

Which means, of course, that my arrival in Boston is not at all that certain. There is the money, $320, for one thing. Now of course, I spend a lot of money on broads, but I have periods of generosity followed by periods of extreme stinginess. For example, I don't even eat so much any more, one small meal a day. Now by the end of April, of course, I may be in my expensive mood again, but who knows? At any rate, can you do me a favour and send me the address as well as the telephone number of the hotel near you (the one with the TV) where I would stay, so that I can make reservations at the last minute, say, three hours before I arrive? The guy at the Harvard Faculty Club will give you the number, I am sure, or the secretary at BU. If I go, I usually ring the hotel at the last minute, even from the airport, jump on a plane and come. Long-term planning is too rationalistic for me. (Do you know that last time, when we met in Minneapolis, I was already at the airport and had my tickets and rang the Minnesota Center to tell the secretary that I was *not* coming, and only her failing to answer the phone forced me to come.)

You can still advertise that the *Devil* is coming, for the Devil chooses strange shapes to confound the innocent. For example, he may choose the shape of a rationalist from London with a Hungarian accent and *not* the shape of an anarchist from Berkeley with a Viennese accent. [. . .]

One thing is certain, however, that you should write a nice contribution to AM. From an *argumentative* point of view our "theories" are equivalent. They are *not* so from a psychological point of view. Nor do we have the same *basic normative judgements.* In the "Battle of the Ancients and the Moderns," *I* would side with the ancients, *you* would side with the moderns. I believe that Galileo cheated and had to cheat, that is; you believe that mob psychology plays only a very minor role in science, while I think that it is everything and reason hardly plays any role in science. I prefer happiness to truth, you—well, here I am not so sure, but I am sure that you will say publicly that you prefer truth to happiness. I want to reform the sciences so that they make people happier regardless of whether they become more truthful; you would resist. So you see, if we stick to basic n. j. [normative judgements] we have lots of things to talk about. We could even invite Habermas to write the introduction to the Suhrkamp edition (I prefer Suhrkamp to Vieweg, even if they should pay less, because Vieweg is for intellectuals, Suhrkamp is for pseudointellectuals).

I just talked to Grover [Maxwell] on the phone and he says that your affairs are now being handled by Roger and everything seems to be OK. Poor Grover. It seems he has lost interest in philosophy, he has lost his sense of humour, he smells *evil* everywhere, he has found out that Herbert Feigl has forsaken him—why are these Lefties such an utterly humourless bunch? Once, in a lecture, a guy replied to this question: because they have serious things to do, and there is no time for laughter. To which I replied that if they are not interested in laughter they can shove the revolution up their arse, as far as I am concerned (my words). To which *he* replied that, of course, eventually there should be laughter for everyone, but that *now* the times were too serious. To which I replied that once you stop laughing you will have a very difficult time starting again, and you may never again be able to start. [. . .]

I have told Dreyfus[148] to contact you in Boston and not to be afraid of you. Incidentally, I forgot to tell you that in addition to being a Merleau-Ponty, Husserl, Kierkegaard etc. freak, he has just finished a book proving that computers will never be able to do what humans are doing. And, I repeat, he is *very* enthusiastic about you and very popular with students here. I once attended a lecture of his, found him very clear. Craig turned this into an objection against him when we were about to hire him, saying that he was falsifying existentialism by presenting it in such a clear manner (he meant this quite seriously).

Next year: well, I still do not know where I shall be. I may be in Beirut. I may have my sabbatical, in which case I shall either study at the Warburg Institute or else, and this is another possibility, make a trip around the world. Have just been invited to New Zealand for summer 1972 and shall accept. What happens after this I do not know. [. . .]

Be well,

Paul

148. Hubert L. Dreyfus was Feyerabend's colleague at Berkeley.

21 July 1971

Dear Paul,

Many thanks for your undated letter of July 15th. Do you mean that you had a heart attack, as you wrote to me, or do you really mean *hepatitis,* about which you wrote on the very same day to Jack Smart? I hope that both reports are false because for the next 15 years at least I cannot possibly cope with your literary heritage; moreover, I want to appoint you as my literary executor, and it is therefore in my interest that you survive. On the other hand, I wish to die at the age of 68, since I shall retire at 67 and I need a year to see that my successor is somebody loyal to the party line. So please keep healthy and take care of yourself. [. . .]

The Hempels invited us over for supper the other day together with the Zahars. They are such sweet people that I would not want to disrupt his peace of mind with my papers or anything intellectual.

Alan Musgrave writes that the whole of Australia and New Zealand is up in arms against the United States and that they don't want to be defended either from the Chinese or the Russians. Actually, I would not mind if the Russians were to busy themselves with Australia and Asia during my lifetime, and come to Britain after my death, since they always rehabilitate their enemies when they die or they are killed; they may even erect a statue in my honour. Just imagine a statue of me to the right of Marx and a statue of you to his left in Highgate Cemetery. [. . .]

I am now in a rush because the fire revealed that part of the wall is missing in my bathroom, and I want to put some bricks in to fill the gap. Write soon and reassure me about your health,

With warm greetings,

Imre

Tuesday

Dear Imre,

[. . .] Having translated [your] "History,"[149] I had to revise my criticism and have found a completely new line of attack. Result: *I am a better rationalist than you are.* The main argument is clear to me, but now I have to make it clear to everyone else, and especially to you. Every day I write about 15 pages, first in longhand, then on the typewriter (my fingertips are sore and bleeding), then I go to bed satisfied, but the next day I tear it all up again. Needless to say, I shall also have to rewrite the Imre chapter in AM. [. . .]

During my lectures here in Berkeley I have already had the third witch giving an account of witchcraft—she is the pretty daughter of Anton Le Vey of the Church of Satan in San Francisco.[150] But *nothing* brings these people to life *in this world.* It is depressing.

I read Noretta's [Koertge] thing on me and thoroughly enjoyed it. She is one of the few people who is both intelligent and endowed with a touch of lightness in everything she does. Quine is regarded as *The Bible* here (i.e. his *Word and Object*). Absurd!

Be well and don't be surprised when you suddenly see me on your doorstep.

Paul

Love to Gillian.

149. Feyerabend refers to his translation of Lakatos's "History of Science and its Rational Reconstructions" (1971a) into German.

150. Anton Le Vey, born in 1930, was one of the most peculiar characters of the occult movement which flowered in California in the 1960s. He is remembered as the founder of the Church of Satan, which numbered among its members the film star Jayne Mansfield and the director Kenneth Anger. Le Vey can be considered at the roots of the contemporary satanist movement. The occultist ideas of Aleister Crowley (1875–1974) left strong traces in his thought.

18 August 1971
London

Dear Paul,

[. . .] Let me tell you first of all how much I appreciated your cable and in particular how much I appreciated your generous attitude concerning the delay in my contribution to our joint immortality. Frankly, I do not now know when I shall finish my MAM. If I am very bright and lucky, it will be Christmas. If I am my mediocre self, it will be February 1st. If I am at my stupidest worst, it will be March 15th, and that is absolutely the latest (on March 15th I am going to Boston to preach against you).

I hope that by now you got my card telling you about the change of address [. . .]. From this you should not infer that we are actually living there. We live in hotels and boarding houses until the builders move out, at least from part of the house. They love it so much that they don't want to leave it. [. . .]

Alan Musgrave is here; he is going to Bucharest to refute you and me. Suppes paid him 800 dollars for the task. He has not been to Penn [State] yet. He has not plucked up the courage. Like me he believes in induction, like you he believes in proliferation, and like both of us he doesn't believe in Popper but he does not dare say so. I predict that he will end up with a paper in which between the lines there will be a big criticism of Popper and big praise of us, but "on the lines" it will be the opposite. So he will have three new enemies, Popper, me and you. (Forgiving as you are, you may still like him after the crime.)

With warm greetings.

Yours ever,

Imre

P.S. I just went down to the bar for a cup of tea and I saw Peggy Marchi and Lynn Lindholm. I fainted and now I am in an ambulance.

22 August 1971
Berkeley

Dear Imre,

For a moment I was worried, but then I realised that your head must certainly be harder than anything the first, second or third world has to offer, and your letter finally proved my hypothesis. It may be that 5 Turners Wood [the address of Lakatos's new house] is the *only* obstacle you cannot overcome, but then there are always hotels, boarding houses, or the beds of students (though the last are not too clean these days). It is nice to hear that you are ready to resume the fight and I am considering scrapping my final reply in favour of the enclosed picture. We shall see. It is also nice to hear that now that *we* have made our contributions to KNOWLEDGE, you in the third world, I in the second, the younger generation is going to pick up the mess where we left it in order to add some distortions of its own (for 800 dollars, that should not be too difficult). I only wonder how Alan [Musgrave] will manage to hop back and forth between the first, second and third worlds without getting dizzy. I get dizzy, too, but then I am accustomed to it, and I enjoy it (besides, to tell you the truth, I stick to the first world in which the other two do not leave a single trace, but then I am not honest as Alan is, and as you occasionally want to be). Will Alan be up to the acrobatics, especially with the ghost of Sir Karl watching from above (or from below, or wherever the ghost of Sir Karl is dwelling these days)? [. . .]

Mara is well, could be weller, and maybe you will make her weller when you come in fall. Lots of other people whom you do not know but who know about you are anxiously awaiting any new bulletin saying you are well, too. One of them even wrote—did you get the letter? *I* am well as well, though my head keeps banging not into cars but soft bodies. It doesn't make me a Stalinist but it doesn't turn me into a rationalist either. [. . .]

Be well, give my regards to Alan and to Gillian and

Write again

Paul

26 November 1971

Dear Imre,

I have finished my contribution to the volume where important contemporary philosophers give their considered opinion about what philosophy is supposed to do. My thesis: abandon scholarship, construct a new myth, and present it with the help of a movie. Title of my paper: "Let's make more movies."[151] That should teach them! [. . .]

I have given up my incommensurability paper in disgust. The stuff bores me to death—you are perfectly right! And I have got stuck in the Feigl paper not because I don't know what to say but because I lack motivation saying it. I have also sent for application forms for the British Stanford job. I think I would prefer administration to teaching, and I also think I could be an excellent administrator. For one, I write good letters (remember how surprised you were at my letter to Lord Annan? And my letters got 7 graduate students jobs last year, when I was what was called a placement officer here, i.e. when I had to find jobs. And no meeting which I chaired lasted longer than 15 minutes). Well, we shall see. At any rate, I could be in England again. During Xmas all I did was watch TV. They showed British singers, British actors, British orchestras etc. etc. while here in San Francisco there is nothing, absolutely nothing for miles. I shall also inquire for a job in Germany when I am there. Jack Smart, incidentally, who is Professor in Adelaide, applied for a *Readership* in La Trobe to get out of too much administration in Adelaide, *and he got it.* So, some people step down to be able to lead a more pleasant life. The mood I am in now may leave in a few days, but I shall not, and I shall stay here at least until the end of January and then leave. [. . .] One of my former TAs [teaching assistants] wrote to me from Cambridge, England and says that my name is taboo there. Spinner writes to me from Mannheim where Hans [Albert] is discussing his *Traktat* in his seminar and tells me my name is taboo *there.* A guy called Brown wrote to me from Nuremberg saying he read my stuff, agrees with me in all points, but cannot say so in public or else he will never get his Ph.D. So my name is taboo there, too. Now I wonder how I could draw some financial rewards from this infamity of mine. [. . .]

And now, be well, and cheer yourself up by writing a brief note to me and thus cheering me up. And give my love to Gillian.

Paul

151. See Feyerabend 1975e.

Saturday [November 1971]

Dear Imre,

So I found you out at last! After your last letter which sounded as if you were lying on your deathbed, I rushed to the telephone to talk with you once more before you expired, and what do I find? You are gallivanting about on conferences instead of sitting at home writing MAM. Disgraceful! I now have a theory about all these sad letters of yours. You are not really ill at all, nor is your house in such a bad shape, but you have no idea how to attack AM and so you procrastinate, invent stories about slothful British workers, drafty rooms, aching heads, while the *only* ache you have is the one created by chapter 25 of AM and by the 12 points at the end of it. Why don't you admit it? Why don't you write a nice confession to that effect, add it to AM and terminate the suspense?

Incidentally, have you read Feigl's attack on the two of us? He already regards us as being one and the same person, and that "Imre Lakatos" are my two first names. So hurry up, establish your identity in MAM—this is your last chance. I liked Feigl's paper, not because it is right but because of the spirit of it. While Sir Karl is moping about, while he is complaining and whining, Feigl is having fun defending empiricism. And this is how it ought to be. I am writing a reply and maybe if it gets finished in time it will get published, together with Herbert's [Feigl] paper, in the *Boston Studies.*

So, will you be in Boston in March? That is when I intend to start on my European trip with Karen, going first to Paris, then to Vienna, then to Heidelberg (looking for jobs in Germany) and then to London (looking for jobs in England—UC Berkeley is getting worse and worse), and, maybe, I will meet you in May in London. I hope I get the money together, right now I have no money at all, I gave it all away to impoverished friends etc., this is the reason why I had to cancel my fall trip to England and Europe. I wish somebody would pay my trip in March, or end of February, but this is too much to expect, nobody wants me these days, so I have to go on my own.

Now, be well, don't work too much, cheer up, and write.

Paul

30 November 1971

Dear Paul,

You didn't "find me out" as you put it, but you caused a bloody quarrel for me with Gillian which didn't do any of us much good. As a matter of fact, I was at a conference during the weekend in Windsor Park and I nearly fainted in the middle of my talk, and at the end I could not finish it. I only went down there because I had already let down Geoffrey Goodwin (who organised it) a couple of times, and I foolishly decided I would get rid of this obligation, and I turned up in my cervical collar and didn't do anybody any good. Your conclusion that I only pretend to be ill in order to conceal the fact that I cannot knock out your bloody nonsense is of course absolute and utter rubbish. The only thing I am now grateful to you for is that I don't talk to Gillian anymore and that makes life easier for me.

I am going to see Blackburn on Monday and then we shall sort out what, given my doctor's reports and prognoses, can be done, if anything at all, about MAM.

I still hope to be in reasonably good condition in March (the only travelling I am allowed is by air) and then I shall be back in London on May 1st. I am sorry that you are going bankrupt because I have no doubt that your association with Karen will lead you straight into it. [. . .]

My illness has turned me into a nasty creep, and unless I recover by the time you are in London, you will see it. I am now organising a movement with the slogan "All sick people in the world unite," and we shall kill all healthy bastards or restore health equality. Looking at Mary who is taking down this letter I can see she must be very healthy because she is looking very frightened. I shall spend the rest of the week in the hospital because they now suspect that the diagnosis was wrong and that it was the collar which bent my spine which was originally straight. And they are returning to the old hypothesis that I have a brain concussion.

Of course, if after reading all this you still don't know what the truth is, there is a simple remedy. Get on a plane and come and have a look.

Yours ever,

Imre

5 December 1971

Dear Imre,

Rereading your letter I realise with some apprehension that you may have been serious which means that you took *my* letter seriously. Let me assure you that there is not a single serious sentence in it. When I finished my telephone call I thought this was a great opportunity to tease you and in this way, perhaps, to cheer you up. Alas, the whole damn thing misfired and the final state is now worse than the initial conditions. Which just shows that one should not tease people whom one has not seen for a long time for one never knows the way they have gone. And there is still a month or two that I have to spend here before I can leave, for it needs that much time to make me fluent again (moneywise).

I am very restless, believe me, I am wandering around in the hills and wish I were elsewhere. I dream of London, and of Vienna, even of Oslo and when I wake I am still in my little bedroom and *Naturphilosophie* is still living on my desk, unfinished, even unstarted. [. . .] The whole intellectual business bores the hell out of me and I wish I were the owner of a restaurant or something. You are much luckier than me in many respects. You have found your profession, your interest and so on, while I just stumbled onto the philosophy of science and often ask myself in amazement: "what am I doing here?" [. . .]

Be well, and for heaven's sake, don't lose your sense of humour!

Paul

9 December 1971

Dear Paul,

Thank you for your two kind letters, which arrived yesterday and today. I haven't quite lost my sense of humour, although I agree it has diminished lately. The trouble is that since my accident at the end of July I haven't been able to do any work, I have constant pains and for the second month now I am wearing this cervical collar which makes me look like a monster. Also, the house is in such a state that I have to enter into a legal dispute with the builders, who have ruined my study to the extent that I can't get any books, manuscripts or reprints out of the boxes and so I can't even *try* to work. I need not say that this combination of events leaves me a bit nervous, to put it mildly. Since my trouble is that I had an accident hitting some arthritic vertebrae, there is a diagnosis but no proper prognosis. I might be completely alright in two weeks' time if I could have a week's rest or it might last for a year or forever.

The doctors firmly recommend that I take sick leave for at least three months, which I have up till now refused to do because, as you know me, I am not the person who can easily take sick leave. On the other hand, I might be compelled to do so very soon.

[. . .] As you surely understand, all this is very distressing. What the outcome will be I don't know yet, but I shall probably take sick leave for two months and then hopefully be able to go to America in mid-March, since if I don't I shall end up in complete bankruptcy because of this bloody house. Altogether I feel completely miserable and constantly irritated, and I am sorry that you became a victim.

Whatever the case, I am looking forward to seeing you in London in February (the earlier the better). With warm greetings,

Imre

15 December 1971

Dear Paul,

This is only to tell you that I am alive but only just. I do not remember how much I told you of the story since it took me a long time before I took it seriously. It was first diagnosed, as you may remember, in Minneapolis with a slight brain concussion, and Paul Meehl frightened me with the conjecture that I might have a blood clot in my brain. When I came home, it turned out that I had had an arthritic spine for many years, and that that was hit in the accident, but they thought that wearing a cervical collar for three weeks would set it right. After three weeks, I took the collar off and came off the drugs and it was then that I went to that bloody conference and it turned out that that was the worst weekend I have ever had and I had to go back to the hospital, put my neck back into the collar and go on the drugs again. The diagnosis is now all right, I hope, although it is not excluded that they might still discover something inside my head. On the other hand, I have got completely Feyerabendised in the sense that I am cancelling all my arrangements and appointments; I have not seen my students for months and I have not given a single lecture at LSE since last February.

I am afraid that during this process I built up an outstandingly materialistic explanation of your anarchism. With your old accident and its consequences, with the pains which you get and the drugs you take, you cannot be anything other than an anarchist, since you cannot live up to any definite promise. The difference between you and me is that I still preach law and order for healthy people and that anarchy should be the only privilege for you and me. I hope you will see my immense intellectual superiority from this argument.

Also, everything is going well with the house for the moment and I live in complete chaos. I had lunch with Robin Blackburn who promised me an edited version of AM to be delivered to me within a few days' time. The doctors now promise that within a few weeks' time I shall be able to work again and although I might have to wear this bloody cervical collar when I am travelling in cars I shall otherwise be tolerably normal. However, I agreed with Robin Blackburn that the printing of your book should go ahead and if by the time they have finished printing, my PS is not there, the book must be published without it since all the world is eagerly waiting for it and I have no wish to hold up this publication.

I am afraid I may already have written this same letter to you a few days ago, but since I am at home divorced from my files which are anyway in disorder now, my letters may become repetitive.

Stanford University is establishing a new campus in Britain. They want an academic director. Don't you want to apply? I enclose the advertisement. I am also sending one to Alan Musgrave although I don't think he would be interested because he is now the cleverest man in New Zealand and he would hate to come to a country where the position would no longer be so distinguished.

I very much hope that I am completely back to normal very soon because I *must* go to Boston in March if I want to avoid bankruptcy and selling the house and Gillian together.

With warm greetings to you and all your girls.

Yours ever,

Imre

31 December 1971

Dear Paul,

I am still wearing the cervical collar and I have just been told that I must take at least January as sick leave, possibly also February. However, my doctors claim that I shall be all right by March for the American journey if I do not travel around too much in America but sit on my bottom in the Harvard University Club. Unfortunately they already promised that I should be all right for Christmas and therefore I am depressed because I do not trust them anymore. I may get a proliferation of neurologists sitting around my skull and I shall judge them by the standards of the methodology of scientific research programmes.

I was delighted that your volume I of the Italian edition of your collected works is out. If you have collected works you must be a great man. Doesn't this disappoint you? [. . .]

I am most impatient for your assassination attempt on Giedymin. You may have noticed that in one place he refers to me and I think that there, in one line, I have killed your incommensurability. Have a look. On the other hand, if you send me your paper I shall publish it immediately and also you will cause me to have a good day which is I hope an argument which may still work with you.

It would be nice if you got the job in Stanford on Avon.

I am still collecting my thoughts for the paperback edition of AM.

The most important question engaging me now is the date of your visit. I would then arrange for Szabó to be here too if I can. However, to succeed, I must know your timetable far ahead. Please send a cable or something with the date of your arrival.

With warm greetings,

Yours sincerely

Imre

1972

8 January 1972
London
STRICTLY PRIVATE AND CONFIDENTIAL

My dear Paul,

Just read your AM chapter on me. I shall carefully compare it now with the old version which Gail has allegedly deposited in my office and shall send you a comment on their relative merits.

I am now greatly grateful for your depicting me as *God* and yourself as the *Devil.* I also return the compliment: for me you are the only phi-

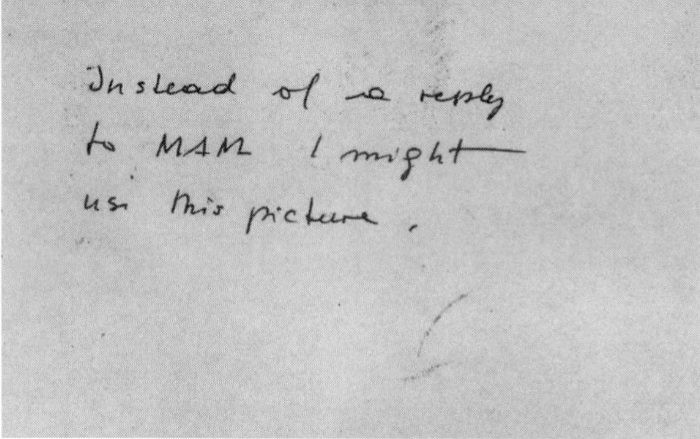

Instead of a reply
to MAM I might
use this picture.

Figure 3.5 "Instead of a reply to MAM I might use this picture." Postcard sent by Paul to Imre.

losopher worth demolishing. But there is *one* trouble: I *can* take you to such little pieces that only an electromicroscope can discover you again. Will you be very hurt? I shall, of course, say that the Devil is immortal and shall analyse your possible retorts, refute them in turn, suggest how you might answer back, refute etc. I shall carry the process to w. Now, you *might,* of course, invent a transfinitary series of arguments from $w + 1$ on.

However, before I start demolishing you, I want a solemn promise that you will not change the text *relevantly to my demolishing work;* but you will have the last word (AMAM). Of course, I shall not catch you out on small slips: I shall shoot directly at the columns holding your philosophy. The shots will already be fired on March 20th.

What is happening to your Fulbright? Watkins and I and 73 girls are dying to have you here for a full year.

Accept the invitation to North Carolina, *please.*

Baron von Weizsäcker and Jürgen Habermas have a joint institute in Munich to Save Mankind. I have been appointed for four years on the governing body and once a year I shall now go to Germany to check whether they progress or degenerate. Weizsäcker is too dignified for my flippant nature but with Habermas I get on superbly as long as we keep away from philosophy. My first official act will be to ask for the secretaries' nude photographs in order to assess some of the external factors of this research programme.

My main problem is now a crack which has developed on the wall of the downstairs loo. It is not dangerous but has to be covered. I am now trying to get a big blow up picture of Sir Karl and put it there in a Victorian gilded frame.

According to London astrologers the earthquakes will start in the Bay Area on August 26th and will last for a period of four years and 319 and 1/2 days. [. . .]

My Copernicus lecture went down well. I spoke for an hour and Jerry Ravetz commented. I proved my point and now it goes to print. (The night before the lecture Elie [Zahar] changed my mind and now we have turned Copernicus into a Progressive Figure, but we emphasised that he never spoke Polish. This upset Jerry Ravetz and the discussion centred mainly on this. I also compared Copernicus with Popper and Rheticus with Bartley, a homosexual, who, when he found out that Copernicus slept with Ann and does not quote him in the Preface, tore up the Preface (this is how Osiander comes in), moved to Hungary . . . (to be continued).

Love,

Imre

URGENT:

When do I speak of "Science in the last 200 years"? I do not remember. Can you give me the reference? I surely meant 300, *not* 200.
IMPORTANT:

what do you think of the title:
LAKATOS AND FEYERABEND
FOR and AGAINST method
or,
Feyerabend: *Against method*
Lakatos: *For method*
and, if you wish,
Feyerabend: *Lakatos's research programmes. An obituary.*

20 January 1972
Berkeley

Dear Imre,

I am glad to hear that our disagreement, which has been degenerating for the past two years, has moved out of its degenerating trough and that I am going to be annihilated after all—at least temporarily. For every little piece you cut me into will turn into a fully fledged AM and so in MAM you will face *thousands* of AMs rather than one. I would be grateful if you could let me know your opinion on this [last version of AM] compared with my earlier version on you *not sub specie veritatis* but simply from the point of view of the book: which chapter, taken together with your reply, would make a better book, which chapter you would prefer to reply to. Of course, I am constantly working on the whole mess. I have just rewritten the last five pages of the chapter on you, but the content is not changed. Right now I am in the middle of the incommensurability chapter. Incidentally, let me warn you that there are no "columns holding my philosophy," and so you cannot shoot them down. Unencumbered by weighty considerations, my philosophy flies in midair. [. . .]

Well, I may come to Brighton after all, especially with the danger of the earthquake here—but I shall have to talk this over with you *very* carefully. North Carolina—you know, I may have to go to Vienna in between Boston and Pittsburgh, my new orthopaedic shoemaker (the old one has retired and *no one* makes the right kind of shoes in this country) made my shoes too narrow. I cannot wear them, or only for a few hours, or else my feet get infected. I have already had a little trouble of that sort. (Besides, there are two chicks in New York I want to see). Your reconstruction paper arrived, I am looking at it, you will have it in a few days (Replacing "rivalisierend" [rival] with "konkurrierend" [competing] everywhere. OK!) Mention a translator to me in connection to *your* papers, NOT in connection with mine, I devoted much more attention to you and also did a better job on them. [. . .]

Every month I sit down once and hit my head against the wall for not having stayed in London at UC [University College].

I met von Weizsäcker in 1965 and gave a talk at his seminar. (There was fog, so I took a taxi from Berlin to Hamburg. Hübner[152] was with me, he was on his way home. Weizsäcker paid for the taxi.) I quite liked him, though he is somewhat ponderous. But he seems to be kind, and he has a very pretty niece. After my talk we went to a beer cellar where we talked endlessly, especially the niece and I. He was mainly responsible for my being invited to Berlin, for he recommended me most strongly (Bunge, who was the second one on the list, was not treated very kindly by him). And, most of all, *von Weizsäcker is responsible for my jump into anarchism* (I have never told anyone this, but it is true). My "spiritual development" always depends on theatrical episodes, *never*—thank Behemoth—on argument. The situation was as follows. I gave my Brownian motion routine, trying to show that it is good, given a highly confirmed theory, to invent an alternative and to work on it. Then we talked for some time about some concrete problems in the quantum theory, the infinities of the field theory were among the problems, as well as displacement and invariance principles, and so on. It became all very concrete. In the end, I returned to the point of my lecture and asked von W.: well, don't you think it is a good thing to invent alternatives etc. etc., and while I was saying that I felt a little ridiculous, for there were all these concrete problems, circumstances etc. etc., and I had started talking as if they did not matter. It was this feeling (which is now always with me) that was the starting point of my interest in the concrete case and the belief that new methods may have to be invented for each case, rather than adapted to "reason," or making them "reasonable" in some overall sense (right now it occurs to me that my feeling may also have been influenced by my Wittgensteinian upbringing). I still remember how foolish I felt with my general question. So, if I had not met von W., there would not have been any AM.

(This incident, incidentally, was much more important for me than the pleasant surprise I got when Sir Karl, then Prof. P., started his lectures on scientific method (in 1952) by saying: "I am Professor of Scientific Method; but there is no scientific method . . . ," which I liked, but that had no further consequences, for then I still became a faithful

Popperian for some years.) Habermas I never met, but I may meet him soon, for he has been invited to come here to Berkeley for a quarter (but he has not yet accepted). If you see von W. give him my best regards, and you may tell him the above story. Von W. puts many people off, but I think he is OK. He is like a very serious graduate student with bits and pieces of humanity underneath. Also ask him whether he needs a typist or a supervisor for all his secretaries. I am willing. I have asked Spinner to get me a job in Germany, seriously; all I want is a little job at a little university in a tiny little town with a tiny department and a BIG salary and lots of time, so that I can visit you in London.

Thank God MAM is in the making. Now at last I shall know whether my anarchism is worth anything. I mean that seriously. So far I haven't run into any argument (or joke, or aside etc.) I found worth considering. Maybe you will make me realise that the Devil, after all, is an angel, and the servant of God.

Let me have your thing on Copernicus, if you have written it down, for I want to incorporate it into a footnote of AM. Also tell Elie [Zahar] to send me *everything he has written on special and general theory of relativity.* His *general* supertheory of RP [research programmes] does *not* interest me, however. [. . .] My mailbox is slightly repaired (I did it myself), but it still leaks. But the rain has stopped, so now everything is OK. No use having a new mailbox before the earthquake. You have a crack in your loo, I have ants again all over the place and the basement is flooded. Result: the LSD (which I keep in the basement) has been wetted through and through.

Enclosed two ladies from Hawaii, either of them might make you come out here for some time. I know one of them (guess whom).

All the best,

Paul = Lucifer

And remember, my name is *Lucifer,* so it is *I* who bring the *light, not* you.

Gillian: is Imre good to you? If he is not, just write, and I come and beat him up, or burn him with a little hell fire.

152. Kurt Hübner was professor at the University of Kiel. He is a historian of science, philosopher of science, of myth, and of religion.

26 February 1972
Berkeley

Dear Imre,

Yesterday I had 102 degrees, today I had 101 degrees, and tomorrow I hope to be well again after a severe cold. There are lots of happy bacteria crawling around on my body, but they will not be happy much longer.

[. . .] I have added a long footnote on Hübner whose paper on Kepler is really a masterpiece.[153] There are now *three* historians of science in the world: Hübner, you, and me. Did you read his paper on Duhem plus Mach? (Problem: history *was* once taken seriously by philosophers of science such as Duhem etc.—why is it no longer taken seriously?)

Today I am very weak, still fever, impatient, I want to leave but workmen have torn a hole into *my* wall (a leak, and water came through) so, apart from my illness there is the illness of my house and the latter may keep me here for a considerable time.

Am a little depressed, especially because I miss Karen (when I am ill I see no one). Ah well. In Heaven we shall all be reunited.

Be well,

Paul

153. For Hübner's view on Kepler, see his *Kritik der wissenschaftlichen Vernunft* (1978).

5 May 1972

Dear Imre,

Damn the *Naturphilosophie:* I do not have your patience for hard work, nor do I have two secretaries, a whole mafia of assistants who bring me books, check passages, Xerox papers and so on. If anarchism loses, then this is the most important reason. The examples which I find are in books which *I* have found in the stacks myself, which *I* have *carried* myself, which *I* have opened myself, and which *I* have returned myself. The counterexamples which you find are books which have been found by others, carried by others, copied by others etc. etc. The papers which I write are my own handiwork from beginning to end. The very bloody version has been written by myself, never have I asked a secretary to do my dirty work (I *could* ask them, but I just can't bring myself to do it). Maybe I explain this predicament of mine in an additional footnote in AM. To restore justice.

I do not have anyone to discuss any of my ideas with and on top of all this I now have a rather severe hepatitis which turns me almost into a vegetable. I had it before, when I was in London (this was one of the reasons why I left: if I cannot be at the top of my form, I'd rather not deal with girls) but now it has become worse, because I had to take pills against pains and my liver now seems to be much more sensitive than before. Had I not been felled by my liver again, I would have come to London in about three days from now to see her again, everything was prepared, but now I sit here yellow like a lemon and mad at the world. I daresay, I do envy you and your mafia. There are very few people with whom I can discuss philosophy, it may well be that you and your mafia are the only ones. [. . .]

I now spend most of the time putting elaborate messages on my telephone answering service (which Ravenaugh left behind). One message is (in thick French accent): "This is the society for the preservation of the California rhinoceros. We are happy to hear that you are interested in the survival of this magnificent creature. Sorry, all officials are out at the moment, for a new rhinoceros has been found and must be catalogued. If you want to leave any message, please do so immediately after the beep." Or (in thick German accent, with a woman screaming in the background): "This is Dr Frankenstein. I cannot answer the telephone personally, for I am now engaged in a most interesting experiment. Please leave any message—of scientific nature only—after the end of this message." Or (with heavy breathing in the background and two different voices—I do all the tricks with change of voice and two tape recorders): "But, Doctor, is there no help for him?" "There is nothing that science can do for him." "But if he dies they will read the testament and we shall be ruined!" "Be calm, my friend (satanic laughter), there are enough messages on the answering service to last us a lifetime." "Like this one?" "Like this one, precisely." And so on.

So, you see, in between jaundice, *Naturphilosophie* and that magnificent broad in London, I am slowly going batty. God help me. Maybe I shall come to London and throw my yellow body at her feet. [. . .] So, and now back to *Naturphilosophie.*

Be well,

Paul

19 May 1972

Dear Paul,

According to the chart containing your movements which you gave me before you left you have been in New Zealand since the end of May. There is no date specified and also no exact address. This of course frustrates me in my desire to write you a long letter because I am afraid it will not reach you. So please do let me know your dates.

I am sorry that you felt unwell and could not come to London before New Zealand; I hope you are better now. Unfortunately, my recovery too has its ups and downs.

[. . .] Yours confidentially,

Imre

(dictated on the telephone)

June 1972
University of Auckland, New Zealand[154]

Dear Imre,

I think I deserve a salary as your agent in the Southern Hemisphere, for I make everyone here read your papers, especially *Proofs and Refutations* and *Falsification* etc. The former is being Xeroxed at the department here with the result that they won't have any money left to buy any books for the next 10 years. Which means that we should send them a complimentary copy of our book as soon as it is out. Everyone knows your work, everyone knows my work, no-one believes that I am the guy who wrote the Wittgenstein essay and "Explanation etc."[155] which are all law and order, and some insinuate that I assassinated the real PKF in Hawaii in order to get the 1,500 dollars he was supposed to earn.

I am *very* tired because my liver is again acting up which is a pity for my desire to lay the broads here (and there are some fine specimens walking around on campus) is considerably reduced. I am now reading Captain Cook's notebooks which are assigned in philosophy one, which worries the chairman no end, for there are supposed to be questions at the end (a set exam) and he cannot see himself asking questions about Captain Cook, who was certainly not a philosopher. The weather here is marvellous—New Zealand is the same latitude south as San Francisco is north—sunshine, I sit on the balcony reading and can watch all the broads drift by—my balcony is in the middle of campus, so I call to them, and tease them, and invite them to have lunch with me in a loud voice.

On the whole, I am reading, reading, reading; as a matter of fact I am doing all my reading for the next 53 years, for there is nothing else to do here. I am now reading Edgar Snow's *Red Star Over China,* a marvellously interesting book, and some stuff on myth. Tonight I may either go to the local theatre or to a movie. Also, I must prepare my introductory lecture for 200 people. In the first lecture I told them about my three marriages as an exercise in critical thinking. All staff come to my lecture on philosophy of science where I blast science, philosophy of science, philosophy, logic and praise Lakatos. So by now they think there must be something suspicious about you too, for everyone thinks I am crazy. Standard question: "By what criterion etc. etc.?" Standard answer of mine: "Any criterion you like," which makes people very unhappy.

Anyway—next time I shall tell them about *Our Great Forthcoming Book* and I think everyone will buy it here, for everyone wants to know how a reasonable guy such as IL is going to get along with a madman such as PKF. And that is all for today. Be well, and *Write.* Remember, I am in *Exile* here! (*You* will be, soon: McGovern will take care of that! Unless your friends in the CIA save you.) Greetings to Gillian, Helena Sheiham whom I seduced last night in my dreams—she was very willing, and to John who wrote me a very nice letter.

Paul

154. "I went to Auckland twice, in 1972 and 1974. The university had established a chair, for which I had applied and had been accepted. This time I had excellent reasons: I wanted to get away from the northern pollution. I was more careful than before: I kept my Berkeley job and simply added Auckland to it. I enjoyed the scenery and the people; I even enjoyed my job. I had an apartment on campus. Waking up in the morning I heard the students going to class; sunning myself on the balcony at noon I saw them returning. The library was just a few steps away. I spent many hours there, looking at books and making notes" (Feyerabend 1995, 134–35).

155. Reference is here to Feyerabend 1955b and 1962a respectively.

15 June 1972
London

Dear Paul,

I was absolutely aghast at your complete lack of financial sense. You say that you deserve a salary as my agent in reducing the further impact of *Proofs and Refutations* and thereby reducing its sales considerably. I would also like to tell you that Xeroxing the paper cannot be done without the permission of CUP which usually charges a fee and passes it on to me. However, in order to extract you from this legal trouble, which may land you in gaol, I kindly offer to you that if you send me one hundred dollars for illegally reproducing my *Proofs and Refutations* I will forget about it. I anyway had a sleepless night last night because Arne Naess in his new book, of which he only sent a review copy and not a complimentary copy, quotes you twice as many times as me (which, if I remember well, still leaves me among the three or four most quoted philosophers). [. . .]

I cannot see what on earth you are complaining about having a balcony in the middle of pretty girls in a sub-tropical climate; not having to do any administration, but reading; and in addition having the opportunity to praise me.

All the girls are terribly upset here about your having seduced Helena Sheiham in your dreams. She is a bloody left-winger and has no logical merit whatsoever. Are you sure that in fact you didn't see McGovern seduce her? You seem to be more mixed up than ever. [. . .]

I failed all the exam papers in "Marxism" as "subject" and I am going to make myself a bloody nuisance. Otherwise, my neck has improved dramatically in the last two weeks, although it is not yet painless and allegedly it never will be, but I am quite content that after nine months of hell it is like it is now. However, I am tremendously in debt and I can only break even by refuting you once a year for the next ten years. Also, you must try at least to reply to me each year, because otherwise people will think that I am repetitive.

I am basing all my summer plans on the assumption that you arrive here on August 22nd and stay here until about September 12th. I cancelled my journey to Boston in September/October because I decided that I cannot hold up your *Against Method* any further and if I am not ready by August I am going on working on it until I finish it and with your stimulating help no doubt I shall finish it, in the very worst case, in September. I needn't say that this is a loss of $5,000 and I am open to suggestions as to how much of it you will help me to recover.

With warm greetings.

Yours never,

Imre

Wednesday, June 1972
University of Auckland

Dear Imre,

The letter I wrote before lunch, this note I am writing after lunch which I took on the top of the poshest hotel in town where all the filmstars passing through stay. It is one minute from where I live, the view is excellent, outside you see Auckland harbour and the sea, inside the Polynesian waitresses, marvellous creatures, the food is very good, exotic fruit, on the whole I like it better and better here and I have occasionally been thinking of staying (if they still want me, that is). This is a small university, nice people, the library open to all, everyone, students and unstudents, can walk into the stacks and read, so I often sit and read in the library myself, for it is nice and sunny. I am about one minute from the library, I have taken lots of books home, I read, read, read, I am so uneducated, and time is short, for in a few years I shall be senile and shall not be able to learn anything anymore. I have given some very good lectures here, but about all sorts of things, theatre, stone-age astronomy, matters of life etc. Everyone reads Lakatos, everyone thinks Lakatos is the greatest philosopher, pretty girls bury their pretty heads between the pages of *Criticism and the Growth* etc., and if one wants to lay them they want to discuss Lakatos first. More and more I am now thinking that our quarrel is a *schulphilosophische Spielerei* [a little philosophical scholastic game] and the only reason I think it must go on is because your *vocabulary* is so seductive.

My liver is much better, only a few pains every now and then, so I am really happy and in a good mood for a change and I am glad to hear that you are feeling better too; there is nothing as infuriating as having to look at the world with a tired brain and a narrow field of vision. [. . .]

And I am looking forward to zeroing in on the reviewers in the joint introduction to the paperback edition of AM. Don't forget to send me a Xerox copy of the typescript of my part of AM with all your suggestions, for I want the thing to be terse. I have the tendency to meander which is fatal in matters such as these. It should be a brief statement with examples, not a sermon which never ends. So cross out, modify, send me the result. I still want to rewrite the last section but don't worry, it will not change anything in *your* stuff. The idea to bring in scepticism is *absolutely marvellous* and in my *Schlusswort* [conclusion] I may not attack you at all but elaborate *on just that point.*

You say that my sceptical conclusion can be evaded by appeal to commonsense only—I think that is OK with me. It finishes Karl who wants to give a *theoretical* solution, and brings us back to Hume who said the same. I still think there is a bug somewhere, but so far, so good. Really, if all I do is make scepticism bookworthy again, then this is more than I ever dreamt of. [. . .] After Helena I dreamt of Honor Blackman who lives right next to me, and was Cathy Gale in the Avengers (TV programme—do you know it?) and Pussy Galore in the movie Goldfinger. She is a marvellous woman, intelligent, attractive and no philosopher. Besides, she does not like McGovern. How is that for a progressive problem shift?

Be well,

Paul

Friday 23 June 1972

Dear Imre,

I have read one book by Foucault *(Madness and Society),*[156] and thought it was a marvellous case study of the different treatments madmen have received through the ages: first they were sent off in boats into the ocean in order to cool off *(Narrenschiff),* then they were put into the poorhouses and were (a) given alms and (b) when the Puritans said that poverty was the result of unwillingness to work, were regarded as criminal elements; and so on. These kinds of things I like. I did *not* like Foucault any longer when he became *general* and started to speak about the human condition (language etc. etc.) as such. For then I found him boring. I really don't know whether I should say yes and review him. Can't you keep the book you want me to review until I come to London so that I can have a look and make up my mind then? Surely, there is not a big hurry about it. If I *do* review him, I want to say a lot about his concrete studies also, for I am sure the worm which is now big and fat can be found there already. So let us wait. [. . .]

I would also like to review plus *annihilate* Sir K on the *Third World*—boy, would I like to make mincemeat of that pompous mountain of intellectual jelly (the book, not Sir K), however, not without having said what there is in it. So all this has to be talked over.

I like to hear that I am an Important Figure, but I also *don't* like to hear it. It makes me nervous. Because then I think I should do this and I should do that, write and write, make things clear, correct here, and correct there, make things "perfectly clear" where I have been misunderstood, and for the life of me I do not want to be dragged down into the sewers by tendencies of this kind. I want to live a quiet, peaceful, lazy life, casually debating things with people like you where publication, at least from my point of view, is like the publication of private letters, so one does not have to worry.

Reading Hacking I get gratified—it is *nice* to see one's name in print—and nervous, for should I not now write the Definitive Summary of My Thought? On the other hand, I think being confined to "obscure volumes" is much better for somebody like me, it raises the mystery—and so on. But I really do not want to think about myself, but just live. You are much better off; your life, your future, is constructed on a much clearer pattern while I sometimes think I have not even grown up. This is a great advantage, because you can do anything, but it is also not an advantage, living among people who say you cannot and who (still!) treat you as a responsible person.

Anyway—about the Foucault, let us wait until I come to London (which may be in two weeks for two days—there are 12 days for midterm break, and I might just travel). [. . .] Don't forget, I want to find a flat and buy it and I have also to have a closer look at Sussex in order to find out (a) whether I really want to go there and (b) whether I can afford to go there (my retirement pay would be £90—not much to live on, unless I can live in your boiler room and earn some income by rolling you around in your wheelchair and doing your garden). So all these things have to be thought over and once I come, you will have a lot to do (unless I am again felled either by my liver, or by my heart which latter always has a strange influence upon my brain).

Incidentally—collective excursions with girlfriends, I am against. I *always follow* my girlfriends rather than expect *them* to follow *me.* So I would go with them to their friends rather than take them with me to my friends (and interests). *This is a Matter of Absolute Principle* not to be eroded by anarchism. I also read, in Hacking,[157] that for Foucault "individuals matter less to knowledge than the discourse in which they participate" *which is entirely against my grain, absolutely and forever.* The individual comes first, *even if this means the breakdown of understanding. So I really Must review Foucault*—keep him for me. And, I shall review him together with Sir Karl because both these bastards seem to go against the sanctity of the folly of the individual (keep this letter, and put it on the *Foucault-Third World* file for I may use the phrases in it in the review).[158] As a matter of fact, it is now quite clear to me: I shall review Foucault and Sir Karl together as enemies of the individual (and, naturally, of Good Science). [. . .]

Be well,

Paul

156. Foucault 1961.

157. Hacking 1972.

158. Feyerabend did not in the end write a review of Foucault's book. He instead reviewed *Objective Knowledge* during 1973 and part of 1974. His paper, entitled "Popper's *Objective Knowledge,*" is pivotal to many of the letters which follow. It was first published in *Inquiry* in 1974, and was reprinted with corrections in Feyerabend's *Philosophical Papers,* vol. 2 (1981b), chap. 9, where one reads in the acknowledgments: "An earlier version was criticised by Imre Lakatos. I have adopted most of his suggestions" (168).

30 June 1972
London School of Economics

Dear Paul,

Thank you very much for your kind letters which I received yesterday and today. Incidentally, I just spoke to Roy Edgley and it turned out that they are anyway still applying to have a chair in Philosophy since the continuation of Corbett's chair is not automatic; but I told him that I am willing to guarantee that you are a man of good character (since I suppose they won't question your genius). I hope that they will believe me although I have to confess that my reputation was not enhanced lately by the LSE authorities finding out that I am forging signatures for all sorts of people to avoid correspondence and that Mary forges my signature for me. Incidentally, this letter may also be signed by Mary and may even be written by her, and since she opens the mail from you you cannot find out the truth at all without coming to London.

I am preparing the Kraft file for you so that you should not complain. I was also delighted by what you said about Foucault and I hope

I will get a review of some of his books from you. He is a man of some talent who, however, got completely confused very soon. It is very difficult not to be confused outside Anglo-Saxon countries just as it is very difficult not to be shallow in Anglo-Saxon countries. So, anyway, I am also preparing for you a Foucault file, although I would like to confess to you that I am less enthusiastic to give you any book by Karl Popper to review since I am afraid that you will object out of sheer guilt feeling; also, as I wrote to you, I think, I got a vote of no confidence passed on Mary Hesse in the British Society for the Philosophy of Science after she took over from Heinz Post who resigned after a month of presidency since he could not stand me any longer, and now the new president is a man called John Watkins who would not let a guy like Feyerabend review a man like the great Sir Karl anyway. Moreover, Karl destroyed the second set of galleys of his OUP [Oxford University Press] book on the third world and now he won't be able to get to the front of the queue for printing for quite a long time, so God only knows when the book will be out. I know that in the name index I appeared thirteen times (his research assistant told my research assistant confidentially) in the first galleys, but in the second set of galleys my name was deleted completely. I now understand that in the presently prepared set my name occurs once which is one more mention than you get. On the other hand, I have just received from Pittsburgh *vol. 5*[159] which mentions your name 35 times while mentioning my name zero times. It is all about quantum physics; I know that it is all absolute nonsense but I cannot prove it because I don't know anything about that bloody subject. Would you consider reviewing the book and saying that except for the 35 mentions of your name all is madness? [. . .] You are most frequently quoted by van Fraassen who is an operational fool.

[. . .] I started work too hastily and now I have some neck ache again but I am through with all my exercises next week and then Mary will tell everybody that I am on holiday, I shall not answer the telephone and shall do nothing but write the first animadversions to you.

Grünbaum wrote a paper against me and he just sent the first 83 pages of it. I am afraid I shall knock him out in three pages and I only pray that he won't understand it or he will think that I am wrong.[160]

To return to Sir Karl, I shall have absolutely nothing against your writing a discussion note against him for the *BJPS;* in fact I shall relish it. But I can't seriously risk my friendship with John Watkins and Alan Musgrave by letting *you* review him.

[. . .] The builders are back in my house and now all the windows are painted white outside together with the window panes so one cannot look out. I also acquired for the bay window a table which was made in 1760 but which got so wrecked that big clumsy pieces of wood which you can only see when you turn it upside-down hold it together in the inside. On the other hand, I got it relatively cheaply. I got a pair of Regency marble columns for the cast-iron lions so that they can look out of the window.

It is now completely established that the bad floorboards don't do any serious harm to the carpets but they can only be put right by lifting the carpets, removing all furniture from the house, putting down hardboard on the floorboards, putting the carpets down and bringing the furniture into the house. It is also established that this will do a lot of damage to the furniture and it is doubtful whether in the end the floorboards would be any better; and the only thing that is definite is that it will make me bankrupt. I need not say that I shall have it done because now I feel so compulsively about it.

I am not going to write to you any further because otherwise I shall never arrive at Pyrrho and Feyerabend.

Send some sunshine to London. Here there is none.

Yours ever,

Imre

159. Colodny 1972.

160. See Grünbaum 1960, 1966, 1971. For Lakatos's contribution to the discussion, see his 1974b. For his reply to Professor Grünbaum, which was published posthumously, see Lakatos 1974c.

July 1972

Dear Imre,

I now have a somewhat better idea about the shape of the MS. So far I have discovered, on average, about four to five changes per page. In all cases the style has been made more abstract and more nondescript. *I* try to write very much like I speak, and I do that intentionally. Therefore I use a lot of colloquialisms and a concrete way of talking. I also repeat parts of a sentence for emphasis, or try to make one particular word stand out from the background. All this has been eliminated and a uniform grey sauce is running across the pages instead. *I can't stand reading it* and I most certainly shall not permit publication in this form. Take only one example, where I remember the original text quite well: having finished my sermon in favour of anarchism with "and reason withers away . . ." I start the next chapter, the chapter on you, with the sentence: "This would be the end of my essay were it not for the fact . . . ." In the New Left Revolutionary Version this becomes: "My defence of epistemological anarchism could rest here . . ."—stilted, full of nouns, and with the phrase "defence could rest" which I cannot stand and would never use.

So, one thing is clear: *there must now be a clean copy.*

Now, if I had nothing else to do I could restore my text within a month. Things being as they are (travel, discussions with you about your translation, Lady Rohan McCullough etc. etc.) it will take me about two months. I also want to rewrite the chapter on incommensurability and take your own suggestions into consideration. Result: restoration plus subsequent change will not be completed before December, and then the clean copy must be written. So starting to print my part before December is *out of the question* (please tell this to the New Left, I have written them a letter but not with this timetable). As it is, this does not matter too much (except for the fact that I have lots of useless work to do)—for you will not finish MAM before March, as you wrote in one of your letters. Incidentally, do not use *their* copy when reading me, use mine. There are many places where their changes have changed the sense as well (this happens especially in the chapter on Brownian motion) and there are places in which the text now doesn't make sense at all. It is a depressing business and I almost felt like giving the whole business to CUP except that I would have the same problem with them (I had the same trouble with the University of Minnesota Press and I held up their publication until I had restored my original text). Anyway, please tell the New Left:

(1) that I am restoring the text;

(2) that this is going to louse up their clean but dreary MS;

(3) that therefore a new clean copy will be needed;

(4) that printing my part is out of the question until I have seen this new copy.

(You can show them this whole letter.)

Be well, and commiserate with me.

Paul

19 July 1972
The University of Auckland

Dear Imre,

If the worst comes to the worst, we have 8 days together. Now, let me suggest how to spend them. First day morning: my flat business in London; afternoon: Sussex. There remain seven days. Now I suggest that you send me (1) your MS of AM with all the cuts, changes etc. suggested by you and (2) as much as you have of the clean copy of my translation with your comments in the margin and suggestions for change, and dictionary. I shall be here until August 8th and I can already start working on the stuff here, or I can do a lot in Hawaii where I shall spend about four days on the way back. I shall go over the translation, consider your comments and make up my own dictionary. I shall also consider your cuts for AM and your suggestions for change. So by the time I come to London we shall not need more than two days to discuss *what remains.* You seem to prefer doing *every*thing orally; I prefer to do as much as I can by myself, and by letter, and to remove only the last stumbling blocks orally. There still remain five days. Now you may have finished MAM before I come. If there is still enough time to send it to me I shall have had time to read it and to make my first informal comments. I shall also have made a sketch of my answer. One day for discussing both. There remain four days to chase after girls—and this if the worst comes to the worst, that is, assuming I arrive on September 1st, 11.30 am. So the worst is still very good, provided you do as I ask you to and send me as much of the material that is already available so that I can work on it (send Xerox copies, so that loss is not fatal). [. . .]

And now I must run to get the mailman again. Be well, send stuff as soon as possible. I'll let you know the exact date I leave about one week in advance, so that you know exactly where I shall be at what time and can bury me in material. I really would love to have lots of things to work on in Hawaii, for the air there is very good for work. All the best, I shall soon write again.

Paul

26 July 1972
London

Dear Paul,

Thank you for your letter [. . .]. I am afraid you are completely out of your mind. This neither surprises me nor disappoints me. I would be a complete liar if I suggested that I was angry with you even in the slightest. I do not think that randomness is an illness which can be cured, so I just take you to be the Random Philosopher which actually suits your image perfectly. Let me first comment on your so-called plan for the 8 days which we are going to spend together in London. Thursday morning: you will sleep and have a rest. Then you may have lunch with me and go off to Sussex. That is fine. On the second day we will obviously chat all day and gossip. You have been vague about what you call your "flat" business in London [. . .]. All that we can do is to spend a full day looking around London checking out the sort of place which you like, getting ourselves acquainted with property prices so that you will be totally exhausted at the end of the day, you will sleep and hang around next day and we shall go on gossiping and talking about your plans the day after. [. . .] Then you will meet a girl and disappear for two days and I don't know what. There will be no substantial time at all to do any serious business whatsoever.

Now let me come to *Against Method.* This is the first letter in which you require me to Xerox the copy on which there are my corrections. Since I thought that we would discuss those corrections on your arrival, I gave Mary the virgin copy to be Xeroxed so I am afraid that more or less simultaneously with this letter you will get a virgin copy without my comments. Do look through it and try to make your cuts on your own.

I have all sorts of drafts of MAM which you can inspect on your arrival but for all sorts of reasons I am reluctant and inhibited to send it. You can inspect them on the spot. The most it seems that I can hope for is that you will leave with me a final copy of AM for the printers (this will take us two days to prepare) and I am afraid my MAM will be completely ready only by Christmas. This of course is another two months' delay but I told you that this is why I cancelled my 5,000 dollar journey to Boston and I swear to you that this is the real reason. (I do not expect to get any gratitude from you for this since it is a purely selfish affair: I have been out of intellectual work for a year and I find it difficult to get back into it especially now that the fire upset my house). I have no intention whatsoever to write something which is not up at least to my old standards. Therefore, I think that the realistic time for sending my reply and your final reply to the printers is about March 1st. Naturally, they can and should start printing your part in October in order to minimise lost time.

As far as the translation goes, I am pretty mad because what I think we could have done *together* in about five days' hard work you will be able to do without my presence in your Berkeley boredom in two weeks and then I would have to check it over in two weeks because I shall be frightened that you Feyerabendised me, which is, of course, much worse than bowdlerising me. Anyway, since the fire I have decided that you are not as bad as the fire so that if I can take the fire in my stride I can take you in my stride. There are only two things which upset me. One is that you don't sit down to the Pittsburgh vol. 5, read it instantly and review it the next day. I promise you solemnly that I shall somehow blackmail you into doing that. My reason is simple. I want to uphold the reputation of your technical competence because strangely enough it has become one of my vital interests to keep up your reputation; and your reputation rests on the mixture of technical competence, intelligence and style, three things which very rarely mix.

Let me say again that I am leaving London in the early morning of September 9th and returning on September 24th. If you are still here at the end of September that might be terribly useful. This could be easily done if you forgot about Vienna this time.

I almost forgot that I am angry with you about something else: you have not written to Alan Musgrave. He is complaining to me constantly and he says he will commit suicide if you hold him in such contempt. Although a falsificationist, he is still a human being.

Bill Kneale joined Mary Hesse and full war is going on between John Watkins and myself on the one side and the outside philosophical world on the other. Bill Kneale joined Mary Hesse because I happened to say that the Philosophy Section of the British Academy is a bunch of idiots whose names I do not care to know and the British Academy itself, of which he is vice-president, is a bunch of phonies, and also that I shall consider it as an honour if I can die without having been elected as their Fellow. It was only then that I discovered that poor Bill Kneale regards it as the acme of his career to have become vice-president of the British Academy and I kicked him where it hurt most. After this clash, it occurred to me that Lionel Robbins was an actual president of the British Academy so I thought I should do an experiment and I went to him and told him that the British Academy consists almost exclusively of common fools. Somewhat to my surprise he said that he had lately come to suspect as much. Perhaps when Kneale turns 73 (the same age as Lionel Robbins) he also might see the light; for the time being he is only 62 and he is simply too young and immature. Let us hope he will grow up.

My finger still hurts but I got completely used to it and I don't write. Incidentally, as you may know, there are no newspapers in this country and there is nothing to upset me in the morning which is absolutely wonderful.

Be good and come as soon as possible.

With warm greetings.

Yours ever

Imre

July 1972
The University of Auckland

Dear Imre,

I daresay you are taking your fire with very good humour. I am sure the cause of it is one of two things. Either it is the incendiary material you are producing, viz. the first few lines (or already paragraphs, or pages?) of MAM or else I have friends in high places ("high" because according to Book Enoch the souls of the damned are in the third heaven, around the North Pole). And I must have friends in high places because I spent my last few lectures arguing that the devil *does* exist, and all his demons *do* exist, and that people such as you who irrationally try to deny their power are just mistaken (now all the *first year* students are reading Lakatos on the Devil, or at least trying to, and they are very disappointed because they only find polyhedra and such stuff. "Read between the lines," I say—to no avail).

I agree with you about Hempel. He is one of the nicest people around, and you can talk with him. I think it is a little unfair that you don't permit him to read your stuff, for you are robbing him of peace and quiet in his old age, and yourself of a very promising pupil. Students who have attended his lectures recently tell me that he has practically given up all his firmest beliefs, the only logical positivist who has seen *all* the difficulties of this position and is honest enough to admit it and intelligent enough to formulate it in a clear way—but then he says:

"I don't know what to do *now*"—and here, reading *your* stuff would be just the right intermediate stage before transition to ME. [. . .]

Please, send me a Xerox copy of the retyped copy of AM the New Left sent you, Air Mail parcel to here, so that I can make some final corrections and cut the stuff down. I asked them for a copy, but they don't have any. [. . .] Also, hurry up with MAM, for I want to have my last word before I start haunting you from the beyond.

Best regards to Gill, greetings to the house.

Paul

2 August 1972 [Cable]
The University of Auckland

PROFESSOR LAKATOS

NEW LEFT HAS BOWDLERISED *AM.* I DO NOT ACCEPT MANUSCRIPT IN ITS PRESENT FORM. LETTER FOLLOWS. ARRIVING ENGLAND AUGUST 27 AND AGAIN SEPTEMBER 24 TO OCTOBER 6

PAUL

4 August 1972

Dear Paul,

I received your sad cable that the "New Left" people let you down. I wonder whether you are interested in an alternative and would publish it with the New Right Publishing House which I am starting up with John Watkins and Enoch Powell.[161] We also invited on the Board David Stove and Spiro Agnew but they have not yet replied. I shall also be sounding out Jack Smart to see what he thinks of it. If the enterprise comes to nothing then Gillian recommends CUP [Cambridge University Press].

I am behind with my work because the workmen are hammering everywhere, the carpets are up and nothing is moving. It also turned out that the house has not been treated for woodworm and worms are all over the place. Because of all this I am delayed in elaborating MAM although it is coming. For the moment I am occupied full time with the character assassination of Mary Hesse whom I charge with indecent public behaviour and I just wrote to the British Academy that they should provide their fellows with boyfriends and girlfriends respectively according to their needs so that they need not intrigue instead. I have never understood why the welfare state does not provide brothels.

Karl had his 70th birthday with Lionel Robbins and Ernst Gombrich. Of his old department only John Watkins was invited but he could not come because he had a yacht race. You may also know that we have a dock strike which cannot be settled because Ted Heath also had a yacht race.

Hans Albert is on the verge of suicide. Allegedly somebody told him that in Kiel you will describe critical rationalism as a "mental disease," and he thinks that will be the end of Reason in Germany. I told him that though you are AN EXTREMELY GREAT MAN, that you will not bring nazism back single-handedly, and I also suggested that he persuade you to tell those German idiots that since they are doomed to mental disease anyway, they should prefer critical rationalism to their sort of Marxism and hermeneutics. However, you must tell Hans that his neo-primitive rational criticism must stop.

I just got a very sweet letter from a German Ph.D. student asking me why it is that in 1968 I quoted "all swans are white" as a paradigm of a scientific (refutable) statement, and in 1970 I said that "all swans are white" is such a trivial banality that it has no place in science. The chap says that my answer would be of crucial importance to him. I told him that the reason is that I have changed my mind. I am afraid he will look up in his dictionary for what secondary meanings the words "change" and "mind" have, since great philosophers in Germany do *not* change their minds. So instead of solving his problem I may have caused him a new one, but such is life.

I shall be trembling on August 27th, which is I see a Sunday, to listen for the telephone when you arrive. If you get hold of some magic formula to get rid of worms and regenerate burnt curtains in discontinued ranges you will be most welcome. If you do not have any magic formulas you will come back to the critical rationalist fold.

With warm greetings.

Yours ever

Imre

161. Enoch Powell was at the time a member of Parliament in Britain, despised by many for holding racist views on immigration.

The University of Auckland

Dear Imre,

I just got what you call the "virgin copy" of AM. Alas, it is no longer virgin. These bastards from the New Left have changed my style, inserted explanatory passages where I make brief, dogmatic remarks (thus turning a series of barks into an extended whine), they have omitted things, added others, changed words, and it will take me *years* to restore the original text. I must first restore my original style, *and then* I can start improving. No hope that my MS will be ready for October.

I don't know how much harm has been done until I have had some occasion to compare the thing with my master copy at home. At any rate, the style is no longer me. And I insist that the style remain me even if it is then different from what some desiccated publishers think a real style should be. I had the very same trouble with my essay *Against Method*—the editor of the Minnesota Press changed whole passages, removed personal remarks etc. etc. Only *by accident* did I discover this, and then I held up production by two months in order to restore the original. So tell the boys that it will take me some time before I can give permission to print my part. New Left indeed! *A couple of bourgeois sissies!* Wherever there is a change it is in the direction of an academic style, of a pedestrian style etc. etc. And I won't have it! So, don't worry: you have all the time for MAM while I do some archeology etc. to restore my original text. God, I am so mad!

Anyway, I just thought I'd let you know. So, *don't read the MS,* for it is not me anyway and there is no use discussing what is not mine. Wait until October, when I am back in Berkeley with the master copy and can restore the MS. For I won't spoil my stay in London with this kind of stuff. Besides, there is time.

And now, be well.

Paul

August 1972
The University of Auckland

Dear Imre,

It took me two hours to restore the Preface. Footnotes had been incorporated into the text, long sentences broken up, changed, terms changed, things omitted and the result: a more academic document. New Left indeed! I thought I was done with my part of AM except for the changes resulting from my discussions with you and rewriting the chapter on incommensurability. No such luck! Fuck them.

[. . .] Back to the restoration of AM. Feyerabend philology. How ridiculous.

Be well.

Paul

August 1972
The University of Auckland

Dear Imre,

The bowdlerisation is not so bad as it looked at first sight; many pages, whole chapters have been left untouched by the trembling hand of the anxious publisher, but there are other places where one change follows the other, paragraphs are broken up, short sentences are expanded by qualifications, repetitions etc. etc. and nice and colourful words are replaced by boring ones. Only one example: the wonderful words "Truth-freak" which every Berkeley dropout would readily understand, they have replaced with "crank" which means something very different if it is not the complete opposite (a "truth-freak," in case you don't know, is somebody who loves the truth so much that he becomes a freak of nature, somebody like Sir K, for example or, rather, as Sir K wants to be seen by others). At any rate, I cannot simply read the stuff and see what *I* can do about it *myself,* I must first uncover myself from underneath the New Left whitewash and this I cannot always do from memory, but I must consult my copy in Berkeley. So keep this in mind when you read AM. As far as I can see, chapter 2 is quite bad. But the correctors may lose heart and it may become better later on. I *always* have this trouble with publishers.

So now I am off for my morning walk (one hour, up into the hills, every day, to keep the old bones nimble). See you soon—at least this is the present state of the randomiser. [. . .]

Be well,

Paul

7 August 1972

Dear Imre,

Here is the last word on the Great AM Affair (which, I am sure, is a source of great amusement to you).

It is now quite clear to me that the "New Left" are a bunch of constipated academics who have hardly anything in common with either Marx, or Lenin, or Mao. I was always surprised by the arid character of most articles in the *New Left Review.* Now I know that this is not an accident. It is *intended.* And where the author has too much juice, the translator will take care that respectability is restored. Now in my own case, I would distinguish two different things. There are certain extreme cases, and I think that "truth-freak" belongs to them. These cases are rare, but I would still insist that no change be made. I am writing for dropouts, or potential dropouts, not for colleagues, or future colleagues (you being a lonely exception). On the other hand, the New Left won't even permit me to write something like: "this means putting the cart before the horse" (their version: "this inverts the natural order of things"). But this means that they want me, or, for that matter, anyone else, to write the most dreary and abstract kind of English. And I won't have it.

I am seriously considering taking the MS away from them and giving it to CUP [Cambridge University Press]. At any rate, I wrote to Robin [Blackburn] (yesterday) that their version of my manuscript is "worthless," that I am going to revise it, that this will take me until about November, that I shall then give it to a typist in Berkeley and that they will receive a clean copy in December, which they must not change in any way, except with my express permission. This is how things are right now. So, read my version, forget their version (and keep my version, for I shall need it to continue correcting in London). End of reports concerning the Great AM Affair.

Incidentally, I have added the following footnote to the title (to the word "anarchistic" in the subtitle, that is):

"When choosing the term 'anarchism' for my enterprise I simply followed general usage. However, anarchism, as it has been practiced in the past and as it is being practiced today by an ever increasing number of people has features which I am not prepared to support. It cares little for human lives and human happiness (except for the lives and happiness of those who belong to some special group); and it contains precisely the kind of Puritanical dedication and seriousness which I detest. (There are some exquisite exceptions, such as Cohn-Bendit, but they are in the minority.) It is for these reasons that I now prefer to use the term *Dadaism.* A Dadaist would not hurt a fly—let alone a human being. A Dadaist is utterly unimpressed by any serious enterprise and he smells a rat when people stop smiling and assume that attitude and those facial expressions which indicate that something important is about to be said. A Dadaist is convinced that a worthwhile life will arise only when we start taking things *lightly,* and when we remove from our speech the profound but already putrid meanings it has accumulated over the centuries ("search for the truth"; "fight for justice"; "passionate concern" etc.). And he is prepared to initiate joyful experiments even in those domains where change and experimentation seem to be out of the question (example: the basic functions of language). I hope, therefore, that having read the pamphlet the reader will remember me as a flippant dadaist and *not* as a serious anarchist."[162]

Am I right in suspecting that this footnote brings me even closer to you?

Be well, and *beware of the "New Left"!*

Paul

162. See Feyerabend 1975a, 12.

14 August 1972

Dear Paul,

I received your dramatic letters of August 5th, 6th and 7th. I am slightly amused that pretty Branka did not care to tell you that she has rewritten your work. Just to calm you down, I can tell you that *Criticism and the Growth of Knowledge* was completely rewritten by a young woman at CUP in 1969, and especially your paper and my paper. I think that the woman who did it even mumbled something about making the style uniform and I think that her standard of the ideal style was Margaret Mastermann's.[163] For three months I thought that the book was being printed and I thought the galleys would arrive any time. After three months, since I had not received the galleys, I rang them, and it turned out that had I not rung they would have sent the whole lot to the printers without my seeing anything before the galleys. However, I found out that the book had been edited during these three months, so I ordered them to return the ms immediately and I solemnly burned the edited copy, that is the three months' work of the unfortunate woman, and told them to print the original as I had sent it to them. The difference between you and me, of course, is that I also demanded her immediate dismissal. Indeed, she was dismissed two weeks later but CUP says that that was not the cause. I never bothered to find out the truth.

This of course does not mean that I would want you to demand the dismissal of Branka Hoare. Before I made my demand I took great care to find out that the girl sub-editor was not pretty.

I am now going to contact the *New Left* and my present plan is to keep my time completely clear for my *final corrections* of your paper between January and March. I say *"final corrections"* because I set you in historical perspective and in historical perspective such minor details of what you think about optics, Aristotle, Galileo, Lakatos etc. and what particular words you use are of absolutely no importance. The message will only be that you are Pyrrho's successor and the only one after Newton. I am afraid I shall need quite a lot of space to explain this simple sentence to the average reader, but that is what I am going to do and all that I need is to insert or delete a few footnotes at the end. But I warn you not to change the image which I just specified. If you do so, I shall distinguish Feyerabend_0, Feyerabend_1, Feyerabend_2, Feyerabend_3, and for the amusement of the New Left readers, I shall call you the Harold Wilson[164] of philosophy.

I do not mind your Dadaism, but when it comes to moral theory even Pyrrho I have cut up into pieces. The primitive Dadaist_0 certainly would not hurt a fly—as you put it. The question is what the Dadaist does when he is in a position to hurt either one fly or another but is bound to hurt one. Will he commit suicide? You may remember that before I started off on research programmes, I discovered that I had to substitute *acceptance* and *rejection* of theories by *preferring* one theory to the other. And you may also remember in my inductive logic paper the thesis that the relation between hypotheses and theory is a ternary and not a binary relation;[165] and this of course also applies to ethics and politics. For instance, I would not like to hurt the North Vietnamese, but if not hurting the North Vietnamese hurts the South Vietnamese I am faced with a problem which *I* am willing to face but *you* are not.

[. . .] I am absolutely flabbergasted that you do not visit Jack Smart and Alan Musgrave. This means that you prefer reading books to pleasing people which surely makes you a serious scholar. By implication this makes you a hypocrite (don't worry, everybody is a hypocrite, only one hypocrite is bigger than another).

For your entertainment, I enclose a draft of a letter which I wrote to Kneale but have not sent off. As you can see, it is already being corrected by John Watkins and it will take another week or two to make up my mind about the final text. I think I wrote to you that Mary Hesse wants to have the British Society for the Philosophy of Science subordinated to the British Academy, and in addition wants to run an international conference connecting philosophy of science to history of science without Kuhn, without you and without me. If she gets her way that means I really have a bad neck. [. . .]

I am playing with the idea of turning up in East London on about October 30th when I am invited to comment on Toulmin. I have not been abroad for a very long time and I am itching to go. On the other hand, I do not know how I would stand up to the journey. Also, I would like to see some of my friends and enemies again.

I do not remember whether I have written to you that Grünbaum wrote a 101 page paper against a talk of mine at Penn on crucial experiments. He is the sweetest person on earth but this paper is both mean and stupid and I do not know whether to write a cruel answer in which I cut him into little pieces or to write a treatise on the psychopathology of a sweet man and have it published after his death or mine. I would not dare to do anything before. The third possibility is not to reply. I am still hesitating. Incidentally, reading these 101 pages I am now completely convinced that Kuhn's psychological incommensurability thesis has something to be said for it. The old simply do not understand the young. If you conjoin this statement with the true proposition that Grünbaum was born in 1924 and I in 1922, this proves the theory of special relativity.

Enough for now, and rush to London.

Imre

163. See Margaret Mastermann's essay ("The Nature of a Paradigm," 1970), in which she describes twenty-one different uses of the notion of (the Kuhnian) paradigm.

164. Sir Harold Wilson, elected MP for Labour in 1945, became President of the Board of Trade and was later elected to the party leadership. He won the General Elections of 1964, but was defeated in 1970 following the breakout of the so-called "Irish crisis" in 1969, during which the British Army occupied the six counties of Ulster in Northern Ireland. He returned to government in 1974 and later accepted a knighthood.

165. See Lakatos 1968a; "Editor's Note: The Value of Novelty," this volume.

August 1972

Dear Imre,

A long letter from you is always a welcome event and it is even more welcome now, for I have been ill and have had some unpleasant experiences. As a matter of fact, I may have to have my body shipped over in a coffin in the manner of your late countryman, Count Dracula (incidentally, do you know that Dr van Helsing, Dracula's bitter enemy who finally managed to destroy him, was an early Popperian? After Dracula had bitten five of his protégées right under his nose he says (*Dracula,* p. 135): "We learn from our mistakes."

I did not know it was Branka Hoare to whom I owe my transformation into a dignified scholar. Well, whoever it was, I am still trying to dig up PKF from Vienna/Berkeley from under the trappings.

[. . .] *Dadaism:* I am wrong, you are right, but I shall leave the footnote[166] there anyway, for I don't mind being wrong here and there. I am a scholar because I preferred reading books to visiting Smart? Do you know the books I read? *Murder in Edgware, From Leicester Square to Piccadilly* (on which Hitchcock's new movie *Frenzy* is based), *Love with a Dagger* and lots of Ibsen and Strindberg. [. . .] The conference won't help [Hesse]. There are hundreds of conferences going on, and they pass like a fart. They smell badly for a minute or two, and then nobody ever remembers them again. I got Adolf's paper on you about refutation and I agree it is long. But it is not the generation gap. The gap between the two of us is even bigger, and yet we seem to get along. [. . .]

Paul

166. Feyerabend is referring to the footnote at the beginning of AM that he talked about in his letter of 7 August 1972.

8 August 1972
The University of Auckland

Dear Imre,

Two waitresses have been dragged out of the sea, unhappy undergraduate students have staged a protest march, happy graduate students and faculty are preparing for a Big Party, at least 34 dogs have run away from their owners and have been found in a state of utter despondency, three possums have died of diarrhoea and why? Because PKF is leaving town. In two days, to be precise. [. . .]

See you soon. If you get lonely, why don't you ring Lady R and wait for me *together?*

Love,

Paul

18 August 1972
London

Dear Paul,

[. . .] Robin Blackburn just rang me on his return from China. Branka is in the hospital after childbirth and is terribly hurt that you don't like her corrections and Robin was very hurt when I told him that you found them of utterly bourgeois character. They got into a real panic and in order to prove that they are true revolutionaries and indeed terrorists, they decided to call Branka's young boy Attila. I am afraid that this is the last effort of the New Left to look revolutionary. Otherwise, I conveyed all your messages and told them that, sentimental creature that you are, I do not think that you will take away your book from the New Left Publishing House and I also said that they should not panic because the book will be out in 1973 unless they go bankrupt before then, but they won't go bankrupt because they have a very good sense of bourgeois publishing business.

[. . .] I am afraid that yesterday night I had a conference in my burned-down place with John Worrall and Colin Howson and we decided that it is our moral duty with people like Grünbaum to adopt a Feyerabendian style and it is our duty with people like Feyerabend to adopt a Grünbaumian style. This actually follows from Taoism. You should know that recently I took an interest in Taoism and I am now writing a book on its superiority to Buddhism. Unfortunately, I still cannot get on with it because I have not yet read Needham's six volumes on Chinese civilisation which I have now ordered. Moreover, I have not solved a very funny puzzle in logical probability which an argument by Grünbaum presented me with. In order to solve it I have to fiddle with silly formulas before going on to Tao. I am expecting you to take the next plane to London.

Yours,

Imre

13 October 1972
Berkeley

Dear Imre,

Here I am, back in the California desert (mental, that is), barely made it from the airport, all the roads out from the airport were flooded by rains and were closed a few hours after I arrived. The flight was quite nice. Lufthansa gave me first class and there we got roses, and something to eat every five minutes. I just arrived in time at Hamburg airport, was almost late; reason: Stegmüller. I gave a rousing talk before all the German philosophers, they sat there in their suits, with their ties, I walked up in front of them in my pullover and my bellbottom trousers talking to them about incubi, succubi, the problem of the cold member of the devil (sensed as such by all witches), the question whether the devil could create seeds or whether he had to steal them from some man to inseminate women; I also talked about von Neumann and the end of reason in quantum physics, about the double language model, prediction etc. etc., and came to the grand conclusion that all philosophies of science in existence were ever so many mental illnesses (separation from reality, formal gestures etc. etc. being some of the most outstanding features of mental illnesses).

By then I had only two points left of my itinerary (which I wrote on the blackboard), namely: "What is to be done?" and "Sales talk." Under "What is to be done?" I wrote down "Lakatos" and "Ich," and as there was no time left I simply crossed out your name and surrounded mine with a halo.

In the "Sales talk," however, I urged everyone to read *Criticism and etc.,* promised them my impeccable translation and told them that most likely they would fall for you as you argue very seductively. The talk took place in a large hall full of sunshine, so I started by saying: "Ladies and gentlemen, I must start with an apology. Most likely I shall get lost in the course of my talk and *ich werde meinen Faden verlieren* for the light of the physical sun for me always outshines the light of reason . . ." The whole assembly was laughing, and laughing, and laughing. I also got in a few jibes against the New Left and their Puritan morality.

And then, discussion. Hardly anyone said anything (they just looked stunned), so I walked up to Stegmüller, quoted something from an article of his into his face and asked him: "Why do you write such silly things?" So Stegmüller started talking and he did not stop. [. . .] While he was talking I considered various answers in your style, but I just cannot hurt a guy who is constantly ill, who suffers from perennial hoarseness, who has a charming wife who is sitting at his side, looking unhappy, so I was very polite. One thing is very certain: he is *very* upset by you, me, and old Sir K. [. . .]

And now, here I sit, the rain has stopped, it is clear again, and I wish I were elsewhere. Fuck it!

All the best, give all my love to Gillian and be well yourself, my mentor (that is what I called you in Kiel a lot).

Paul

3 December 1972
London School of Economics

Paul,

[. . .] Roy [Edgely] rang me to ask me whether it is true that you were sacked from several universities for your irresponsibility. I answered him that *you* sacked several universities for *their* irresponsibility. To be serious, I was very solemn and I told him that you are terribly serious and there is always lots of gossip about interesting people. Does he believe everything he reads about Elizabeth Taylor? He did not notice the joke (i.e. that Liz is a *bore*). Altogether, I think, we are now on good terms because I stopped making obvious jokes and he does not notice non-obvious jokes. [. . .]

Love,

Imre

December 1972

Dear Imre,

Just finished my *Kiel paper* (65 pp). Last line: "We see our principles are both rational and irrational as well as humanitarian. What more can one demand?"[167] You'll get a copy as soon as I have arrived at the final version which will take me another week.

Enclosed a report which shows that proliferation is well and alive in California.

On the other hand, Reagan wants to increase teaching time for faculty to 9 classroom hours per week and he may fire the President of the University of California. This is one reason why I want to leave.

Another reason: this is a desert. For weeks I don't see anyone worth talking to. The only person now is a nice girl, Anita. Otherwise—books. Not a very interesting life. London is much better. So is Germany.

I hope I can get next fall off, for then I shall come to Europe, mainly Germany and England and try to find a place there. I am dreaming of a nice German university in the woods with 27 students, all of them studying Lakatos and leaving me alone except for me telling them anecdotes about you.

Here *Criticism and the Growth of Knowledge* is the basic book in two of my courses. Recommended, not required, because I don't *require* things. But even without this, your fame is assured.

It is even more assured in Germany where historians, Marxists, organisers, sociologists prefer you over the "unhistorical" Sir K.

[. . .] Radical translation looks more and more mysterious to me the longer I look at it. But this is what everyone here is talking about as well as interesting things such as "assume there is a hippopotamus in the icebox" (it may be a rhinoceros and the difference may be essential, but I have now forgotten). I am being taught all this by a guy who wants me to be on his thesis committee.

I hope you are well and don't yet have to beg on street corners. If you are starving, tell me and I shall come and feed you. Best wishes to you and Gillian.

Paul

P.S. In my introductory lecture into philosophy I talk about drama, science, witchcraft, Aristotle, Brecht, and I invite visitors. Two weeks ago a local warlock addressed the class, one week ago a very clever astrologer. This week it will be the beautiful daughter of Anton Le Vey of the Church of Satan (I have to pay her $25, for Satan is poor in the US). Why don't you come some day?

167. See Feyerabend 1973.

mid-December 1972
Berkeley

Dear Imre,

Congratulations for bringing Reason to Bielefield, but what an utterly distorted sense of values you have! Instead of visiting Hübner who is the most gracious host imaginable, who has a beautiful and intelligent wife, a ravishing and intelligent daughter, whose colleagues are all interesting to talk to and whose house lies directly on the Kiel canal, not forgetting the hired help who would have woken you up every morning, instead of doing this you return to London to defeat Mary Hesse! I think fate pointed out to you the absolute insanity of this choice by having your side defeated. Anyway—do not refuse Hübner again. If you do not visit him in the near future, visit him in the fall of 1973. (I may come with you then if I get my grant.)

While you were battling irrationalists and frigid women libbers I was battling ants. They are all over the place: in the kitchen, in the bathroom, they swim in the bathwater when I am having a bath, they are in my soup, my tea, my coffee, they crawl all over my bed and there seems to be nothing I can do. I spray ant repellant. The result is: *I* throw up, the *ants* continue appearing. Right now I am sitting in the middle of a cloud of ant repellant, I am feeling sick, all the windows are open, rain comes in through the doors and through the windows, and so do the ants. How I am longing to have them replaced by a single problem like Kautsky the Rat. At least he was *one;* they are *many.* We shall see. At least they do not eat paper, so they may kill me but the last version of the Imre chapter of AM will remain untouched.

I have given up on Gail the typist and regard the last version she was typing as lost (I don't have a second copy). This version contained three completely rewritten chapters and lots of other changes. Well, I have to start all over again. Tell Robin that this will take me some time, so the March deadline is off. At any rate, I have started with *your* chapter which is about to emerge in *a completely new form* (so, you see, I am battling with you all the time); similar to what I write about you in the German paper, but, hopefully, *better.* After that comes *incommensurability* [. . .]. There will be lots of pictures in this chapter. All this takes me quite some time, for I am a slow writer. But the new and final version of AM is gradually appearing in all its glory. If I survive the ants it should be ready in March or April, i.e. when we meet in Boston.

Then comes [. . .] *Sir Karl.* [. . .] I shall work hard on this review, I want to make it quite fair and objective so that my final *subjective* chapter will have lots of "authority" behind it. Of course, you will get all the versions. [. . .] This is a funny book, indeed. Well, it will all be said in my review. My answer to his dig at *me* is (a) that my acknowledgements are friendly gestures and cannot be used as historical evidence, and (b) what you say about Duhem. But all must be said *right*, it must not come out in a *self-righteous* manner, nor must the review be too full of these things. There are lots of things to *describe*, to *criticise* (especially his historical remarks) and when all that is said and done there will be a lighthearted section on these personal things. I really want to make this a good review, for I want to send it to Karl when it is published instead of a long overdue letter.

Last night I had pains, the night before last I had pains, then I threw up, then there are the ants—these are the circumstances in which I praise you and criticise you. Are the gods against this? Do they want me to do something else? Are they giving me a hint? Well, right now I feel sick again and must lie down. Then, on with AM!

Be well, best wishes to Gillian and tell her that I took her picture just when she was imitating you. This is why she looks like a monkey.

Paul

P.S. One of the girls to whom I am talking about you and who is reading you, and me against you, thought that you spelled your first name "Emery." She was quite surprised when she saw the real thing. She may be even more surprised when she sees you.

20 December 1972

Dear Sir Paul,

(1) I read the script of your address in Kiel which completely convinced me that I am *obviously* right and you are *obviously* wrong. (Formerly I mistakenly thought that my complicated arguments were valid and yours invalid.) Cannot say that details are to follow, for there *are* no details to follow when things are obvious. My MAM however will tackle this obviousness thanks to thoughts born under the banner of Democratic Control of Intellectual Parasitism. For (1) intellectual parasites must be *selectively* killed and (2) there must be a socialist law for *selective* murder and (3) Lakatos's law is the best we have.

(2) I am in a panic. I shall address the History of Science Society on the 500th anniversary of the birth of Copernicus. January 6th, black tie. My message: Copernicus was a charlatan and thus Newton showed that the sun was *not* at the centre of the planetary system, let alone the Universe. Copernicus's programme was eliminated *before* it even had a chance to supplant Ptolemy's. [. . .]

P.S. COPYRIGHTED (part of MAM):
The two schools of thought about science:
(1) WE NEED A SUPER-ELITE
Sociologists of science (King: Robert Merton)
Philosophers of science (King: Imre Lakatos)
(2) WE ONLY NEED A KING (Paul Feyerabend)
I VOTE FOR PAUL FEYERABEND
Signed: Imre Lakatos.
P.P.S. [. . .] In the August 1972 [number] of *BJPS,* Spiro Latsis applied my criteria to economics and found Milton Friedman degenerating. Friedman replied in a 17 page letter to Latsis and shouted that he has been "talking *Progress* all the time."

26 December 1972

Dear Imre,

Quickly, before my demise, my last letter. A layman has predicted that San Francisco will be annihilated by an earthquake strength 10 on January 4th (it's plausible: there is a solar eclipse on that day which increases tides) and, naturally, I believe him.[168] I am now sleeping with my shoes on, a suitcase is packed in front of my house, tomorrow I shall have the fundament strengthened (I looked at it and found it quite weak, a few matchsticks holding up all my thinking and writing)—but in case anything happens, you are my literary executor, which means that you will have to find someone to decipher my *diary* which contains a lot of uncritical episodes. I am glad to hear that Gail is on the horizon—I have almost finished my new Imre chapter and shall send it to you with the next mail. After that, *please compare the new version with the old one* (i.e. with the Imre chapter of AM which you have) *and tell me which is better.* I shall choose the one which is better—if I am still alive by that time. You know, Aristotle said: Plato is my friend, but the truth counts more (or words to that effect). My principle, of course, is: the truth is this and that, *but* so and so is my friend. In your case, however, friendship and truth *go hand in hand,* for your work is indeed the most interesting thing since Mach and Duhem (and Hilbert and Gödel in mathematics), and considering it, is bound to lead to discoveries whether it be true or false *now.* This, incidentally, is one of the most serious objections against "my" view: accepting it does not lead very far, even if the view should happen to be true. But accepting *your* view leads very far, even if it should happen to be false (which situation, incidentally, is one more argument in favour of *my* view).

Tomorrow, workmen will start working in the basement here to strengthen the foundation. This earthquake bit has really got me frightened. While writing and rewriting AM I always keep one copy with me and put one outside, close to the goldfish pond, in case the house burns down. I cannot sleep during the night and have the most silly dreams (last night I was nominated Urine for 1973, whatever that means).

Did I write to you that John told me that Peter Hempel and I are being considered for the Islam section of the Finland conference (considered, not yet invited; the whole thing now goes to a vote). Which, of course, is in perfect agreement with the fact that I am the foremost expert on that great Islamic scholar *Lak-el-tussi* and the great tyrant *Al Poppuni* and am just translating the latter's book on "The Third Heaven, a moving account," or, however the title, which is somewhat complicated, should be translated.

[. . .] Best wishes for the New Year. Maybe AM and PM *(Pro Method)* will after all see the light of whatever it is that gives light, in 1973.

Paul

168. Apparently it was because Feyerabend was frightened of the "Big One" that he returned to Europe in 1990.

27 December 1972

Dear Imre,

Enclosed the new Imre chapter of AM. Compare it with the old chapter (of which you have a copy) and tell me which is better.

I wrote the thing during Christmas, when everyone had left for their mothers', brothers', sisters', fathers', aunts' and so on and so there was peace in my house. To compensate for that peace I was invaded by a very severe case of jaundice, the most severe one I have had this year. I cannot stay up more than eight hours during the day, and I am tired all the time. I tried to take a walk, but I had to return immediately: I started perspiring and hardly made it home. On top of all this comes that blasted earthquake prediction. So one of the things I did was to drag almost all my books into the basement over the steep garden path, without any help. Maybe that contributed to the jaundice. So in between the liver, the earthquake, the changing of furniture, the consulting of architects about the foundation of the house (it is a little too weak and needs strengthening) your chapter came into being and it may not be too well written. Let me know what you think. Right now I am just going to write this letter and then go to bed.

Be well, greetings to Gillian, I shall try to give you a ring before committing suicide, if I don't reach you and I depart, it will be *your* fault (I have already started exercises in haunting; my time will be Fridays between 1 and 2 in the morning; I shall start with Gillian, but tell her not to be afraid—quite the contrary: she is going to enjoy herself).

Paul

27 December 1972

Dear Imre,

I *knew* it—an *internal* account of Copernicus is impossible in your methodology which forces your historiography into a degenerating trend ("Copernicus is great" being a basic value judgement), and leads to its elimination. (See also chapter 1 of Kepler's *Mysterium Cosmographicum* for argument in favour of Copernicus.) *I* am not in any difficulty for I don't share in the external-internal bit. For me, Galileo was a coward for whom the motion of the earth was acceptable only when it was thrown at him by a von Neumann—and for Galileo, Copernicus was the von Neumann of the hypothesis of Philolaus. In addition, Copernicus *explained* why for the upper planets the motion of the epicycle = the motion of the sun, and for the lower planets motion of the deferent = motion of the sun. Also non satisfactory was his account of the brightness of Mars *and* of Venus. I shall use the footnote of AM to turn your account of Copernicus into refutation of you, *so send manuscript of your account of Copernicus!* (Besides, the Copernican research programme was the movement of the earth—"the Earth is a star in the Third Heaven"—and NOT the rest of the sun. Making the sun move was a progressive problem shift within this programme.)

I am in mortal fear of the earthquake here. *I agree that intellectuals must be turned into obedient slaves* (*and I have said so in writing*—"On the Improvement of Science and Art"),[169] but that has nothing to do with rationality, only with practical politics. In practical politics vis-à-vis intellectuals I am in complete agreement with you and you can quote me to that effect. But this has *Nothing Whatever to Do* with your theory of rationality.

Be well! And here's for a long and prosperous *and irrational* 1973.

Paul

169. Feyerabend 1967a.

1973

1 January 1973

Dear Imre,

I just heard that while my house is earthquake-safe, the ground on which it stands is *not,* which means that the house will stay whole but will run down the hill at an increasing speed. At any rate, the earthquake predictor who has gotten everyone in a frenzy now says that he made a calculating mistake—but I am not reassured. The smallest tremble, and I am running to the door. I am a nervous wreck, in addition to the liver I now also have a cold. Nevertheless, I am proceeding with AM; your chapter, new version, is finished and I have already sent it to you. It is stylistically not very satisfactory, and I am just rewriting the last five pages to make them better. Tell me what you think of it; and especially whether you think it is better than the old version—older, that is, from the point of view of the Book, and publicity, and the way you are going to write your answer, NOT from the point of view of the Third World. Can you do that? I think it is much clearer than the old version, though the ending is not very clear. I hope I can finish all this before the earthquake strikes. For now, all I have to do is to reread the MS (whenever it arrives), and to rewrite incommensurability. Then the whole catastrophe is finished.

[. . .] Your exchange with Sir K—well, it is a pity. But your decision to break off the exchange was the only reasonable one.[170] I shall say a few things about such matters (not about your affair with him) in my review, I really shall.

I didn't get the humanities research fellowship for fall 1973, so I shall have to stay here. But I shall come to Europe over the summer. I shall come to Alpbach, certainly, but I am more and more inclined to come as a private visitor, rather than as a spiritual leader for, to be honest with you, after I shall have finished AM and have had my dis-

Figure 3.6 *"Reading the Black-Book* = AM." Picture sent by Paul to Imre.

cussion with you in Boston I shall be sick and tired of anarchism and want to do something else. And the something else will be the finishing of *Naturphilosophie* which is entirely non-anarchistic.

Be well, and a good new year.

Paul

170. See also letter, IL to PF, 5 February 1970.

15th Nisan 375
(30 January 1973)

Dear Imre,

[. . .] The little girls are out as well. First it was the ants, then the earthquake, now it is the little girls. They are bigger than the ants, more difficult to kill, much louder. I don't know what I prefer. I shall have to talk with Gillian about this.

I now see Alpbach as follows. First day: you are surrounded by 21 pimply youths who demand to be instructed in the straight path of Reason and who promise to write your name with their greasy fingers on every piece of paper they can lay their hands on while I walk off into the hills with Gretl. Second day: you are exhausted by three hours of talk, exhilarated by the tremendous response you got—three more pimply youths leave other seminars to come to you—and frustrated, for all you see of the girls is their backside, for they are walking off with *me*. Third day: everyone is convinced that you are right, you give a triumphant talk in the biggest hall in Alpbach, 70 year old matrons come up to you, throw their arms about you, breathe garlic into your face and say: *"Professor Lakatos, Sie haben mir aus der Seele gesprochen"*—but I am nowhere to be seen, nor is Gillian. Fourth day: you retire to bed with a headache, I visit you with Gretl, Gillian, Inge, Laura. Fifth day: you give up the seminar and leave it to Stegmüller who has been imported in a hurry to replace you, and you are happy again. Not a single pimply youth, not a single ageing matron to hang from your lips (figuratively) but lots of juicy broads to hang there (literally). And this will be the proof that anarchism is better than Law and Order.

Cheers!

Paul

15 January 1973
London School of Economics

Dear Paul,

What on earth happened to you? This is the first time since 1967 that you put the date on your letter. I find this most irregular, although a very pleasant surprise, because, until now, I used to put the rational reconstruction of the date on the head of the letter and then file it carefully, so that Gillian will be able to sell our letters after our death to make some money from them.

I am chasing like mad that bitch who left half of AM with me but not the other half, and I am blowing my top.

The only good news from the department is that everybody is learning German to read your recent stuff. I also understand that all the

Marxist students in the School now learn German to read Marx and Feyerabend.

You drive me nuts with your news that you are investing so much in your house in Berkeley. Are you still serious about Sussex? Although I note from my records that you haven't answered my query about Fulbright. What happened to Fulbright? Also, why don't you apply for a Guggenheim, which gives you more money? Or are only American citizens eligible? I think you should marry a girl who mothers you and looks after your finances.

I changed my mind about the letter to Karl, and I sent instead the letter enclosed.

[. . .] Have you written to Hans [Albert] that you are definitely coming to Alpbach? I repeat, I am willing to give all the formal lectures, as long as you are willing to say no to what I say.

With warm greetings.

Yours ever

Imre

Berkeley

Dear Imre,

It is raining. I have just stopped singing, the last aria I sang (with great conviction) was *La Calunnia,* and I have it on tape. It sounds *very* good. One of these days you'll get a tape with a singing anarchist on it.

Anarchist? More and more I am convinced I am a rationalist of a kind, that is, I am an anarchist not just for anarchism's sake, but *for a reason* and the same reason will soon convince me to give up anarchism. So it is quite possible that I shall change my dedication to you: " . . . friend and fellow *rationalist*" and so confuse everyone.

Your letter didn't make any difference (here in Berkeley). To Searle etc. it just showed that crazy Paul Feyerabend has crazy friends. And that may well be true. Incidentally, the affair is not yet finished. Last time the secretaries wanted to come to the lectures but to spy on me. I said no. Now Hans Sluga and that drip Dreyfus are giving my TAs the third degree ("Does Feyerabend come to his lectures? What does he do?" etc.) and I have the suspicion that they sent somebody into my lecture with a tape recorder. Unbelievable. All the while I am talking about Brecht, Homer, Festinger, Evans-Pritchard and one of the textbooks is the *Malleus Maleficarum.*

[. . .] Yesterday I played ping pong with Karen for a full hour, I am getting *very good.* And when you come I'll play ping pong with you in return for a visit to Merton.

All the best

Paul

Dear Imre,

Please send me your Copernicus paper. My LIFE depends on it!

Love

Lucifer = he who brings the LIGHT!

P.S. I have already started AMAM, and I have refuted two arguments you are bound to use against me and all modifications thereof. How do I know? I sent 3 incubi and 4 succubi through the crack in the wall of your loo *(don't* put Sir Karl on the crack, for that would trap them and *they* could then be forced to stay with you AND Gillian *forever).*[171]

171. Feyerabend refers to "the big blow up picture of Sir Karl in a Victorian gilded frame" which Lakatos mentions in a previous letter (IL to PF, 8 January 1972).

25 January 1973

My dear Paul,

I am very upset by your sloppy letters of unknown date. (As your literary executor, I would expect you to put dates on your letters.) In your latest letter there is a picture under which stands *Figure III, Reading the Black Book.* It was obviously a photograph taken from an earlier Popper seminar before I arrived at the LSE. However, you do not give the reference, and I cannot get hold of the engraving. In spite of what you indicate, the audience there is not listening to your AM, but to the *Logic of Scientific Discovery,* and the picture looks as if it was taken in about 1956, so please do send me the reference instantly.

Just this moment I received a big envelope without even a compliment slip, and there is the second copy of your *Against Method,* which I shall be taking home, studying and comparing your different versions and advising you accordingly. I started to give my lectures again after 21 months. There was a packed audience and I delivered my final refutation of everything that you can ever say in your life. I also pointed out that Lucifer denoted the chap who brings false light, *while I am shrouding them in the darkness of truth.*[172]

This is all because I have to rush to a day-long meeting of the Royal Society on the planetary motions. There will be one lecture on each separate planet, with all their bloody anomalies. It will be a joy to my ears. I shall write to you immediately afterwards (the day-long meeting will be followed by a reception at which one has to turn up in evening dress, and is given by the President of the Royal Society. I have no doubt that Mary Hesse will be there).

With warm greetings.

Yours fondly,

Imre

P.S. I just met Lord Robbins in the Senior Common Room, and had the following conversation with him:

Lord Robbins: I just read *Criticism and the Growth of Knowledge.* It is like a planetary system, with your essay as the sun and the other essays as the planets orbiting it.

Lakatos: I just read your recently published autobiography, Lionel. It is like a galaxy of super stars.

We were both very pleased with each other.

172. Lakatos is referring to his lectures at the LSE, first published in this volume.

The first day of the end of your life (intellectual, that is).

Dear Imre,

I agree, I made a grievous mistake in the caption of the historic picture that I sent you. "Reading the Black Book" is *of course* "reading the LSD [*Logic of Scientific Discovery*] at the LSE," and I can identify the date: it is the first Tuesday in March 1953, I was there (you can see me in the background with a radiant face—at that time I was one of the foremost BlackBookReaders), John Watkins was there (you can see his stern face in the middle) and, of course, the Evil One himself, Sir (then still Professor) K. How could I have ever mixed up this foul event with the innocent joy created by reading AM? Innocent joy it is, but not for you, for the final version of the final word on Reason i.e. YOU, has just been finished and is being Xeroxed today, and if it does not get lost or buried in an earthquake you will be able to read the notice of your demise in a few days. Enjoy yourself as long as you are still unaware of the impending end of your intellectual carrière—*ecco la fine del suo cammin,* straight from Otello. So this is the first day of the end of your life as a rationalist, though your *influence* will increase and spread—I have taken care of that. And now, on to Greek pots and incommensurability.

Be well,

Paul

Saturday May 15th 1978
(26 January 1973)

Dear Imre,

I am, or rather, I was not aware that all my letters are filed away for HISTORY. Hans Albert does this—as a matter of fact his cellar is full of letters from you, John, Habermas, Sir Karl, me and others and it was very amusing to read my letters of five, six years ago. *I did not recognise myself at all.* As a matter of fact, I am now applying for a Guggenheim to give me time and opportunity to study my letters in John's, Hans', your and other files. *I* have no files at all and I keep letters only *by accident,* for example, when I put them into a book as a bookmark and forget to take them out again. Some of the most surprising letters to *me* got preserved in this way. All the others, letters by Carnap, Popper, Schrödinger, Born (a letter where he curses Sir Karl, for example) went the way of all paper. Maybe I should not do this, but I cannot see myself surrounded by files. Anyway—I am not a scholar, and that settled that.

As far as I am aware, that is also going to settle the question whether I shall get that salary raise I told you about. A big commotion is going on here, I have to fill in lots of documents about earlier appointments, written work etc. etc., much of it I have forgotten and it is such a waste of time. It would be *so nice* to have 900 dollars more *per month,* for then I could come to London every year *at my own expense* (I would just take a quarter leave of absence without pay) but I rather doubt I shall get it. But, you never know. So write them a nice and impressive letter, in case they ask. I am not too good in making propaganda for myself.[173]

Now business. Item one: *Alpbach.* If *I* accept Alpbach, I shall of course run one half of all the seminars which, I think, may be about four or five mornings (3 hours each). I do not say that I shall give half of the "formal lectures" for I do not think that a formal lecture is the right way of advancing knowledge. I have already a very definite programme, and there is lots of Plato in it, also Aristophanes, all in all an answer (*my* answer) to the question: where did we go wrong and slide into rationalism? There will also be a tiny little bit about the question whether *science* is run in a rational way, but this will be secondary, for my interest is now mainly *why we should swallow science at all* not only as an aid for technology etc., but also as a view of the world. Nobody I know has given an answer to that. *You* reconstruct science and make this the measure of everything else, but you do not say why this should be so. Why not reconstruct the Church fathers such as Lactantius of the flat earth and make *him* the measure of everything else? At any rate—this is my problem and I want to find out why and when the scientific ideology eventually took over.

This, then, I would do if I *were* to accept. *But I hesitate.* On the one hand, I am convinced that raising these questions in the seminar in my own way, with you present, would help me a lot, on the other hand I am only a part-time intellectual and *very* part-time, and so I'd rather not think at all during the summer. Now you suggest that you run the seminar, give *all* the formal lectures, and that I just "sit and sleep." But Imre—to come to beautiful Alpbach, lock myself in a room full of *Erkenntnisfortschritt* and sleep? *If* I am in Alpbach, and if I am *not* running my half of the seminar, then I shall wander around in the hills, try to pick up girls and avoid seminars, lectures etc. *like the plague.* So, I think, the best thing for me to do is *to come as a visitor,* to have breakfast, lunch and dinner with you, to be with you when you are (a) free and (b) not talking about *Erkenntnisfortschritt* with some pretty girl or some ugly but "rational" boy and otherwise have a good time. I have not yet made up my mind, and I may come and talk about Homer, Plato, and the Sad Downfall of Unreason in the fourth century BC (and even earlier)—but it is much more likely that I shall come as a visitor and behave as described above. Of course, I shall also comfort all your opponents such as Stegmüller, in case *he* runs the seminar with you. There will be no need to comfort *you,* for you will triumph anyway.

[. . .] *AM:* I am working on your chapter—it is now *much much* better—and on incommensurability, and I am waiting for the MS to make my last changes there. Tell her to send me what she has done *and forget about the rest.* Or, rather, ask her *to give you* everything she has now, keep the carbon copy of what she has written, and send me the master copy, i.e. my original MS together with what she has typed.

And let her do this even if she has written one chapter only. Send the stuff to the *University,* my mailbox is too small for big things like this, and it is falling to pieces again (letters are OK).

[. . .] Do you know the Marx brothers—of course you do, we saw one of their not so good movies in London. Well, if I get that raise, we shall go to the theatre *every week,* I shall invite you *every day* and shall give you the education you lack.

All the best, love, write again, and give my love to Gillian if you happen to remember who she is.

Paul

173. Here is how Lakatos accomplished his duty:

Professor H. L. Dreyfus
Chairman Department of Philosophy
University of California
Berkeley California 94720

Dear Professor Dreyfus,

Professor Paul Feyerabend

You asked me to write about Paul Feyerabend who is being proposed for promotion. I am delighted to be able to testify that in my judgement Paul Feyerabend is one of three or four contemporary philosophers of science whose name will still be remembered in the next generation and who has made exceptional contributions both to general problems in the philosophy of science and in several particular fields, especially philosophy of quantum physics and several aspects of the history of physics and history of ideas in particular. I think one can safely say that for philosophers of science between the ages of 25 and 40 the philosophy of science would be very different without him. His influence extends far beyond the boundaries of the Anglo-Saxon world and in German-speaking countries he is regarded as the most stimulating philosopher of science alive.

I should also say that sociologists of science received very considerable stimulus from him and at least in Europe his influence in the sociology of science equals, or surpasses, Merton's.

Perhaps I should add, if I am allowed, in a lighter vein, that I have been trying to refute some of his ideas for quite a few years now, but apparently without much success.

Yours sincerely,

Imre Lakatos

10 February 1973
Berkeley

Dear Imre,

Most amusing, most amusing. This is indeed the case of Achilles and the tortoise. Both run from A (Sir Kappa) to B (Sir Pi); the tortoise (Sir Lambda) is ahead of Achilles (Sir Epsilon Omega) and will never be overtaken by him unless Achilles gives up his hankering for Rationality.[174] Remember when people judged an action according to whether it pleased God or no, and then they had to go through the strangest contortions to interpret the pleasure of God until some clever man suggested leaving God out altogether: *"Wenn man dem Hund den Schwanz nicht lassen will, dann schneide man ihn mit einem Griffe ganz ab"*[175] says Boltzmann, my hero.

So—forget about rationality and find out what it was that made everyone accept Einstein's research programme and abandon Lorentz's. "Everyone," this means a few bigshots in England, Germany, France, for the rest are content with the Lorentz transformations and E equals mc^2 tacked onto it; that is, they are content with some purely formal tricks and would not even know the difference between Einstein and Lorentz.[176] That Lorentz turns out to be not ad hoc at all, but progressive, pleases me very much and cheers me up on an otherwise rainy day. "That rationality must lie in the *extra merits* of Einstein's theory," says Sir Epsilon Omega [Zahar] and I am interested indeed in what these merits are going to be. And if there are any such merits, were *they* the reason why Einstein was finally accepted over Lorentz? These are two different questions. And I would not examine "merits" abstractly, but I would proceed in two steps: What were the merits *for Planck; for Born; for Minkowski.* Maybe the final acceptance was due to merits for *influential people* who, via their *influence,* did the rest. I can well imagine (though that may be entirely contrary to fact) that a mathematician such as Minkowski was not too pleased by a complicated theory such as that of Lorentz and pleased by a simple procedure such as that of Einstein which he could immediately translate into his own lingo—and his influence ("great mathematician!") did the rest (that may have been part of [Einstein's] success story in *England,* where Eddington brought with him all the glamour of the Cambridge mathematicians plus, of course, the circus—identical in function with the telescopic circus of Galilei—of the sun-experiments).

So remember, you two: you may find that Einstein was better off than Lorentz in 1910. But this may not have been the reason why he was accepted. [. . .] Could Lorentz derive $E = mc^2$? Did he do it before Einstein? Was there ever even the possibility of a crucial experiment? (Kennedy and Thorndike came so much later! They don't count. Etc. etc.) At any rate: I smell another Galileo-Forman case. Good luck to you rationalists. Without you, we anarchists (that is, *I myself*—there does not seem to be anyone else) who are such lazy people would never find any material for our pamphlets. *Why don't you take Elie to Alpbach?* You see, he could do the *tough physics* part, while you take care of the ideology. I would be most interested to drop in for five minutes now and then.

All the best, greetings to the unwitting gravedigger of reason, Elie, and my best wishes for him to become a real Achilles—but I am sure that by the time he becomes Achilles, the tortoise turns out to be, and always to have been, an apparition from the fifth dimension (which, incidentally, would also refute Adolf Grünbaum).

Love,

Paul

174. Understated is the double play in the references: Sir Lambda, Sir Pi, etc., ironically stand both for the characters in the dialogue of *Proofs and Refutations* and for the real people: Popper (Sir Kappa), Feyerabend (Sir Pi), Lakatos (Sir Lambda), and Elie Zahar (Sir Epsilon Omega).

175. "If you don't want to leave the dog its tail, you cut it off in one fell swoop."

176. These underlined sentences were underlined by Lakatos himself with exclamation marks on the original of Feyerabend's letter.

15 February 1973

Dear Paul,

Thank you very much for your letter of a Friday in February 1973. I enclose a Xerox copy of this letter with two exclamation marks on passages which I can assure you have been answered months ago by the great Zahar. Your letter is a wonderful example of how the methodology of scientific research programmes asks questions which have never been asked before. In order to satisfy your curiosity at least partially, I am sending you a copy of Elie Zahar's paper which is now being printed under separate cover (part 1).[177] I am also sending you, with some reluctance, a copy of my talk which I gave on the five hundredth anniversary of Copernicus's birth. I am reluctant to send the paper for two reasons. (1) it was based on a tape recording, and I have not checked it; (2) the first part of the tape was messed up and only partially reconstructed; in this part, right in the first few sentences, *I started with my tirade against your approach, but, alas, the tape recorder did not cooperate.* In the printed version, which will be published jointly by Zahar and myself, *there will be an introduction about you and lots of technicalities at the end to tie up loose ends.*[178]

[. . .] I am starting to worry about March 20th.[179] I propose that we have our last circus of mutual teasing, get *Against Method* and my

Postscript out of the way and start a new life by emphasising what we share instead of emphasising what (no more) divides us. The trouble is that I am afraid your epistemological anarchism will get complicated by all sorts of piecemeal concessions, so that, approaching truth, we will lose a lot of fun. I do not think that we can drag this game out much longer, and I am very keen to have our fun and start some new music.

In particular, I seriously want you to put down your bloody position as untenably as possible, in no more than ten typed pages in form of thesis. Then I am willing to set out my position in maximum ten pages, which will be *equally* untenable and idiotic, but it will be fun. After that we had better return to caution and common sense, and grow into respectable scholars who have a less exciting balance between caution and fun.

With warm greetings.

Yours sincerely,

Imre

P.S. I am extremely puzzled as to whether my last sentence was written in by Sophie (my new secretary) or dictated to her by Watkins. I think I shall *switch* to mad philosophy of *maths* and, having finished AM, *you too* should switch. After that we need a radical change for some New Fun.

177. Zahar 1973.

178. See Lakatos and Zahar 1976. The paper, "Why Did Copernicus's Programme Supersede Ptolemy's?" which Lakatos wrote with Zahar in 1972–1973, had been prepared for a conference held on the 500th anniversary of Copernicus's birth at the British Society for the History of Science on 15 January 1973.

179. In this as well as the following letters, Feyerabend and Lakatos are referring to the conference they held together on 20 March 1973, at which they both put forward their points of view, which were later to become the contents of AM and MAM respectively.

24 February 1973
Berkeley

Dear Imre,

Two days ago I sent Bob Cohen a letter, in reply to a letter of his, telling him how I would proceed on March 20th: I am going to write *20 Theses on Anarchism,* have them copied, hand them out at the door of the lecture hall before our encounter, and simply discuss them in my talk, thesis after thesis (this should not take longer than 20 minutes, and then you have all the rest of the time to yourself). Yesterday I got your letter where you ask me to do precisely that. You will get my theses in about 2 weeks. (Mail, I notice, is very slow these days. Your letters take about a week to arrive and so, I guess, do mine.)

I am still walking around with this heavy equipment on my feet and my foot is still swollen. In addition, so is my liver, and I may cancel my lecture today. Incommensurability gets longer, and longer, and longer, and I may drop the whole bloody thing.

I read with amusement what Important People say about the Important Things in the Important New Book of Sir Karl's. My review is *slowly* taking shape and it will have no resemblance whatever with anything you sent me. As regards style (Sir K's), all I can say is that *anyone* looks good when compared with the average paper or book in the philosophy of science. But if you compare a thing with Brecht, or Pound, or, within the profession, with Lakatos (do you know the chap?) then things are *very* different. For example, one notices then how bloody self conscious the whole thing is (style-wise).

I won't ever hear about the effect of your letter of recommendation here. These things are *confidential* and I am not on good enough terms with any of the guys above me to have their confidence. Scriven, whose confidence I sometimes *think* I have, is too proper to tell. Besides, he has been ill (hospital, heart attack, kidney trouble), but he is already up and about.

[. . .] Saw Groucho Marx the other day on TV, he is very frail and very weak, but still in great spirits. *These* are the people I admire, not Sir K., but then, I am only a part-time intellectual (and *totally uncommitted*).

Be well, and write. Give my best wishes to Gillian, if she is still alive. If not, drop a rose on her grave for me.

Paul

Monday, end of February 1973

Dear Imre,

Never have I prepared an event in such great detail as I have prepared March 20th. Usually I wait until the last moment and then, a few hours before the talk, I write down a few notes in my hotel, memorise them, and speak on that basis. My two triumphs, i.e. in Kiel and in Cincinnati, originated in precisely this way. Hübner was aghast when I told him I had no manuscript. The last day before my talk he went out with his family, I stayed home and watched TV and while watching Sarah Leander (I was madly in love with her when I was 12—in movies, of course) sing a few songs, I wrote three pages of notes on wrapping paper. When Hübner returned I showed it to him and said: "Here is my talk," and he looked *very* doubtful. Yet things went perfectly.

Now, look what I have done for you: I have sent you a whole chapter on yourself, I have written THESES ON ANARCHISM (enclosed), and I have also made an hour tape, complete with arguments and music in case my feet get bad again and I cannot come. I shall send the tape to Bob Cohen and you can listen to it when you arrive, whether I come or not. At any rate, you will be *threefold prepared* and should be able to make *mincemeat* of me (usually I am very bad when I prepare too much—like a boxer who overtrains before a fight). *Has any opponent ever made your task that easy?* NO.

I hope that your eyes are better, I had some similar trouble some time ago, and it is great bother. Right now my feet are somewhat better, I temporarily abandoned the walking machine and I hope I don't have a relapse. But my liver is bothering me, I am yellow and *very* tired. I wanted to ring you yesterday, but I got absorbed in writing the THESES and I did not notice the time until it was too late: 11 pm your time. By that time, you were bound to be in bed, dreaming of Mary Hesse, no doubt. *I* cannot go to bed that early, Susan and Anita arrive at 10.30 to watch a movie on TV: they *did* arrive, then they left and *I* spent another hour getting rid of smoke and ashes (I have to speak to Sue and Anita about the role of females in society). These are the pleasures of human company.

I have got the Xerox copy of your copy of AM and I am revising it. There are lots of mistakes. I added a footnote on the dirty footnote Sir Karl made about me in his *Objective Knowledge.* It runs as follows:

> In his book *Objective Knowledge,* Oxford, 1972, pp. 204f, Karl Popper quotes me in support of his contention that the idea "that theories may *correct* an 'observational' or 'phenomenal' law which they are supposed to explain" originated with him. He makes two mistakes. The first mistake is that he takes my references to him as historical evidence while they are in fact just friendly gestures. The second mistake is that the idea quoted occurs in Duhem, Einstein and especially in Boltzmann who anticipated every philosophical observation of "The Aim of Science," in *Ratio I,* p. 24f in note, and of its predecessors. For *Duhem,* cf. *Objective*

> *Knowledge,* p. 200, where Popper concedes priority to Duhem in a footnote and then suddenly identifies himself with Duhem! For *Boltzmann,* cf. my article [on Boltzmann], in the *Encyclopaedia of Philosophy,* Paul Edwards ed.[180]

So far my footnote. Have you ever read my article on Boltzmann? I think it is very good. And Boltzmann indeed quite explicitly introduces the idea of a correction of observation laws by theories (even mathematical laws are for him not exempt from revision) [. . .].

With love,

Paul

180. Cf. Feyerabend 1975a, 35–36, footnote 2.

End of February [1973]

Dear Imre,

Today I felt like Christmas with all the presents I got from you: your letter, my letter (Xeroxed), Elie's paper,[181] your paper on Copernicus[182] and, lo and behold, a pale copy of AM. It has every reason to be pale in contrast with your juicy stuff. I loved your paper on Cop., it is the first interesting thing I have read on him amidst all this junk produced in connection with his birthday (poor chap!). I especially liked the way in which you explain Galileo's return to the simple model which was a great step back from the point of view of "saving the phenomena" but which you make appear rational. I have to mention you on that. (Other example: when Schrödinger did not get anywhere with the relativistic wave equation—because he neglected spin—he made 'a step back', used a non relativistic wave equation and started wave mechanics. Dirac says this: never give up an idea when it looks beautiful. You would give a much better explanation.) [. . .]

Elie is definitely Sir-material, the way he writes: dignified, in complete command, not one word missing, not one word too many. This is a fine trick he introduces to make Lorentz progressive but as I like the story, I like the trick. As a matter of fact, it is the best, the most interesting fairy tale I have read in a long time. Now all I want to know is how he is going to turn the Big Bad Witch into the Beautiful Fairy—I want to hear the Einstein story. Please send it to me as soon as it is ready. I envy you all for each other's company, you have a big and juicy idea-factory going there at LSE. If I get my raise (IF, mind you) I shall come over every fall at my own expense as an external graduate student to cheer up Sophie and listen to all of you.

As you will see from the last version of my chapter on you, I have returned to the *Commentariolus* stage of AM—all the epicycles have disappeared and the story is even more radical than before; that is, I may have returned to the *Philolaus* stage. At any rate, you will have fun demolishing it. The *20 Theses on Anarchism* will be even more simplistic and more radical. Then we shall discuss AM in August to prepare it for publication, to make it most efficient and then, as you say, we shall turn to other things, hopefully *together.* As far as I am concerned, anarchism is just another passing stage in my life (after Dingler, Mach, Reichenbach, Wittgenstein, Sir K).

Be well, and send me stuff here on my island so that I don't completely drown in irrationalism.

Paul

181. Zahar 1973.
182. Lakatos and Zahar 1976.

2 March 1973

Dear Paul,

[. . .] On studying your *Against Method* I am increasingly worried. To my mind you have a basic weakness in your position which is at least as bad as mine. To be fair to you I am telling you right now where the contradiction lies. If you were consistent, you would have the courage to be a sceptic. For the first time to my knowledge, you now say that epistemological anarchism cannot be equated with scepticism. If so, I shall prove that epistemological anarchism is *doublefaced.* One face is the face of a sceptic, the other is the face of a Kuhnian authoritarian. I am terribly sorry about this, but you either return to complete scepticism or I shall show that you are inconsistent. All that I can promise is that I shall do it with a light touch so that you will be killed and most people will believe that you are being praised.

Gillian says that unless you write to her and tell her that you are at least going in person to Alpbach, she will not allow me to go.

With warm greetings.

Yours ever,

Imre

March 1973
Berkeley

Dear Imre,

I have thought it over, and I do not think you have discovered an "inconsistency." What does the sceptic do? He says there are no reasons to believe in anything, all views are equally unproven *hence* he will not make any positive assertions and he will keep close to customary forms of life. To me this is not going far enough. If he does not know anything, then he may well do what he wants to do; that is, he may engage in propaganda, he may defend the status quo, he may oppose it—"anything goes." "Anything goes," on the other hand, clearly does not mean scepticism. It means: *anything* goes, therefore also law and order, argument, irrationalism etc. To amuse himself, and not really expecting any result, the epistemological anarchist may for short times or for long play the thinking games of the rationalist to throw him in confusion. For example, he may "show" him ("show" now interpreted in the way in which the rationalist interprets it) that anarchistic moves (moves the rationalist calls "anarchistic") have led to results the rationalist loves—and he may thereby throw him in confusion (being an anarchist he may not really expect this to happen but, lo and behold, it happens, most of the time). So, Imre, I don't think I am inconsistent, I rather think the classical sceptic is inconsistent, going on as he does *as if he had* some reasons for doing so, while as a matter of fact he can now do anything he likes, including defend the status quo. You see, the sceptic *argues* (for example), *despite* his sceptic stance. I can do this by pointing out that it may (or may not) throw people in confusion. At any rate, it is a marvellous way of sneaking into the minds of rationalists.

"Mistress," like "rational," has many meanings. *I* meant it in opposition to "master."[183] *You,* talented wordbender that you are, meant it in opposition to "wife." We shall have a marvellous argument (in your sense) about this.

Wish me luck. I am writing the thesis on 200 stencils to send to Bob for distribution before our fight.

Love,

Paul

183. See the conclusion of Feyerabend 1970c: The shift from a rational conception of scientific method to the position of "anything goes," "far from being undesirable, changes

science from a stern and demanding *mistress* into an attractive and yielding courtesan who tries to anticipate every wish of her lover" (229, emphasis added).

March 1973
Berkeley

Dear Imre,

Enclosed my most recent version of THESES ON ANARCHISM which is much better than the version I sent you a few weeks ago. Why don't you hang it on the notice board of LSE? Especially as you are in favour of the truth triumphing? For if these theses have the truth, then obviously reading them will make the truth triumph. And if they are falsehood, then being written in this marvellous fashion they will make the truth shine forth much more brightly (falsehood, well presented, makes Truth shine forth brightly).

[. . .] All the best, give my love to Gillian.

Paul

18 April 1973
Berkeley

Dear Imre,

[. . .] I am convinced that research programmes are good for history: why? Because they are connected with a *historical* method of evaluation while paradigms etc. etc. are not connected with any method of evaluation at all. Research programmes make you look for things in history which are much more interesting than what historians or philosophers look for. They are excellent for history, much better than anything that went on before (except, of course, Hegel) BUT THEY DO NOT SOLVE THE PROBLEM OF RATIONALITY, that is, even the greatest success in history does not refute anarchism.

I haven't heard anything about Fulbright, NSF, nor about my advancement. Somebody in the department is still reading my collected works. I don't know who. Also, I have been criticised by the committee on courses for running my courses in too lax a fashion, and it has been decided never again to let me have more than 50 students (I now have 900) except if there is evidence that giving me more than 50 students turns out to be of advantage to the department. Considering this bit of criticism it seems very unlikely that I shall get the advancement.

Roy wrote to me that the chair in Sussex will be announced soon—and I don't quite know what to do. The salary is not very much compared with here and I have somehow become accustomed to the climate. *I'll think about it between now and the advertisement.*[184] [. . .]

And now I am going to bed. I am very tired especially as I had much to do with administrators in the last few days. They are not my cup of tea, and I am not theirs. Did you get the missing link to AM which should now be complete?

All the best, love to Gillian, I shall write soon.

Paul

184. Feyerabend was appointed to this chair and taught at Sussex University in Brighton in 1974–1975.

20 April 1973
Berkeley

Dear Imre,

I just found out that the increase of money per month for my advancement is $178 and not, as I was wrongly led to believe, $900. I also found out that I won't get it. The Dept. is not going to make a case for my advancement. Reason: to make a case the Dept. must have positive evidence in three fields: teaching, research, administration. Now, as Michael Scriven has told me, all they know about my teaching is that I come late to class, cancel classes, and don't give the required final paper (I ask for a research paper instead). On this evidence I am an extremely bad teacher, not fit for advancement. Now, for a lousy $178 per month which, after taxes, is about $4 per day, I am not going to be punctual. So I shall withdraw my application for advancement and never hand it in again. The only thing that upsets me a little is that nobody seems to talk about *what* I say in class, *how* I present the material, only latecoming, wrong kind of examination etc. etc. Anyway—I shall not make a Mary Hesse out of it (though, right now, it is difficult).

Another two months and I should be finished. Then comes *proof of non existence of Sir K* (which follows from *proof of non existence of third world* as he is a chimera of the third world). I shall certainly let you have the review before I send it to *Inquiry.* You *always* make good suggestions.

Can you please have the illegible pages of AM retyped and the grammar checked by someone at LSE (at my expense) so that the New Left need only print it? But leave in the Berkeley colloquialisms such as "truth-freak" and "lay something on somebody" (which is somewhere in between "put something over on somebody" and "force something on somebody") and so on.

I am *certain* I won't have anything ready on quantum theory within the next two years. But then I shall start writing a new edition of "Problems of Microphysics." Of course, I start sniffing around, and I may come up with a little note now and then, but nothing comprehensive.

For the secret Greek meeting[185] I shall have to be chairman or something, no talk.

Research Programmes are the best thing *history* has had yet, incomparably better than anything else that went on before. Should be applied also to the history of the arts etc. If you find *one* excellent art historian (one, not more, just to set an example) to apply research programmes to the arts, invite him to Greece. Forget Toulmin. I know, there are lots of guys who, feeling uncomfortable with either Kuhn, or you, or me, drift to the Nondescript, and that is Toulmin (with him you don't have to make up your mind on *anything*). The best way to get rid of Toulmin is not to attack his philosophy—here you just sink, and sink, and get swallowed up by the mud—but to show how his philosophy has loused up his *history* assuming he has ever done concrete studies in history. If anyone can do that (anyone of your mafia, or submafia) let him do it in Greece. I.e. explain one particular historical study of Toulmin, criticise it, show how his errors or what have you flow from his philosophy and get rid of him that way. In *history* YOU ARE KING. But, as I said, that has nothing to do with the rationalism-irrationalism issue.

[...] Gardner—most likely he never heard of my paper on Reichenbach which is buried in 1958 *Philosophical Studies,*[186] a journal which was at the time edited by the Minnesota Center and where nobody would expect anything of importance to happen, least of all in quantum theory. But in 1957 I had a debate with Putnam on Reichenbach's *Logic* in the Minnesota Center, and Putnam then wrote a paper in *Philosophical Studies* defending Reichenbach (1957, I believe). I wrote my paper criticising Putnam and Reichenbach and I think I made mincemeat of both. At any rate, after that Putnam followed Finkelstein, and Reichenbach was not mentioned anymore. Even Hooker, who is a kind of completeness-freak as regards bibliography did not know of the paper until I pointed it out to him. I'll send you a copy of it and of the German paper (which contains three short sections on how one can argue about metaphysical research programmes) as soon as I find them.

I had a look at Giedymin 3. First (G1 and G2) incommensurability is absurd, now it is a triviality for the older positivists. I may briefly reply to him, for the gentleman protests too much. In AM I refute his contention that the issue is one of logical reconstruction and show that it is one of anthropological field work, but I omit almost everything else I said about him, as a result of your criticism. Then *I* protested too much, and you were right in criticising me.

I wish to heaven either NSF or Fulbright comes through. If not, I still have Anita and Susan both of whom however give speeches in their sleep and so I don't get much rest when I am with them.

Paul

185. In this and many of the letters that follow Feyerabend refers to the conference in Nauplion, Greece, organised by Spiro Latsis. The conference, held notwithstanding Lakatos's sudden death in 1974, was dedicated to the methodology of research programmes. Lakatos's introductory address was read by Feyerabend, who tried "to be faithful to his way of speaking, venerating Reason and demolishing irrationality." The proceedings of the conference were published by Cambridge University Press in two separate volumes (Howson 1976 and Latsis 1976).

Here is how Professor John Watkins remembers the stormy events that preceded the conference:

> In 1973 Latsis and Lakatos were planning a conference, to be held in Greece, which would give equal time to the methodology of the natural sciences and of economics. One person they were keen to get was Milton Friedman, but he didn't respond to their letters. So they flew out to Chicago to ask him in person. He was impressed ("You guys are serious"), but still declined. But other distinguished thinkers, including Sir John Hicks and Herbert Simon, accepted. For us in the Philosophy Department, 1974 was a drama-packed year. It had now been decided to hold the conference in Nauplion. And then, early in February, Lakatos suddenly died. (He was fifty two, and at the height of his powers.) Quite a few people assumed that this meant the end of the conference, but Latsis decided to carry on. It often seemed that Fate was against our conference. Paul Feyerabend was a rather natural replacement for Imre Lakatos, but he had a well known contempt for conferences. So we made him an offer he could hardly refuse, and this appeared to work. But in early July he airily announced that he was sending a tape of his talk to be on the safe side, as it was rather likely he would not come. I sent the tape back and told him to come in person. And some of the Americans who had been invited were uneasy about coming because the "Colonels" were then in power in Greece. And then the "Colonels" did something which endangered much more than our conference: they organised a military coup in Cyprus to overthrow Archbishop Makarios and bring about "enosis" with Greece. The coup was on July 15. On July 19 a Turkish invasion force set out. War between Greece and Turkey threatened. Foreign tourists got out as quickly as they could. Spiro Latsis's whereabouts became a mystery: he had left London but without arriving in Athens, from whence came anxious telephone inquiries after

him. Listening to the car radio as I set off for my summer holiday, I could only assume that the conference was doomed. But the Greek army commanders decided, it seems, that it would be safer to topple the regime in Athens than to invade Turkey. Suddenly the sun was shining, and not just on our conference. I and a few others went out ahead (our plane was almost empty). There wasn't a tourist in sight; but there, in the entrance to our Athens hotel, I almost literally bumped into Mr Karamanlis, returned from exile in Paris as the prime minister of a democratic Greece. In due course, everyone who had been invited, including Paul Feyerabend, arrived. The methodological mix of physics with economics worked well. For all of us it was the nicest conference we had ever attended." (John Watkins, letter to editor, May 1998)

For Feyerabend's contribution to the conference, see his (1976a).

186. Feyerabend 1958b.

27 May 1973

Dear Imre,

[. . .] What do you think of my criticism of Sir E [Zahar]? I was *really* looking forward to hearing a good case for Einstein being progressive. But it just did not turn out that way. Incidentally, I have the suspicion that *even before* Mercury one could make a stronger case in favour of Einstein being progressive compared with the space-time part of Lorentz (*not* the whole of Lorentz) than Elie did. But it still won't suffice. I may take up the matter myself, for it is a fascinating case.

[. . .] I have the suspicion that acceptance of Einstein before Mercury was an irrational bandwagon effect of mathematicians fond of simple invariances and of absolutist physicists unfond of complex theory. And I *very much* distrust Mercury. The reputation of the Cambridge mathematicians played an important part when Eddington joined the flock, and circus played an important part with the solar eclipse observations at Sobral and Principe (the circus is *very very* similar to the circus about the telescope, skilfully engineered by Galileo). And *never* trust Einstein when he talks about the importance of observations. For any utterance of that kind there are dozens against trusting too much "verification by little effects." The *objective* situation, however, is *extremely unclear.* I wish somebody would write a little masterpiece about Einstein comparable to your little speech on Copernicus. This is really philosophy at its best.

Have you noticed that Bohr in his papers always gives an account of the development of rival research programmes and that this history is for him quite an essential part of the argument? He is the only physicist as far as I know who writes like an Imreist. (Einstein writes like a Kantian.)

Be well, don't work too hard. Had lunch with Karen, Anita, Susan today and then gave a rousing lecture about irrationalism on the grass.

Love to Gillian.

Paul

29 May 1973

Dear Imre,

[. . .] I am glad that you are enjoying the last few months with your research programmes: history (Worrall), propaganda (your BBC talk which I liked immensely for it shows that you have realised the poverty of argument and the strength of a well placed phrase).[187] All this of course will be over once AM is published. Or, rather, it will be over for those people whom you or Sir Karl have convinced of the importance of argument and of the existence of a third world. In the first and the second world you will be as popular as ever, and perhaps even more so. So we have here a nice objective test for or against the existence of the Third World: if you slowly lose your followers, then *you* are right, and the Third World exists and argument is important. If you *keep* them and *continue* to be successful, then success is just a bandwagon effect and there is no third world. I have the suspicion that you prefer the second alternative. I also have the suspicion that this is what you will get. This means, I shall be right in the Third World, but nobody will give a damn. Which is as it ought to be.

I almost feel inclined to send a short note to *Encounter* or *The Radio Times* (or how it is called) criticising your talk and arguing for state interference in science, quoting the Chinese as an example. You know I am for the separation of state and science and for state interference with scientific chauvinism, wherever it shows itself (argued in detail in some of the footnotes to your chapters in AM). Lysenko went wrong not because the state should not interfere, but because it was a totalitarian state that interfered.

If I find some time between now and October *I shall worry Worrall*[188] by finding evidence that Newton disregarded, that was taken care of by his opponents. I am sure such evidence exists. I am sure there was a lot of ad hoc adaptation going on. Etc. etc. And even if there was not, *"eine Schwalbe macht noch keinen Sommer":* Newton may be OK, Lakatos-wise, but Galileo is not, Einstein is not, Bohr is not, and so on and so forth. Time for me to return to physics. [. . .]

In June I am giving the first final examination in my career. I was ordered to do so. It will be a disaster. Of the 900 students only about 30 know about it. Sample question (this is a general course of introduction to philosophy):

What is precession?

What is the meaning of the 'Age of Aquarius'?[189]

Give an account of the doctrine of form-criticism and its application to the interpretation of the gospels.

How was the coldness of the Devil's penis explained by St Thomas Aquinas? What do you think of the explanation?

Discuss 3 of Sextus's [Empiricus] 10 modes.

Love,

Paul

187. The BBC talk was published posthumously as "Introduction: Science and Pseudoscience" in Lakatos's *Philosophical Papers* (1978a).

188. The word play here is self-evident. Feyerabend is referring to Worrall's paper on Newton, "Thomas Young and the 'Refutation' of Newtonian Optics: A Case-Study in the Interaction of Philosophy of Science and History of Science" (1976).

189. It was trendy in the 1960s to maintain—and many people asserted openly—that the spread of values linked to the so-called counterculture (such as love, creativity, interior peace, world peace) was actually the sign of the beginning of a new era, corresponding to the astrological sign of Aquarius. This interpretation was upheld by the song from the musical stage play *Hair* entitled "The Age of Aquarius." Feyerabend is also possibly referring to the fourth section in his article "The Theory of Science—A Form of Madness As Yet Unexplored?" (1973), called, precisely, "Methodology in the Age of Aquarius."

4 June 1973

Dear Imre,

I am glad to hear that you liked my letter on Sir Elie. Actually, all I said in my letter was that part ii was *not* Lakatosian in spirit although it could have been made so.[190] The question of the excellence of the methodology of research programmes has not yet arisen, at least not in the case of Lorentz vs Einstein (though this is a case where it *might* arise). If I have time I would *love* to write a short note on Elie i and ii, from your point of view. But for that purpose I would have to look up a few things about Lorentz and Planck, especially Planck's first relativistic paper which, if I remember correctly, is on the proper transformation of force and momentum. I just *might* be able to do it by the end of July.

[. . .] Your Copernicus paper is marvellous except that you use the same trick Elie is using in his Lorentz-Einstein business. He calls the "research programme of Lorentz" that *part* of Lorentz which Einstein deals with, and he acts *as if that part were all there is to Lorentz.* When comparing Copernicus with Ptolemy-Aristotle, you again talk only about the kinematics of celestial paths and so you can easily talk about progress, for you omit that part of Copernicus that regresses with respect to the previous cosmology, viz. *dynamics.* In the *Commentariolus* Cop. of course does not talk about dynamics but then you should warn the reader that a research programme which is progressive with respect to *part* of its rival does not need to be progressive with respect to the *whole* of its rival—or, at least, *you need a few more stipulations to deal with that case of part vs whole which occurs almost always:* almost always a rival of an established programme is a rival of a part of the programme only, and does not say anything about the rest. But would you call progressive a cosmology with respect to another cosmology when it says something, for example, about the shape of the earth that is progressive relative to what the rival says about the shape of the earth, but says *nothing else,* while the rival talks about paths of planets etc. etc.? And in the *De Revolutionibus,* where Cop. *does* talk about dynamics he does so by using ad hoc hypotheses (and besides, his kinematical programme had started degenerating by that time as well, as you point out). Is there any way I can publish a note on your Copernicus when it comes out? A note written according to your specification, of course.

I am sick, my liver hurts, I have to go to the bathroom every two minutes, yesterday I had arrived at page 160 of *Naturphilosophie,* today I am back on page 140, destroyed the rest—this is how I work. Karen looks splendid, Susan looks splendid, Anita looks splendid, Anita's mother looks splendid, so does her grandmother, Kay looks splendid, I look lousy. I know the problem of trees: in the winter some of my trees froze, I had to have them removed, now I see my neighbours and their houses and I don't like what I see, I had new trees planted but who knows when they will be tall enough to cover the catastrophe.

Back to *Naturphilosophie!* Be well, don't work too hard, be nice to Gillian and tell her that when I come to Alpbach, I shall be even nicer (or before).

Love,

Paul

190. See Zahar 1973.

Wednesday, birth of Susan

Dear Imre,

I have nothing to say, but there is no reason not to write a letter (there is not even any reason not to write a book under these circumstances). I am slaving away at *Naturphilosophie* with frequent breaks at the TV—*Naturphilosophie* already bores me. Later today there will be a picnic in the garden with Susan (the birthday child to be honoured by the picnic) and Karen, Lisa, Robin (a girl), Anita, Ruth and, of course, myself. In the evening Anita's mother is going to cook for us all. Tomorrow we are going to see Marcel Marceau on what seems to be his last tour.

Today in my lecture I defended the Church's stand for law and order against the anarchist Galileo and convinced everyone that Galileo was a bastard while Bellarmino was wise, scientific, humanitarian etc. etc. Some people looked rather upset. Sir Elie's part two arrived yesterday. I have not yet looked at it but I am looking forward to hearing the reasons that make Einstein progressive over Lorentz, especially in view of the fact that [Einstein] took only a *minuscule* part of the Lorentz programme, viz. its transformation properties ("of Lorentz"), while the rest was run to the ground by the quantum theory (starting with Bohr's thesis on the electron theory of metals). In addition, he wiped out any hope for a good theory of solid objects. To take an example: assume we have a flourishing theory of the constitution of matter, of the movement of particles etc. etc., when somebody introduces the *phenomenological* second law and nothing else. Is this progress? Especially as the investigation of the micro-motion will now be upheld? I don't think so. The best one can say is that the guy starts a new subject that partly overlaps and clashes with what is already here, but is much poorer in content within the domain of opinion.

Anyway, I shall read Sir Elie and give my considered opinion. I might even include it into AM as an appendix with the title: Zahar's proof of the degenerative properties of Einstein (I have an inkling that I can use Sir E for *my* purposes). We shall see.

Had a look at the *Time* article. Well, there is reason and unreason. *Their* unreason does not suit *me.* Am pleased to hear that you are reduced to *moral* arguments against AM. To me this shows that I have won my case. (After all, you are a professor of *Logic.*) And, of course, you know that for every bit of moral propaganda you offer I have better (im)moral propaganda. For I am a ruthless moral scoundrel. Nixon is not even that, he is simply an *idiot.* He should have had a few Hungarians instead of his German mafia, then all the Hungarians in the world would have known about Watergate, but all would have helped him, and nobody else would have known. Just sent another $5,000 to Austria to escape the vanishing Dollar. Some people tell me that I anticipated radical translation (Quine) by two years in my paper "An Attempt at a Realistic Interpretation . . . ".[191]

Met Davidson today (accidentally, against my will). A little gossip revealed him to be a bore. But then, there are only few people like you and me. Start publishing your stuff, stop writing, except MAM (which should be your farewell performance, writing-wise).

Love to Gillian, and see you soon.

Paul

191. See Feyerabend 1958a; on the thesis of radical translation, see Quine 1969, 1970.

13 July 1973

Dear Paul,

Thank you for your letter which has just arrived. I had the ghastliest time of my life (perhaps this is an exaggeration) in Finland. It was the worst conference I have ever attended. It was a gathering of illiterate people who had no common background knowledge, no common language and were, with a maximum of two or three exceptions, operators inviting each other in the hope that they will be invited. One of the very few people who knew something was Tom Kuhn who, however, was bitter and unspeakably rude. It is little consolation that some big "lead speakers" were completely demolished. [. . .]

The accommodation was awful, a student hostel on a building site, the temperature was always above 30° C and I got a relapse of Bornholm (this is why I missed Sabra's lecture). He is terribly serious and I think he might be better than the others. The Russian team is a long and amusing story; more about it when you arrive.

[. . .] You do not say a word about whether you are coming to be interviewed by Essex and Sussex. What is up?

All fond greetings.

Yours ever,

Imre

1 August 1973

Dear Imre,

I have just returned to life after two awful days of pains and, because I took so many pills, an upset stomach. For two hours I had *Schüttelfrost,* I was perspiring and at the same time shaking with cold. Now I feel a little better, but still very weak. I don't know how it will go on but I may have to postpone my departure by a few days.

I have finished rewriting AM; it needs so many changes, making things shorter, clearer, but there is no time left to do that. Right now I could write an excellent AM if I were given a full year with no teaching etc. etc. for right now my ideas are very clear, and I could present them in a very simple way. But, first of all, I am sick of the subject and want to go on to something else, and secondly there is no time. There are, however, some good chapters, and my chapter on *you* is one of the best, for in this case I had a worthy opponent to argue against and so my intelligence rose to a high pitch.

I refuted point by point a paper by Machamer[192] about my Galileo paper. [. . .] A good philosopher is always a much better historian than a good historian—*that* I have learned from you.

Right now I have started on Sir Elie and it is proceeding nicely. *But you must promise me one thing:* let your mafia read it before you publish it, if they say *no,* then don't publish it. Apart from you they are about the only people whose judgement I trust.

Have you been well? I haven't heard from you for *ages* (I have not been at my office for two weeks, so any mail you sent there remained unread for quite some time). I very much miss a little intelligent conversation. I have lots of intelligent conversation about theatre, literature, general philosophy with Anita, Susan, Karen, who are all very well read and most intelligent but I am sorry to say that I am still interested in science and there is nobody I can talk with about *that.* [. . .]

For the first time in my life I am reading *Dickens* (I may have read him in German when I was nine and read all the books in the local library, but I forgot; I read Zane Grey, Karl May, Marie von Ebner Eschenbach, Redwig Courts-Mahler and other notables; and Edgar Wallace until papa forbade me to read bloody books for they excited me a lot and I could not sleep). You know, there is one great advantage in being illiterate: you have still lots of Big Things ahead of you. A literate man is reduced to the small things. I absolutely *adore* Dickens. Maybe I'll start writing in his style. Yesterday saw a play by Bus-Fekete, very amusing, but very long. And read more Roda-Roda (do you know him?).

Love, and I miss your company, and Big Love to Gillian.

Paul

192. Peter Machamer, professor at the Department of Philosophy and History of Science at the University of Pittsburgh.

15 August 1973

Dear Imre,

From your long silence and the anguished cries of your research assistant (which can be heard right here in Berkeley) I infer that you are working on something. From the content of the essays I get, I infer that you are working on your book on the philosophy of mathematics. From the nearness of Alpbach, I infer that this is what you will be talking about at Alpbach. From the quality of my criticism of your more recent efforts I infer that this is due to the fact that you can't beat anarchism—so you walk around it. Excellent! Especially as your chapters on Euclid and after are so marvellous. Why did you not send them earlier? In *Naturphilosophie,* I talk about Euclid at some length, but I had only Szabó to refer to. Of course, I referred to *Proofs and Refutations,* but I wish you could have used what you say about Euclideanism directly. Anyway—I may add it to the proofs. While you are labouring to get recognition (as if you didn't have it already!) in the select circle of mathematical philosophers and other rationalists, I am making headway among the politicians. In a pamphlet with the title *"Stalinismus und Anarchismus in der Spanischen Revolution"* I am referred to as *"der Amerikanische Oberdada Paul Feyerabend,"* and there is a long discussion of *"das negative Element in Feyerabend und Marcuse."* If things continue like that I shall be Marcuse's successor with the New Left of 1980—not an attractive prospect. [. . .]

Be well, and best wishes to Gillian,

Paul

19 August 1973

Dear Imre,

I have just read Miller's[193] demolition attempt and I don't think he succeeds in any way. His criticisms fall into two classes, viz. (1) irrelevant, (2) failure to distinguish between the *subjective* opinions of Lorentz (and other physicists) and the *objective* logical relations they establish in their papers.

Examples: Class (1), page 8 of Miller, concerning Newton's three laws of motion (i.e. the third law being inconsistent with the electromagnetic theory). This is entirely irrelevant, even if it should be true. Not an iota of Elie's argument depends on it. Miller likes to do such irrelevant bickering. The same is true about Miller's accusation that Elie did not include ether into his basic postulates (Miller, page 9). This is simply an oversight that is easily rectified. Also, on page 27: "Nature of molecular forces is entirely unknown" does not refute Elie who only needs their transformation properties, and never says he needs more.

Class (2). This reveals a *didactic* disadvantage of Elie's paper, viz. he only describes the objective relations between statements in the various papers of Lorentz, and never mentions Lorentz's subjective attitude to these relations to which all historians are liable to direct their attention. He is of course right, but only Lakatosians will notice this, other people must be *taught* that Lorentz's timidity (e.g. letter to Rayleigh, Miller page 10) has nothing to do with the matter. The letter to Rayleigh reveals what Lorentz *thought* about his efforts, *not* what his efforts had already achieved. This is constantly mixed up by historians. Elie keeps to the objective side but without explaining to all ignoramuses why he neglects such things as the letter to Rayleigh, and so he can be accused of superficial history when as a matter of fact he does much deeper history than his opponents. Of course, Elie wrote the paper for you and your Mafia, and you understand what he means, but the trouble is: nobody else does, and everybody else must be educated. So, I would suggest that *somebody* (either Elie, or Elie and Worrall and you, or all of us) give *a point by point analysis of a single paper by Lorentz, splitting it up into an objective and a subjective part, and explaining, as if talking to children, why the paper is split up that way in every particular case.* Really you, Elie etc. must realise that you are surrounded by children, and that you must explain every single step.

Same remark about page 15, construction of Lorentz's transformations "by a purely geometrical argument." This is a *Hilfsargument* that sets the stage for the molecular force etc., even if Lorentz *at that time* did not realise this. If "the majority of physicists" by 1904 regarded the Lorentz-contraction as ad hoc (page 30), then this simply shows that they didn't know what ad hoc hypotheses are, and Elie has explained in sufficient detail why it is not necessary to mention them—but he should have added, again for the benefit of the children, why he has omitted them etc. etc. So there is not a single argument in Miller that knocks Elie out, but Miller and co. won't realise that unless you make it clear to them how you subdivide every Lorentz paper into subjective and objective elements, and why you subdivide it that way. Elie part ii *won't* solve the problem, only a detailed and simple and patient *educational* paper will.

Anyway—have fun in Alpbach. [. . .]

Paul

193. See Miller 1974.

[no date]

Dear Imre,

Yes, if somebody else makes up my speech in Greece and reads it, it is OK with me. Only I don't want to promise going back to physics at this point. It is a long way, it needs time, and when *Naturphilosophie* is finished I shall take a *long* rest from thinking or whatever I have been doing the last ten years. I am glad to hear that Lakatos-party labels are spreading: for example, that Jensen is progressive. Big deal. Lorentz was progressing when Einstein started, only there are no Einsteins left today, in *any* field (thanks for the details on Eysenck's nose. I read lots of Eysenck about 15 years ago and enjoyed his general philosophy). Everyone who calls himself *progressive,* or is nice and vaguely progressive gets embarrassed when one broaches the Jensen topic.[194] *Everyone feels* one should be against him. Only Anita says—yes, one should be against him, but only if you can argue the case. My reaction? The whole thing is a pseudoproblem because intelligence is not important anyway. It's more important to be able to sing, for example. So, forget about it.

You vastly overrate my reputation when you think I got an invitation from Lévi-Strauss. Don't you realise that you are in the upper echelons of the world's intelligentsia while I am not? Despite the fact that I have defeated you in argument in the great chapter 16 of AM? Which just proved how right I am: it is rhetorics, power play, pressure tactics, intimidation etc. which changes our state of knowledge.

I shall read Sir Elie's paper[195] in about two or three days. Have to get in the right mood. I have a suspicion the Einstein case will be water on my mills and if you don't hurry with MAM I shall include it into AM, thus proving that not only the school of proto-Popperians but even the school of neo-Popperians is falling apart. The only school that is not falling apart is *my* school because it consists of only *one* member, *me,* and I am not yet falling apart and if I do, I have a philosophy that proves that I must, viz. anarchism.

Everyone here reads Quine, nobody understands him. Yesterday in my seminar I had a couple of Quineans giving an account of radical translation. Result: nobody knows what the argument is, but everyone is excited. The argument seems to consist of two parts: (1) given some body of evidence you can always have many different theories which fit the evidence, and: (2) solipsism in its linguistic form i.e. I can never know that when you say "Jensen is progressive," you do not really mean: "Popper is a donkey." Big deal, eh? But the more obscure a thing, the more it attracts the mudflies, which again proves my point.

Yesterday we celebrated Susan's birthday with cakes, fish, drinks, songs; today I am hoarse and have a headache. Better than having a headache from Quine, or Popper. Today I shall prove the excellence of witchcraft and convince the ladies that incubi do exist and to be ready for them. Last time I argued for the separation of state and science which has become a church.

Be well, be not too good, and keep your fingers crossed that I get *Naturphilosophie* off my back soon. It bores me.

Paul

194. See also Lakatos, lecture 6, note 1, this volume.
195. Reference is always to Zahar 1973.

Thursday 11 October 1973
London

Dear Paul,

[. . .] I read out some parts of my paper in our seminar on Copernicus and I think it is bloody lousy and it will take another year to complete it. However I am sending you a copy of the new version. Philosophically it is not sharp enough, and historically it shows very badly that neither Elie nor I have read the *Almagest* or *De Revolutionibus.* I am in no hurry to publish but I want it to be good.

[. . .] I wonder whether you could take a job in England for two terms instead of one—that means from September to March or from January to June—and still retain part of your American job? I think you should think about this more seriously. When the new Director, Dahrendorf, arrives at LSE, I wonder whether we can make a one-term appointment for you here. With the present administration there is no chance but one never knows whether Dahrendorf will turn out to be bright. Anyway, we can talk about this in Berkeley.

With warm greetings from the Department and from Gillian. Until soon.

Love,

Imre

P.S. Many thanks for your Hegel quotation. I am afraid Hegel might very well be inconsistent; I have to confess to you that I now remember that I quoted this passage in a Hungarian paper of mine. So you caught me out.

October 1973

Dear Imre,

Yesterday I cleaned my kitchen. I dragged the stove from behind the corner, dismantled it, cleaned every single piece, cleaned and waxed the floor, washed all the dishes, the walls, same in the bathroom and then I fell into bed tired and dreamt of my time in the military service when I was commanded to clean the toilets ("toilets" that is). Today I started painting the kitchen cabinets. They will all be in different colours, one royal blue, one green, one red, one yellow. The yellow one is for Karen, for she is a yellow-freak. As a result I am partly red, partly blue etc. There is also some colour in the cabinets. The skylight at last is finished and it looks very nice. Now while standing in the kitchen I can see the blue sky and the trees above me.

There is reason to suppose that this splendour won't last long. Today there were two small earthquakes, so there have been four earthquakes all in all since I arrived. The first four earthquakes after years of immobility. According to the most advanced and progressive theory, this means that the Big Quake is imminent. I hope you are wrong in your evaluation of research programmes so that I can live a little longer in my dog-hut. However, if it collapses I hope that the Latsis foundation will build me a new house.

I had a look at the university. The amount of creeps there is beyond belief. Depressed creeps. It doesn't seem much better in England, but somehow British creeps are easier to bear than US creeps. Well, we shall see. I am finished painting for today, so first I write you a letter, and then I'm back to detective stories. Too tired for anything else. As I have said before—this is the reason why you will triumph. You fight for Reason while I am cleaning my toilet bowl. But then I just cannot stand having people do things for me. In three days I shall have finished and if by then the earthquake has not annihilated the results of my work, I shall finish my article for Brennpunkt, then ask the chairman here about a joint appointment (if I don't get a joint appointment, I might come to Sussex nevertheless, not part time but full time, provided I don't get too small a salary. I shall say so in my letter to the Vice Chancellor). And that is all for today.

Be well, see you soon

Paul

Anita read your early Copernicus and loves it. She suggests that I resign from the university and be kept by her. That is another reason why you will triumph.

12 October 1973

Dear Paul,

[. . .] I do not remember if I told you that we have a dozen 1st year graduate students and a dozen 1st year undergraduate students and they are all very bright. I explained to them that we are having a dialectical unity: Lakatosian theory and Feyerabendian practice. This means propaganda for Reason. I am explaining to them that you are doing Lakatosian practice with Feyerabendian theory, which means Reasoning for Propaganda.

[. . .] The second printing of *Criticism and the Growth of Knowledge* is now slowly being exhausted and a second edition will be out next August. I am leaving on Monday for Boston and we are arriving in San Francisco on October 29th. I do not know yet where we shall be staying. [. . .]

My radio talk is going to be printed in *Conceptus,* the student newspaper of the University of Innsbruck.

In a rush.

Yours ever,

Imre

30 November 1973
Day of your departure

Dear Imre,

You depart, and already it looks darker. Even the food tasted better when you were nearby advancing knowledge or plotting the downfall of somebody. I am a little depressed because it seems to me that in the case England and/or New Zealand vs. Berkeley I shall decide in favour of Berkeley and that won't do me any good, that is, neither my lungs nor my brain will be improved from that. And now that Karen is coming to London not even my eyes will be pleased any more. But we shall see. I looked up Kant in greater detail and found that the quotation is a fluke. One might think so, he says, but one should not. Too bad. [. . .] And now I have to wash the floor in the bathroom. It is rather dirty, and I have to dust the rest of the house.

Be well, see you soon

Paul

Monday, early December 1973

Dear Imre,

I am happy as I haven't been for a long time. First, because the beginning chapters of AM sound very good (the trouble, of course, will come later, with incommensurability). Secondly, because the *Hair* girl is here. She is beautiful, she is kind, she is intelligent, she is fun to be with. She did a narration at the San Francisco Symphony. When she came to join me in the audience after her piece, she was almost devoured by the males present—and females and people looked at me quite suspiciously: "What is this bum doing with a superb and beautiful artist such as her?" Yesterday we walked for hours in the hills behind my house, today I am going shopping with her in San Francisco. On Tuesday she leaves for Los Angeles and New York and I shall be disconsolate. She may come to Greece in the summer, and to New Zealand before that—we shall see. At any rate, I told her to ring you and give you my regards when she returns to London and I also gave her special instructions on how to ring you. You won't regret hearing her voice. Well, on with AM and with the defence of irrationalism!

Paul

P.S. I told her quite a lot about you. She asked me about AM and your views, and I described them to her, and I discovered, while I was describing my views, that I was describing yours. We are indeed, basically, of the same opinion. And the questions of a beautiful woman will bring it out.

P.P.S. Her name is Rohan McCullough.[196]

196. "Once, just back from Berlin, I went to the Shaftesbury Theatre to see *Hair.* It was an astonishing show. I felt especially attracted to a vivacious brunette who danced up a storm. I met her next day. Daniel Ravenaugh—pianist, conductor, Busoni scholar, man about town, and my neighbour in Berkeley—had picked her up on the way to our luncheon date. She was Rohan McCullough, daughter of the famous Colonel McCullough whose radio interviews during the war were already a legend. I occasionally waited for her at the stage door. We would talk briefly, and then she was off to one of her many parties. About two years later, when feeling bored in Imre's library, I gave her a call. From then on we met more frequently; we went to concerts, movies, the theatre, and operatic performances. When about to leave London, I received a message from Ravenaugh; he needed my house in Berkeley to solidify an affair of his own and offered a sizable sum if I would stay away. "Well," I said to myself, "it's time for a change," and moved to Blakes Hotel in Roland Gardens in Kensington, my headquarters during the next few years. I saw Rohan in Berkeley where she performed with the San Francisco Symphony, and in Zürich, where she played Nerissa in an Old Vic production of *The Merchant of Venice.* Since then we have met only once, but the love and the friendship remain" (Feyerabend 1995, 130–31).

December 1973

Dear Imre,

White as a sheet am I at the thought that SHE is going to leave today and suicide is not very far from my mind: one full year, one full year before I see her again. I don't think I shall survive. So, just in case I take the plunge, I appoint you and your mafia sole agents for my literary remains: to do with them whatever you wish, publish, change, bowdlerise, suppress. [. . .]

In case I survive, don't be surprised when I won't write for a long long time. For IF I survive I shall do so sitting in front of the television, staring into it day and night, for weeks, and weeks, and weeks and, maybe, after that, I shall return to the chapter on incommensurability. This is the only person after about 15 years I have ideas of marriage about (never told her about them, never shall tell her) so, if she were not so much wiser than I am, a catastrophe would happen very soon. I saw her again at the Symphony, musical voice, melodious delivery, graceful comportment like an angel from heaven, she stood there with the big orchestra, like an angel from heaven she walked with me in the Berkeley hills until we both got mud all over, like an angel from heaven she swallows the food I am stuffing into her and the words I bury her in. I am prepared to do anything for her, become a rationalist, even a naive falsificationist if she should want me (but she won't).

Well, the next few weeks will be void like Hesiod's gaping abyss (*Theog.* 721 ff). May the gods assist me.

Paul

14 December 1973

Dear Imre,

Today I finished my final corrections of AM. 24 pages singlespaced, the last chapter ending with my plea for a separation of Science and State. Now I will put 10 miles of string around the MS and put it in the basement and I won't think about it any more. It is a good book, and some parts of it are really excellent, but it is very uneven. [. . .]

Next comes Sir K. I think I have the review pretty well laid out in my head. It isn't going to be long, but it isn't going to be flattering goo either. "There is no body of informed and serious criticism of Popper's thought," writes Sir Anthony[197] in *Encounter.* Well, he is going to get it. Just my luck to spend Christmas with *Objective Knowledge.* But I guess it is still better than hanging myself.

Rohan is back in London and I am vegetating from day to day. There hasn't been a woman for 12 years who has made such an impression upon me. Well, we shall see.

Speaking of women: Rachel Busch rang me the other day (when

I had my telephone working because of Rohan) and complained about her financial situation. She is happy, she says, but she is poor, she wants to work, she needs a recommendation. Well, I shall write her a recommendation.

Just read through the "Newsletter" of the History of Science Society. Machamer is there all over the place, and so are lots of other creeps. There might be something in rationalism if it can put an end to them.

Anita has complained about the dedication to AM. So I have decided that women are not worthy subjects for dedications in serious books and I now have a new dedication. It reads: "To Imre Lakatos, friend, and fellow anarchist." With your permission, of course.

Clouds outside, clouds inside. I would not mind spending the rest of my life as Rohan's manager. It must be true what she says about me: "Paul, you have no ego at all." But what is the advantage of having one? She also says that I *exist,* that I don't *live,* by which she means that I don't have a life-plan. All these things meant as puzzled compliments. I guess I am as opaque to her as she is to me (so far).

So that is my letter for today. Be well. Give my love to Gillian. Your life is well ordered and [. . .] peaceful. You have a life-plan. It seems I shall die without even knowing what a life-plan is.

Paul

197. Anthony O'Hear, professor at the University of Bradford and director of the Royal Institute of Philosophy, is the author of a book on Popper's philosophy (1980).

Friday [14 December 1973]

Dear Imre,

Just got Hübner's postscript to his Kepler paper with the Grand Inquisitor Lakatos examining dear Johannes. I think he (Kurt) makes two good points: (1) Kepler is not progressive if we consider kinematics *plus* dynamics. This is also the objection I would raise against Lakatos-Zahar on Copernicus (I mentioned it in my little note on Zahar on Einstein).[198] (2) Production of novel predictions cum recapitulation of the OK content of the rival programmes is not sufficient. The predictions must have been brought about in a "reasonable" manner, e.g. by logically satisfactory derivation from the positive heuristic. [. . .] Now Kurt thinks that Kepler produced his predictions in a very messy fashion. I don't know whether this is true. Sometimes it seems that Kurt simply means that Johannes used daring hypotheses, which of course he is permitted to do, but I have not read his whole argument. This point, incidentally, is the same I have made against you (and Elie) re. classical perturbation theory: general relativity gives this perturbation theory only if one uses approximations in a shamelessly ad hoc manner—so one cannot really say that the content of general relativity (truth content) exceeds the truth content of classical celestial mechanics. The matter is still open. Elie seems to agree.

In sum: Kurt ain't that bad and, considering that he had to do everything *by himself,* not being carried along by a tradition, he is very good.

Here the sun shines, and I have finally started working on the Third World. My reason is that fewer people commit suicide out of boredom than commit suicide out of lovesickness. Well, the boredom certainly is there. All I need is that my reason is correct (it does not seem to be).

Do you think there will be any England left to come to in fall 1974?

Do you think AM will be printed, or will disappear as part of energy saving? I would not mind if all the other books remained unprinted as well.

Be well, count yourself lucky that you are past the age where women can drive you to distraction (maybe you were always past that age and thus much wiser than me). But even if I hang myself I think it is better to hang oneself over a beautiful woman than to *live* for no reason at all.

Be well. Am reading a 24 volume History of the Church. Reminds me of the LSE philosophy department.

Paul

198. See letter, PF to IL, 4 June 1973.

19 December 1973
London

Dear Paul,

I just got three letters from you this morning, which of course made my day.

A few short comments:

1. Your new dedication "To Imre Lakatos, friend, and fellow anarchist" would be absolutely wonderful and flattering for me on the condition that you take out the comma between "friend" and "and fellow

anarchist." On the other hand you also get my permission to change your mind and put it in a dedication to some girl.

2. What about Machamer? He wants to write a review of Shea's book on Galileo. I suppose I'll let him do it because I have so few reviews.[199] I met the chap and if he spent five years here I could educate him. But he is not. He says that you scored a couple of points in your reply to him: if he sticks to this then he is not hopeless.

3. Kurt Hübner. Of course he is charming. Of course he is intelligent. But what he says on twenty pages, and is correct, you say in four lines. There is no chance that I publish anything of his in the *BJPS*. It would be an enormous service to the friendship of Kurt and myself if you could persuade him that his style is not really suitable for an Anglo-Saxon professional journal. Of course loquaciousness has nothing to do with professionalism, but this is I think a nice way to put it. If he does not want me to publish his stuff I shall be forever his friend.

4. Rachel Busch. I do not believe that she is happy and I do not believe she is poor and I do not believe that she wants to work but I do believe that she needs a recommendation. What I most remember of her is when in Finland she looked with great care through a wine list and made very expert remarks on different wines of which I had never heard, and at the end, after careful consideration, she ordered the most expensive one. In the meanwhile Larry Laudan was sitting opposite to her, watching her choice with trepidation, and cold sweat came out on his forehead, [he was] obviously counting his pennies.

5. Rohan. I hope she will still exist when you arrive and I am looking forward to meeting her.

6. The funny thing about the present crisis is that it really cannot last. Either the government gives in or the miners give in. Until now there has been no change in the country except for the predictions about the gloomy future. Perhaps this is an exaggeration. Printing workers only work three days a week which means that they can do sixty percent of what they were planning to do. I have no doubt that the end result will be that your book in fact will be speeded up because Robin Blackburn will be intelligent enough to realise that he has to select, and your book will be the biggest financial success by far.

7. Incidentally I put in my Copernicus paper "my friend Paul Feyerabend" and five pages later I put in "the brilliant Paul Feyerabend." Then I thought that the two are inconsistent and therefore I crossed out "my friend." But it transpires from the context anyway. I shall send you the Copernicus paper as I am sending it to the printers, in a day or two.

8. I invited Eysenck for a seminar talk but after careful consideration and after having had a pleasant supper with Urbach, Zahar and me, he decided that he does not dare to come because he may be roughed up again. I think that it was a mistake to invite Elie Zahar, whom he obviously regarded as an Arab terrorist.

I hope you have received my parcel with celestial anomalies.

9. I am enclosing a correspondence between Donald Gillies[200] and me for your entertainment. I hope it will be a relief from the third world. Donald Gillies has got an Italian Bolshevik wife who turned him into an environmentalist in the IQ issue. The last letter is of 6th December and now we have called a cease fire but I am determined to bully him into intellectual submission.

10. Noretta [Koertge] is very grateful to me for having offered to her that she can review on twenty printed pages the great event of 1974, *Against Method.*[201]

[...] With fond greetings for a boring Christmas and an unhappy New Year.

Yours ever,

Imre

199. Lakatos is referring to his service as editor of the *British Journal for the Philosophy of Science.* Machamer's (and Lunsford's) review of *Galileo's Intellectual Revolution* by W. R. Shea would eventually be published in the *BJPS,* but not until 1975.

200. Donald Gillies, then lecturer at Chelsea College and now professor of the philosophy of science and mathematics, King's College, London, remembers that the IQ controversy (see also lecture 6, note 1, this volume) caused a very big rumpus at the time:

> I was strongly opposed to Jensen, Eysenck, etc., but I thought that they should be allowed to speak on the grounds that free speech was a good thing, and, anyway, I thought that their arguments could easily be answered. Lakatos supported Jensen, Eysenck, et. al., and he persuaded Peter Urbach to write a piece defending them using the MSRP. This situation led to quite violent quarrels between me and both Imre and Peter Urbach. The whole thing exploded at a seminar which Peter Urbach gave at Chelsea. The arguments at the seminar grew so heated that it degenerated towards the end into a shouting match. It was the nearest I ever saw in England to a seminar ending in blows! Imre was furious with me, because he thought I had packed the audience with hecklers. Now it is true that I had invited many of my leftie friends to come, and they had arrived prepared with arguments on the side opposite to the speaker; but it seemed to me that this was perfectly fair, and an instance of open discussion. My wife Grazia was also there. Imre always blamed her for having a bad influence on me, and, in particular, converting me to bolshevism. There is some truth in all this, but, as usual, Imre exaggerates. Grazia, on the other hand, had no interest in IQ testing, and it was only on my instigation that she read Eysenck's book. Grazia, however, did think of an argument against Eysenck. Eysenck at one point in the book states that the Irish in America have higher IQs than the Irish in Ireland, and says that this is because the more intelligent

Irish emigrated. At another point of the book he says that the Italians in America have lower IQs than the Italians in Italy, and says that this is because the less intelligent Italians emigrate! Grazia pointed out that this is absurd, since in both cases it was the poor members of the community who were forced to emigrate because of economic hardship. The Irish emigration occurred earlier than the Italian, however, and so the Irish were better placed to get middle class jobs. This, on the environmentalist position, explains the results of IQ tests. Grazia produced this argument at the seminar in Chelsea, and Imre was furious. He delivered a long speech to the effect that this was a purely emotive argument, and that he too could cloud the issue by discussing Hungarian emigrants. I am sure that he blamed Grazia for influencing me in favour of environmentalism, but, as I have explained, this is only partly true. In the wake of the seminar, Lakatos and I exchanged a heated series of letters. Unfortunately I seem to have lost the letters he sent me." (Donald Gillies, letter to editor, June 1995)

201. In fact, following Lakatos's death, the new co-editors of the *British Journal for the Philosophy of Science,* John Watkins and John Worrall, gave the review to Ernest Gellner to write (1974). The decision did not please Feyerabend, who wrote a ferocious reply (Feyerabend 1978c).

1974

2 January 1974

Dear Imre,

I have read your Copernicus paper once again. It is indeed a most beautiful paper. Briskly it moves along, covers an immense amount of material, and is completely free of those long meanderings which you find in "professional" papers. I hope it leads to some interesting discussions and to a new (post-Kuhnian) book on the Copernican revolution. If I had the time I would try to write such a book, exactly like Kuhn, or you, namely: understandable for the wider public, but, of course, from a completely different point of view. It would contain a *history of the change of standards* that led to the revolution (I think the whole Aristotelian physics, and not only the principle "save the phenomena" implies a praise of degenerating adaptations which, for them [the Aristotelians], is OK because it leaves the basic structure unchanged), the foreshadowing of the breakdown of "celestial vs terrestrial" in painting, [. . .] and, finally, a detailed account of the revolution itself and the main reasons for its success. [. . .]

But these are just beautiful dreams (I assembled lots of material while sitting in the Warburg library on these items; it is a marvellous library, don't forget it). Incidentally, the first reaction of the Lutheran

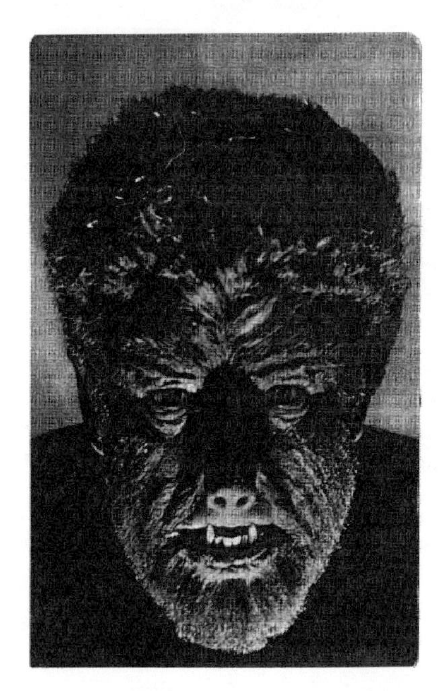

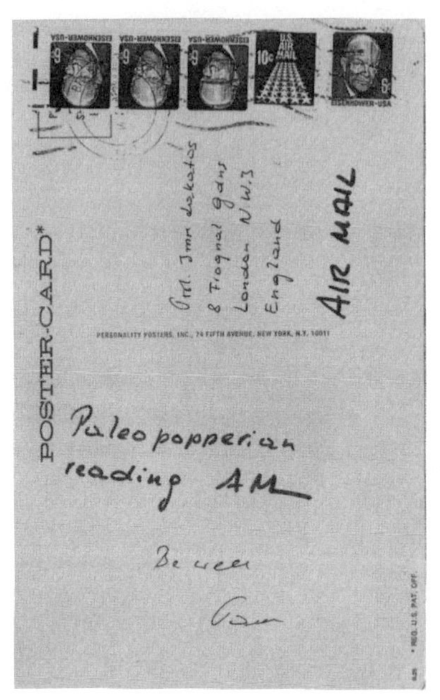

Figure 3.7 "A paleopopperian reading AM." Postcard sent by Paul to Imre.

intelligentsia to Copernicus, whom they knew from Rheticus, produced the following two arguments: (1) [he] contradicts the Bible; (2) [he] contradicts dynamics. Melanchthon's *Initia Doctrinae Physicae* contains these two arguments, and they were repeated again and again in his other writings. Suddenly he repeats them with less conviction, and a little bit more politely—why? This happened long before there was a decent dynamics. I want to find out. Do you know? The same thing happens when I look at the register books of the University of Vienna which contain university decisions and verbatim contents of major lectures and dissertations. Slowly, Copernicus becomes more popular—but nobody mentions any argument (this, again, before Galileo). There were no arguments among the "Great" at that time either. So what happened?

A beautiful day today such as you could not possibly have in England. No smog at all, mainly because for three days there was no driving. With gasoline problems it seems that there will soon be much less smog—I hope. And now, goodbye, until the next letter.

Paul

Contd. I entered the new year on wings of painpills, and now I have a pill hangover. Jan. 1st, a marvellous day, clear, you can see all the way to the Pacific Ocean, I already took my morning walk pursued by a pack of dogs, they have predicted frost and snow tonight. Just five more days and lecturing starts again. This time I have to give finals. So, in one class I let them read Descartes's first two meditations, Mill's *On Liberty,* three chapters, and Castaneda, *The Teachings of Don Juan* (how to turn into a dog, or a raven etc.) and tell them: read this, don't ever come to my lectures, and you'll be OK. In the other class they have to read your essay in *Criticism.* Having read Sir K, I *almost* thought there was something in it, so to make sure, I read Marinetti and St Augustine, and now I have a sense of perspective again: one more week and Sir K will be buried (but *your* third world will remain, it is different from his). I may write a short note on Stegmüller's Sneedisation[202] of you and me, either in German for the Geldsetzer-König magazine, or in English, for *BJPS,* but first I have to bury Sir K, he already starts to smell. Stegmüller is OK. Just think: things change so quickly, hardly has Sir K overthrown poor Rudolf, and he is overthrown himself and not in an orderly way, but by two loudmouths (you and me) and a sociologist-historian (Kuhn). Still, Stegmüller takes up the challenge and writes a 420pp opus. Let us praise him for that! (But my praise, if I ever write it, will be in one line; the rest will be criticism).

Just sent off another "NO" to a conference: to the Vienna Institute for the Retardation of all Knowledge. This Vienna Crowd—a mass of nasty and unhappy creeps, I just couldn't stand it. But why do I say no to *all* conferences? Because I want to lead a peaceful life, because I want to stumble along at my own speed, produce my nonsense at my own speed, watched only by a few friends and knowing the reactions of these friends only (Anita, Karen, Susan—broads first—and you and your mafia). I am a contentious fellow, and if I am at a conference I get worried about "my position in contemporary thought," like Sir K whom I understand very well. I *don't want* to be worried in this way, it is time-consuming, and demeaning. And it is absolutely idiotic. But I have this weakness—so I'd rather stay away. Having a battle with you and Elie or Worrall is an entirely different thing, it is a most entertaining and interesting way of expressing our friendship and I would not want to be without it. Your paper on Copernicus for me is just another personal greeting to which I have to reply with a greeting of my own, viz. my four items in my last letter. But a conference! No, no, and again no. Moving the world spirit? Not for me. I am a private citizen, not a general.

Anyway: in front of me is my reply to you *re* why do I use an elevator?[203] Well, I certainly don't use it (rather than jump out of the window) because I have applied the methodology of research programmes and have found elevators to be progressive over jumping out of the window. For the alternative programme (jumping out of windows) has hardly been developed at all. A fear of heights might well be innate (kittens when presented with the perspective drawing of an abyss on a flat floor don't proceed but stop in terror, even before they have made any experience of heights). *Some* people who jumped out of windows in order to escape Roman persecution and became martyrs did so to enter heaven, and they returned to their relatives and told them about the pleasures of heaven. So, we obviously have to do a little more research about the two programmes before we can use my not jumping out of the window as an argument for anything ("anything goes" is a *theoretical* principle; it does not exclude lack of proliferation because of *fear*).

So now I am going to make myself some beefstew. I have stomach cramps because of too many pills.

Joan McKenna was here yesterday. All her dreams of being a millionairess within two years have collapsed; her collaborators have left her, some have stolen money from her, she may have cancer, is very ill, had to stop smoking [. . .] is very nervous, and now a waitress at an Irish pub. She is very lost, right now, poor thing. She sends her regards.

All the best and love to Gillian.

Paul

202. With this word play Feyerabend is referring to Joseph Sneed, who put forward the so-called "structural" conception of scientific theories (see Sneed 1971; Stegmüller [1973] 1976).

203. This is an argument which Lakatos uses in the course of his lectures to criticise Feyerabend's scepticism: if indeed "anything goes" for the sceptic and any theory is as true (or false) as any other, why does he *in practice* choose to use the better confirmed theory? i.e., why would he choose to use the elevator to descend from the Eiffel Tower instead of trying to land gently after having jumped over the railings? The same metaphor was later taken up again by John Worrall (1989a) in order to criticise the alleged solution given by Popper and Watkins to the problem of induction.

11 January 1974

Dear Imre,

[. . .] I am *almost* finished with the third world. This is the *most confused* thing I have ever read. "Autonomy" now means the one thing, now something else, then something else again, and all the time it is asserted with the same pompous air of infallibility. It makes one puke. The *theory* that is behind this disaster is most interesting, but the arguments—well, let's not talk about it (but, remember, I shall be careful to show in what sense the third world can be retained as a way of speaking, and in what sense its so-called autonomy corresponds to certain features of world 1. For there is something in world 1 that is knowledge and yet independent of our *momentary* whims, and that is: our *petrified* whims).

You say that Sir K just messed up Hume's problem. This is precisely what Schrödinger said, and I was there when he said it. It is a very interesting story. Karl wanted to dedicate the English edition of the *Logic of Sci. etc.* to Schrödinger. He had never given the book to Schrödinger to read and wanted to know, desperately, what he thought of it. Karl was sitting at the Böglerhof, Schrödinger was at another restaurant in Alpbach, in a very bad temper: "This Popper! There he gives me this confused book of his and wants me to consent to have my name on the first page. He says he does something about Hume's problem—but he doesn't, he just talks, and talks, and talks, and Hume's problem is still unsolved." So I tried to explain to him the difference between the problem of demarcation and the problem of induction. "Yes, yes," he said, "I know, he solves the one BUT HE DOESN'T SOLVE THE OTHER and that is just what Hume said, that it couldn't be solved . . ." etc. etc. I was running back and forth between Popper and Schrödinger telling the one what the other had said, it was most exhausting, and the whole story will occur in my autobiography. Of course, Schrödinger had drawn the lesson from Hume that science is a little irrational, and that agreed with his own inclinations—spirit of the age etc. etc.

Now—after I have finished with the third world part of *Objective Knowledge* I shall go on to the methodological part, and I shall be very brief there, for I have no patience left, but I shall certainly mention your appendix to the Induction paper[204] as well as his complete failure to reply to his critics despite his "Full Reply" promise.[205] And I shall maliciously discuss footnote whatever where he says that OF COURSE Newton is refutable and that Einstein had said that relativity is refutable, pointing out that he must mean you (evidence: the index), that he is wrong, and that Einstein *in his published writings* consistently said the opposite of what Popper attributes to him. Example: "It is rather strange that human beings are normally deaf to the strongest arguments while they are *always inclined to overestimate measuring accuracies*" (Born-Einstein letters). Finally, I shall say that Karl has often accused contemporary philosophers of a completely unhistorical attitude, a lack of sense of perspective (philosophy started with Wittgenstein, e.g.), but that he has given rise to exactly the same phenomenon. And in support I shall offer a series of quotations from Boltzmann which contain exactly the same evolutionary view of science *including logic and mathematics* (they may be modified as well as time goes on) which Popper, because of the Vienna Circle, first excluded from "his" critical philosophy, and it took you to put it back there.

My God—reading this book I have learned *nothing whatever,* and I think that is a very bad thing to say about a book (for example, there is nothing in it I can use for my lectures, or tell someone etc.). So be it. Stegmüller will be next, and *he* will be much easier, *because clearer.* I tell you, Stegmüller is OK. He may create his own disaster areas, but at least he looks around at other disaster areas, he tries to take you and Kuhn seriously, which Popper never did. His *results* may be rather puny, but he has the right attitude. A whole volume on Kuhn.

Anyway—Karen these days is very beautiful, quite happy, but seeing her always makes me sad for some reason, and I am saddest when she is happiest because it can all disappear in a puff.

Be well,

Paul

In two days I'll be 50, but I still feel rather stupid.

204. Lakatos 1974a.
205. See Popper 1974b, 967.

10 January 1974

Dear Paul,

I was absolutely shocked by your outrageous suggestion that you may not come to the Greek conference. It is absolutely out of the question.

Greece will be an ideal place for you to rest and you are not requested to come to the discussions at all, or to the lectures; if you just sit in the sun that will be perfect. But you must come. Noretta [Koertge] is invited anyway and she is coming.

The programme you sent me in which you figure with Velikovsky is a PSA [Philosophy of Science Association] conference, and I cannot really complain that they did not invite me since they invited me both in 1970 (with Kuhn) and in 1972 (with Toulmin). The only thing that amuses me is that Hilary Putnam reappears on the scene. He was not there in 1970 nor in 1972 when he obviously had more amusing things to do. Also, it puzzles me how Velikovsky is attached to Princeton University. [. . .]

[. . .] I do not know whether I told you that Peter Clark wrote a paper two years ago on the Velikovsky programme and I said that he should rewrite it. But I told him that he should first get his Ph.D. in a more respectable subject and now of course he writes on the superiority of the phenomenological theory over the kinetic one up to 1905. After that he will finish, with Urbach's help, his demonstration that Velikovsky's research programme is ludicrously degenerating; or, rather, that it does not even add up to a research programme.

I was amused by your suggestion that the scientific revolution was a revolution in the standards. This is of course the story which I encapsulated in the announced title: *The Changing Logic of Scientific Discovery.* A chapter of this book would have been my Newton paper in which I discuss this change in standards in detail.[206] The trouble was that Worrall and Zahar persuaded me that the standards which I ascribed to the seventeenth century were already there at the end of the sixteenth.[207] Bashi Sabra almost beat me up since according to him these standards were there in the age of Ptolemy.[208] Now your letter caused a conversion effect, and now I think that my Newton paper is perfectly correct and can be made consistent with the Copernicus paper, and now I am going to publish it before you get in.

You may be amused to know that according to my research assistant, and to your favorite, Greg Currie, the Copernicus paper is an unscholarly muddle; the first part is sloppy, the second part is knowingly false and dishonest; and the whole thing will be a lot of shame on the department's reputation. Also, according to him, you only praise the paper in order to undermine standards in our department. What do you think? Looking through the paper the other day I thought that he was right. On the other hand I hope that at least some people will not notice that he is right. Such is life.

I do not know what John Worrall thinks of the Copernicus paper; I think he did not see the very last version. He was certainly not very enthusiastic about it. Therefore I was not surprised when he announced yesterday that he withdraws his paper on Newtonian optics[209] which I scheduled for the June issue of the *BJPS.* However, a much improved version of the Urbach paper[210] will be published. Worrall said that he does not publish a paper which is not epoch-making; publishing papers which do not make an epoch is immoral, or so he says. [. . .]

Stegmüller: I should be most grateful if you wrote a review of his book[211] (I mean the second half of the second volume, which deals with you, Kuhn and me). The trouble is that for this you would have to look at Sneed. Do you have the energy and the courage? Of course, you may replace energy and courage by humour and wit. [. . .]

Be good.

With warm greetings.

Yours ever,

Imre

P.S. did I tell you that two weeks ago I was in Birmingham and in the discussion they asked me why I do not discuss critically your views but only explain what they are. I answered that I can prove them. And I did. However that will only appear in MAM.

P.P.S. What is your opinion of Herman Hesse's *The Glass Bead Game?*

206. The reader is already acquainted with the fact that "The Changing Logic of Scientific Discovery" was never written (see "Introduction: A Dialogue," note 63, this volume). Lakatos had, however, worked on the rough drafts of his paper on Newton since 1963–1964, coming back to them often since he claimed they were in need of thorough revision. Worrall and Currie edited the last version, "Newton's Effect on Scientific Standards" (Lakatos 1978a, chap. 5). See also appendix C, this volume.

207. On this topic, see Worrall 1988, 1991.

208. Bashi Sabra, historian of science and professor at Harvard, was among the first generation of Popper's students, along with Agassi, Wisdom, and Watkins (see Sabra 1967).

209. Worrall 1976.

210. Urbach 1974.

211. Lakatos is talking about Stegmüller's *Theorienstrukturen und Theoriendynamik* (1973). It was reviewed by Feyerabend for the *BJPS* in 1977 (Feyerabend 1977b).

17 January 1974

Dear Imre,

What a glorious long letter. And it comes lying on top of a glorious long letter by Rohan. What more can I want—except that having you both here in writing I now more than ever want to have you here in person. Rohan won't come for some time, but you, you must definitely come *in spring.* Visit Merton, this time I might even go with you and sleep in Merton's bed while you two sort out the Reasons Behind Everything (or, maybe, Merton has a TV and I can watch TV while you are talking. Also I might visit Susi Priess in Palo Alto while you are engaged in serious things). *Do come!*

Well, today I prepared myself for this letter: I got up at 8.30, had breakfast until 9, sang until 10.15, a full 75 minutes until I was bathed in sweat (those big Italian arias such as *Ella giammai m'amò* really take it out of you). And now: of course I am coming to Greece, maybe even with Karen or with Rohan. Both seem to be willing. I think I shall first come to London and leave with you, so that I don't get lost (nor do I want to get arrested in Greece as a spy). BUT: I won't be seen anywhere except at your talk on Copernicus. [. . .]

I did read Peter Clark's paper[212] on Velikovsky and told him he was a naive falsificationist. Also I found some of his arguments *very* questionable. For example, he attacks the way in which Vel. uses ancient documents—but every astronomer tries to get as much astronomical evidence out of old records! If Peter Clark ever takes up that topic again I would be very interested in discussing it with him. (I just gave a lecture on Velikovsky in my philosophy course as an example of the viciousness of the new scientific establishment and I also discussed Schliemann who made factual discoveries by taking Homer literally.) I remember Peter Clark's paper: he did not show any *degeneration,* but just waved the flag of naive falsificationism. Besides, I would be interested to see what he has to say about the new data collected by Mariner 5 (I believe) about Venus, Jupiter etc. etc. It is a *very* interesting case.

I am glad to have been the cause of your reconversion to the *Changing Logic of Scientific Discovery.* I knew about the title, but it never occurred to me that you meant what you now say it means—and now, of course, it is very obvious. I know that the situation is not quite so simple, historically, but you should never permit yourself to be beaten up by a mere historian, even if it is Bashi, who has some sizeable philosophical background: having been a historian for such a long time, most likely wiped out this background years ago. At any rate, it would be nice to have some discussion about this some time. Now, whatever the outcome of the discussion, you have a problem: in your Big Paper[213] you explained how research programmes change rationally *given some standards.* Nowhere do I see an explanation as to how standards change rationally *(given what?).*

I have tried to draw a very primitive sketch of such an explanation in my *Beschränkte Gültigkeit,*[214] later on in section 6 [of AM] when talking about what I call the "cosmological criticism of methodological rules"—see also the drawing in that part of section 6. Now, I am very curious: is this similar to what you say in your Newton paper,[215] or do you there just *describe* the change without discussing standards for judging its rationality? You never showed me the Newton paper, you always kept it from me, why don't you send me a copy? My theory is that one wants to find standards which lead to good results ("good" *in terms of the standards*) *in the physical world* described by the best hypotheses favoured by the standards. The same thing is discussed at some length in my Kiel paper which will be out any time now,[216] and also in my chapter on you in AM.[217] Why don't you just add a footnote to your Copernicus paper to the effect that the argument *assumes* that the Copernicans had the same standards which *you* recommend (essential part: novel predictions), that you have had reason to doubt this assumption (refer to your Newton paper), but that top historians have assured you that the assumption is correct, so you just make it for the time being, and refer that problem to a future paper. I think you should add some such footnote. You should also add a footnote that you are discussing kinematics only and that you are aware of the difficulties of dynamics etc. etc., however you want to formulate this.

Now, your Copernicus paper: it is written in an easy style and what Greg [Currie] wants, probably, is more constipation. I didn't find the paper sloppy, I found it was written very elegantly with the left hand and I liked it for that reason: all the arguments are crystal clear, *brief,* that is the main advantage, straightforward in a field where people heap epicycle on epicycle. If you feel uneasy, don't mess up the text, except by perhaps making it even shorter, but add lots of learned *footnotes,* or *appendices.* My only recommendation for the text would be to make it shorter. Maybe we can both work on it in fall, when I am in England. And always remember that you *don't* write your papers for your Mafia, you write them for *the public at large* and so the standards of the Mafia don't apply. Down with Worrall for being so constipated (not publishing his paper—why, does he want to write for eternity, or does he want to cheer up a few of his contemporaries? Most likely his constipation is the fault of his girlfriend—it always is). Besides—what does "epoch-making" mean? Epoch-making *with whom?* And why does he want to make an epoch rather than some individuals he knows? Or does he think that epochs are more important than the few people who are looking forward to having his paper in their hands? Not a very humanitarian attitude (does he sleep with an epoch, or does he sleep with a girl? Maybe the answer to that may be the answer to the other question as well). Anyway, let him know my opinion.

Stegmüller: I really respect him. To take on a trio like US (Kuhn, you, me)—that takes guts, and I'll probably review him. Don't send me his book, I already have it. I shall *not* study Sneed except for what Stegmüller says about him in the book, but I shall certainly show that we cannot be Sneedified. Incidentally, I just hear that all of us (K, you and I) are in the Munich University catalogue printed in Stegmüller's course.

Thanks for proving me in Birmingham. I now feel like Brouwer must have felt when the first formalisations of "intuitionism" came out. Anarchism proved! Well, that happens when one gets 50. But at least it is good propaganda. Hume was absolutely right: reason is the slave of the passions which means that rationalism is a pipedream unless firmly glued to certain passions (this is also what Hume says). So, without fear, propaganda, etc., no rationalism, and no persistence of rationalism. I just read a little of the *Treatise* of Hume—a fantastic book (if you omit the sense-data).

I am finishing up Popper. I have dealt with all his "arguments" for the third world and have arrived at the last one: the infinite number-series. This, of course, opens up the nightmare of formalism vs intuitionism vs Cantorianism vs psychologism etc. and I just have to find a way of getting out of it. At any rate, Popper has no *new* arguments in these issues, all he has are *appeals to old prejudices*—but it has to be said right. I am now on page 15 (1 and 1/2 spaced, typewritten) and the third world is slowly dying. As a matter of fact, it was never there in the sense of Popper (Platonic heaven). But of course, there is a difference between momentary whims and more lasting rules, just as there is a difference between afterimages and the more stable visual impressions one gets when looking at a table, for example. This does not mean that the stable perceptual image of the table is in a third heaven while the afterimage is in the mind. Both are in the mind, but the one is more stable than the other (in the case of numbers, I have found, this is the view of Helmholtz and Weyl and there is not a single argument in Sir K that shows that they are wrong!). Anyway, I shall be VERY happy when this nightmare is over.

[. . .] Funny you should ask me about the *Glasperlenspiel.* I am just reading it and I like it. I am also reading Feuchtwanger, who is one of my favourite writers (*Hässliche Herzogin; Jud Süss* which the Nazis made into a movie).

So, now back to Karl and *numbers.*

Love to Gillian.

Paul

212. This article by Peter Clark was never to be published.
213. Lakatos 1970.
214. Feyerabend 1972.
215. Lakatos 1963–1964b.
216. Feyerabend 1973.
217. Feyerabend 1975a, chap. 16.

17 January 1974

Dear Paul,

[. . .] I do not believe that you wasted your time on *Objective Knowledge.* Proof: since you invented the (TRUE) idea that I use the third world intentionally, I understand myself much better and I solemnly promise to develop this beautiful idea of yours. I shall also embed it in my philosophy of mathematics.

[. . .] Love,

Imre

January 1974

Dear Imre,

So few days have gone by, I have become more depressed, then less, then more. Sir K has not helped. That [*Objective Knowledge*] is bound to be the most badly argued book around and I have a hard time making sense of it. All his arguments for autonomy are circular, *if* they are arguments for autonomy, that is. Most of the time he argues for the reality of a certain domain, and calls *that* autonomy. The reality of World 2, for example, is perfectly compatible with materialism.

Outside it is snowing. I think I shall come to Greece, after all; it would be silly not to, nice countryside, paid fare and, mostly, pleasant company (I am thinking of Gillian, Gretl etc., not of you menfolk). So, keep me on the list. All the best, and wish me luck for the next two months—lots of lectures!

Paul

15 January 1974
Berkeley

Dear Imre,

Do not despair when you suddenly feel 127 years old. I have felt I was 352 years old, and I have recovered. Besides, you have more energy than any one of your young revolutionaries, such as uncle Elie, and you even have more energy than all of them taken together. What makes you feel old is just the general situation of the world, but *that* was never very good. I am now made to feel old facing a new generation in my classes, all of them law-and-order and not because they have thought things through, but almost as if they had been born that way. And the same with my dear colleagues etc. etc. I'd really feel like a creature from another age, or from another planet, were it not for the fact that there are a few people who think like me, such as you, Anita, Susan, Karen, Rohan etc. etc.

But this is how things always are except for those brief periods when the masses, temporarily dragged along by some outer influence, seem to show an ounce of reason. But it is not reason, it is imitation, and it disappears like a cloud on a summer's day (and by "masses" I now mean the intellectual masses and their followers—about the rest I don't want to say anything; but it does not look much better). How I am looking forward to retirement! 15 more lousy years—but I may be able to make it shorter IF I get that Distinguished Professorship in Delaware, IF they pay me a lot there, and IF I don't have to do too much teaching there.

[. . .] Incidentally, don't forget your plan to get me to the LSE via Dahrendorf, if that is at all possible, even if only for a term, or two, or perhaps *half time:* two terms at the LSE one year, one term the next year, and so on. It would be OK with the people here, they would only be too happy to get rid of me half time and I could still get my British salary without British deductions, being a resident of the USA. And, most importantly, I would be close to you.

Anita has left for a hike in Yosemite, Karen has left for a hike in Malibu Beach with her new boyfriend, she (Karen) may come with me to Nauplion, at least for a day, maybe more, send me the last version of Urbach for her to read, you have no idea how intelligent this woman is and yet her intelligence is not one of the competitive type, it is not nasty, I haven't ever seen such a thing. Send me also Worrall, that bastard, where is his Newton? Where is it? There are now about 100 guys in my philosophy of science class, all of them are reading your *Criticism,* I have forbidden them to read mine, they come to me and ask: who is Lakatos? Has he written anything else? What has he written? So I give them the *Proofs and Refutations* article and I shall discuss the Copernicus article with them *at length.* It bores me to preach anarchism, I have given it up, but apart from anarchism there is nobody else to talk about, only you and your mafia. In this course I have lots of science students and *they* take to you like a fish to water (it is different with the philosophy students who have been corrupted by Quine, Davidson etc.; incidentally, I have now discovered that I said everything Quine is famous for, such as radical translation, much more briefly and with much better arguments in 1958, in my Aristotelian Society paper).

Popper is approaching the end and I wish I had never taken him up. My main criticism: having read the book and having criticised it *I have not learned anything.* You can even criticise Popper 1935 *very briefly*—and you have done that. To criticise *you* takes much more thought. And that is what makes you (and Hegel) better than Popper.

Well, I have to cook my beefstew, for today I am staying at home and have to make my own lunch. I shall come to Sussex for *two* terms only (that is OK with Asa Briggs) for if I decide to go there forever I want to make a few more dollars before I leave, so that I can live comfortably.

All the best, give my love to Gillian that poor girl, I hope you are as good to her as she is to you and as forgiving, though I am sure there is more she has to forgive you than you to forgive her. If you feel up to it, I would be *very* curious to know what you think of my two criticisms of your Cop. paper, viz. (1) dynamics and (2) change of method. Just a tiny little line, or let Sir E do the answering. I am bursting with curiosity.

And now I have to start lunch, and write millions of recommendations.

Paul

18 January 1974

Dear Paul,

Congratulations for your fiftieth birthday which I am afraid I missed by a couple of days. I suppose it is on 13th January, isn't it? You amused me no end with your story about Schrödinger. I am terribly amused because if Schrödinger could see that while Karl gives a conventionalist demarcation criterion he does not even touch the problem of induction, then how on earth could he have confused the learned world for thirty-five years? Today everybody, or nearly everybody, believes that Karl Popper has solved the problem of induction or is at least somebody very important in the history of the problem. Of course, you and me and Dick Popkin know that the problem is insoluble and all one can do is formulate it and sit back. And then of course comes the Festinger effect. You have obviously read his book: *When Prophecy Fails* [1956]. This is what happened to Karl: his prophecy failed and therefore he started to convert people to the failed doctrine. And how successful he was!

I just received a letter from Professor Owen Gingerich. I enclose a copy of his letter. He is like Machamer. Alas, a historian can never win a point against philosophers. Of course I will have to change a few sentences but that is all. Of course I am grateful to him that he helped to clarify a few points. I only hope that Rosen will do the same. I heard from Rosen that Galileo's observation of the phases of Venus was sure to be inconsistent with Ptolemy only a good twenty years after his discovery. However, I still cannot find the reference, but I shall find it and it will be there in the final version of the paper. I have no doubt that you will make capital out of it. But you need not worry, there will be a footnote pointing out that this looks very much like a point scored for Feyerabend. But of course, I shall say, the reader has only to think a little to see that it is *against* Feyerabend. That is all I shall say, and the reader who will think and will not see that I am right will be very much ashamed and will not tell anybody.

It was something of a relief that you changed your mind about the

Greek conference but I seriously never expected that you would let down Hans [Albert], Kürt [Hübner], myself, my wretched gang and of course John Watkins and Micky. Also Gillian would not speak to you again. You will soon get a report about a committee meeting which we had the other day without the foreign members. It was held under candlelight since all use of electricity is banned here. Hampstead is full of horse carriages because there is no petrol.

With warm greetings.

Yours ever,

Imre

24 January 1974

Dear Imre,

Here, at last, the End of the third world as an autonomous entity. As a figure of speech, however, it will continue forever and all the things you want to say with the help of this figure of speech remain OK as well. If you think that no major review is necessary, please send the review on to Alastair Hannay, editor, *Inquiry,* whatever the address. If you think I ought to make major changes, let me know. You will not see remarks about the rest of the book. Reason: I just lost my patience. I had lots of notes on falsificationism, on his so-called reply to his critics, and so on, but I decided that, for the time being, I had done enough for (or against) Sir K. [. . .]

Today is an absolutely clear day, you can see all the way to the airport. Yesterday, Susi (not Sus*an*) paid me a visit and we spent the whole afternoon going from one bar to another. Needless to say, I am not very much awake today. Returning home I put the last touches to Sir K and he looked almost human after all the drinks. I am also hoarse from too much talking and too much singing. Did I tell you that I bought myself records with accompaniment but no voice (called *Music Minus One*) and that I am now singing big arias all day? I think my voice is rising so, after a few weeks of exercises I might be a baritone again and sing all these arias of vendetta. In my classes I am talking about Aristotle's poetics in the one, and about chaos in science in the other, with Lakatos-Musgrave giving a firm background of Reason. Two more months, and this quarter is over. You don't know how lucky you are. No lectures, but I: two lectures every Monday, Wednesday and Friday and then, in fall, in Brighton, *12 hours* of tutorials per week. Well, I'll then have you and Rohan to console me.

All the best,

Paul

25 January 1974

Dear Imre,

There seems to be telepathy working between us. You hardly ever write, or ask questions, about novels. I start reading *Glasperlenspiel,* your next letter is about it. You hardly ever write, or ask questions, about psychology. I give lectures on Festinger, your next letter is about him. Pretty embarrassing for a rationalist, I must say, unless inherited, in which case it falls under the Urbach-rationality. Yes, the spread of Popperianism is a strange phenomenon but not strange enough to make me review the two volumes of Schilpp [1974]. Do you really think I want to read through autobiography, self justification, attacks, further self justification and so on? After *Objective Knowledge* almost killed me? Not on your life! His book on physics—maybe, especially as I have already written about the quantum theory contained therein and all I need to do is review my views. But the entire two volumes? No. Stegmüller is another thing altogether, I am having a look at him right now, and I might review him. It can't be too difficult. Incidentally, what you say about Popper *re* induction I can say about him *re* quantum theory: the interpretation he attacks is entirely in his own mind and the solution he suggests has been refuted long ago, in about 1926. Nobody knows that, for nobody goes back more than 1 and ½ years when judging a view. In this respect Popperianism is very much like Wittgensteinianism: in both cases philosophy seems to start with the master, and what exists before him is what he says exists before him. Anyway. This is a good thing to *forget,* and not to dwell upon.

I was very much amused by Gingerich's letter. "I read it as a historian of science rather than as a philosopher of science," he says, which I translate: "I read it as an unthinking person . . . ." And then he tells you about his "gut feeling." His so-called technical points don't impress me at all even if they were right which I don't know. I long ago had some correspondence with him and I asked him how the brightness of stars was determined in Galileo's time. I couldn't get any chswer out of him, for all he wrote was how Galileo *predicted* the brightness of stars. But Bashi Sabra is very impressed by him, and that is why I wrote to him in the first place. I got all his offprints on Cop. and Kep. and I shall write you an "evaluation" of his letter next time.

I am curious to know what you think about my anti-third-world paper. The part of it that is of interest for you: theories, mathematics etc. are all objective but they are not autonomous in Popper's sense. What Popper can show at the very most is objectivity, what he claims to show is autonomy which is *non-physical* objectivity. *This* he has not shown, though he insinuates he has. Altogether, the book is a MESS.

And now I have to run.

All the best

Paul

P.S. Most likely I shall bring Karen to Greece. We shall attend *your* lecture, but then we shall disappear never to be seen again (except at meals).

29 January 1974

Dear Paul,

Thank you very much for your letter and for your paper on Popper. First, let me clarify half a mystery. You say that "there seems to be telepathy between us." You say that you are dealing with Festinger and my next letter is about him. Then you start reading *Glasperlenspiel,* and then my next letter is about it. Alas you are wrong. My letters are as a matter of fact joint letters from me and from Wendy Boulter, my secretary. The question about *Glasperlenspiel* was inserted by her without my knowledge. I only noticed it in the second copy, as I occasionally notice that she adds, detracts and alters the letters, and on the whole improves them a lot. So the telepathy seems to be working between the three of us. Anyway, write to us about *Glasperlenspiel* when you have finished it. You might even persuade me, jointly with Wendy, to read it myself. On the other hand, as far as Festinger is concerned, you obviously seemed to have overlooked Latsis's paper on "Situational Determinism,"[218] a copy of which I enclose. Popper, in his chapter on the Autonomy of Sociology, in which he takes Karl Marx's view, takes a very strong stand against what he calls "psychologism" in the social sciences. Latsis attacked this position and John Watkins's elaboration of it long ago. It was Latsis who persuaded me that Popper underestimates the importance of the Second World simply because he has a sub-conscious resistance to psychology. What Latsis baptised "behaviourism" is nothing else but psychology in economics, which Marx and Popper forbid.

Now a remark concerning Clark's paper on Velikovsky. I was very dissatisfied with the paper as you have known it, and as a consequence Clark has now made Urbach promise to help him in technical details, and as soon as they have both finished their present work, they will turn jointly to Velikovsky and finally delete him from the Third World. That will come sometime in mid-1975.

I am just asking Wendy to prepare a Xerox copy of your letter of 17th January in which you recommend John Worrall to sleep with his girlfriend rather than with an epoch. I suppose he sleeps with both although I find Jennifer prettier than our epoch.

Incidentally, I am dictating this letter at home and sending it with the help of Gillian and Greg to Wendy for typing.[219] I am very ill and dying. I have a sort of bad flu which I did not think I could get since I have obediently been taking a gramme of vitamin C a day for ten years. According to recent reports, even if vitamin C does not completely eliminate these sorts of things, at least it reduces their length: we shall see.

Latsis visited me yesterday and explained to me what messes there are in Karl's *Open Society* and in his *Poverty of Historicism,*[220] with special reference to his so-called "Rationality Principle." He is writing it out now and I think it will be extremely good.[221]

Now let me come to your paper. As I am reading it I shall take this tape recorder and talk into it. The whole thing therefore may be very patchy but might be of some help.

First, your paper starts on a low key. Is not the key very low? You might give some thought to whether you do not want to put some excitement on page 1 which will help the reader to carry on with more interest. Also I have a small semantic complaint: I think "metaphysics" is a better word than "ontology." But words do not matter much.

On page 3 you claim that "Popper isolates three different types of entities." But surely even Popper himself acknowledges that this Holy Trinity comes from Bolzano and Frege. I have not seen it in Bolzano, but I certainly saw it literally, and I think more concisely, in Frege. And also I saw it in Planck. Popper, although he gives references to his predecessors, is extremely parsimonious with his acknowledgements. I think that your review should clearly make the point that the basic idea is not originally Karl Popper's.

On page 6 your third comment is not very good. It is true that Popper assumes the autonomy of the Third World. But it is a *non sequitur* when you say that therefore it is in need of further arguments which *establish* that autonomy (my italics). Surely you can assume the autonomy of the Third World and show that this assumption has great explanatory power without previously "establishing" it? I do not think that you mean here what you wrote and I think that you should improve it so that some vulgar Popperian should not be able to hit back. The real question is whether the assumption of the autonomy of the Third World can figure in the hard core of a progressive research programme or not. You do not have to "establish" the hard core before starting the programme. But surely you know this very well.

I see on page 6 at the end that I have arrived at section 5 of your paper. Until now I did not really notice the numbers and now I have to go back to see what is in the first four sections. I strongly disapprove of your not putting a table of contents at the beginning of your paper as I always do. It makes it much easier for the reader to follow your arguments.

I think that your starred footnote on page 6 should go into the text. I think you should say clearly and loudly (as I shall say at the next possible opportunity), that Popper seems to cultivate rhetoric deliberately, partly to cover up the weakness of his argument and partly, I suppose, for moral reasons, to give the impression that he has an argument when he has none. This is characteristic of his whole life's work. Most of his famous theses are slogans repeated again and again, until the propaganda has worn down the reader and he has completely lost his sense and appreciation for the obvious counter arguments. The starred footnote on page 7 should also come into the text and reinforce it. If you put these two long footnotes into the text you will reinforce the reader's impression that Popper is a very considerable phoney. Do not forget that one of the main points you wish to get through to the reader is that the Great Philosopher is a Göbbels of Reason.

I already mentioned that you do not put this idea of the Third World at all in historical perspective. I do not know Plato sufficiently in order to tell you whether it goes back to Plato or not, but it is a very long-standing tradition. It certainly goes back as a tradition to Leibniz (as I always show in my lectures). I mention this again at this point because on pages 7 and 8 you describe how Popper's addiction to human freedom leads him to complete conceptual chaos, which he abhors because he feels that it is so important from the moral point of view to persuade people that they are free that the poverty of the arguments for this freedom should be hidden. Now I am sure that this tradition has a long history, and for instance Marx's greatest weakness (or at least one of his greatest weaknesses) was that he advocated both historical determinism and then human freedom. For his whole theory it was vital to encourage the working class by showing that they are supported by historical necessity; on the other hand he was afraid they if they take this argument really seriously they will find out that they do not have to make a revolution and suffer for it, since the revolution will come anyway. Therefore Marx is for historical necessity on even pages and for freedom of the will on odd pages, so that revolutionaries should know that without their efforts historical determinism will stop working. This is of course a classical inconsistency in Marx but unfortunately I do not know the literature concerning it. I have no doubt that many people must have discovered this.

I would strengthen your argument at the end of paragraph 2 on page 8. I think you should add that according to the materialists we are miraculously perfect computers which come about once in a trillion years. Once the programme is there one can go on with the miracle, but once the initial momentum gets lost then "there will be no re-emergence of our civilisation for many millennia." This surely used to be Karl's view. Am I not right? On the other hand, you have to say something about how finding again books, libraries, knowledge, acts as the stumbling block of a bad computer on a good programme. The analogy is far from obvious.

On page 9 you do not exploit Popper's idiotic weakness concerning the question whether there is a Third World in the animal kingdom or whether there is none. Pages 116 and 118 in his *Objective Knowledge* are a load of confused knowledge in order to prevaricate on this problem. Cannot you mobilise your humour a bit more than you have done on pages 8 and 9 concerning this point? You may remember that I quote in my paper on Popper in the Schilpp volume[222] his idiotic statement in *Conjectures and Refutations* that amoebas conjecture and refute like men, indeed, like Einstein. I make the point that trial and error can be on the level of stimulus-response. A large research programme is more difficult to explain on the basis of behaviourism.

On page 11 you make a grave psychological error. You say in your third line that "Popper seems to believe" that "abstractness is not sufficient for ejecting an entity from the physical universe." But Popper does not believe that a word can stand for an argument, what he does is to abbreviate "ejectable from the physical universe" into "abstract." This is not an argument but propaganda based on the emotional connotations of "physical universe" versus the emotional connotations of "abstractness." Incidentally, I think I am right in believing that on page 11, line 2, the word "not" should be deleted. However, I am afraid that here it is you that gets into confusion. I, for one, do not think that energy and spin parity belong to the First World. This is a subtle problem which I am afraid you treat a bit unsubtly. Alas, I am unhappy about your pages 11 and 12.

Alas, at the end of page 12 I get to your section 6. Now let me think, what was really the main point of your section 5 as opposed to your section 4 as opposed to your section 3? You have to provide the reader with subtitles. I think that you should give some more concrete and more convincing propaganda for your idea that abstract entities can well be part of the First World. You should drag in here neurons, or some other First World entities which at least will give some more plausibility to your arguments on pages 11–13.

Come off it: on page 13 you really get into trouble. The "Dogmas of Christology" (don't you mean the "Dogmas of Christianity"?) are not irreconcilable with the physical world. In the physical world there are facts, and facts are always irreconcilable. Or so it is once the linguistic theory of logic became a progressive research programme and therefore its hard core attained psychological credibility. (Note: this last remark is one of my discoveries and if you think it makes any sense, then please give me a little footnote.) Deductive logic and its linguistic theory has become a victorious, progressive research programme from Bolzano to Tarski, while inductive logic has remained degenerating. According to me, it is the progressive research programme of the linguistic theory of logic which really created the Third World. For Karl, the Third World is vital in order to uphold his escapism from the problem of free will versus determinism. More of this will be in Latsis's forthcoming paper on Popperian rationality. (Whether the development and progress of the research programme of arithmetic has also contributed to the belief in the Third World is a question to which I have not yet been able to give sufficient attention.)

I am now on page 20 and I feel a bit out of my depth when it comes to philosophy of mathematics nowadays but I thought that, under separate cover, I would send you a paper which concludes that arithmetic can after all be shown to be false if the physical world is one sort of structure rather than another. I submitted this paper to Mary Hesse in 1966, she accepted it for publication in *BJPS,* and then I withdrew it because I was unhappy about it. So now I am sending you a copy to see what you think about it. Incidentally, I sent a copy of this paper in 1966 to Goodstein, the intuitionist, who got very upset and without asking whether I had in fact published it wrote against it a naughty discussion note in *Dialectica,* I think in 1967 or 1968.[223] It was a silly little note.

I am just about to just start on your page 21 but I feel a bit frustrated. You built up an argument and then you missed it out. The question is this: if not for mathematics, would Popper abandon the Third World? *How much does this argument depend on the existence of Bolzano-Tarski logic and of arithmetic?* I think Popper should be asked this question. I certainly think that I have shown that the conceptual world of mathematics is inextricably involved in the conceptual framework of physics, and the conceptual framework of physics is no doubt interdependent with reality. Of course, Popper does not deny that the three worlds are interconnected but he insinuates it all the time. The question is: does he believe that mathematics is a world of its own, independently of the First and Second World, and if so: what are his arguments? You touch on these problems but you do not thrash them out, to my mind, sufficiently. On page 21 I think you are rather facile in attributing to Popper the view that the human mind reacts passively rather than proceeds actively when getting to know the world (the First World); do not forget that Karl always laid stress on his activist epistemology.

My God, this section 7 is terribly long compared with the others.

Now that I have come to your alternative theory on pages 23/24/25; I am getting tired and my headache increases. I do not attribute this increase to your paper, in fact your paper distracted my attention considerably from my illness. I think that at this stage I have to start reading the paper again and then I shall write to you again.

My final comment is that the paper seems to be very good. However, you have to introduce sub-headings. You have to emphasise and separate from others, as much as possible, problems like: Can Popper's addiction to uphold both scientific determinism and the doctrine of free will be saved except for this device of the Third World? (I am convinced that this was Karl's main motive: he invented the Third World to save both determinism and free will.) If I remember well, this was emphasised in Watkins: "The Unity of Popper's Philosophy," which he wrote for the Schilpp volume.[224] I am sure you have a copy but to be on the safe side I am asking Wendy to send you one under separate cover. Then another problem must clearly be separated, namely, how much philosophy of mathematics depends on the Third World. This is a separate question.

Complete collapse. I am going to bed.

Yours,

Imre

P.S. Last effort: What about forgetting about this paper for a couple of weeks or a month and then rewriting it more crisply and with clear sub-headings?

218. Latsis 1972.
219. From here to the end of the paragraph the writing has been lined through with a pen, but remains perfectly legible.
220. See Popper [1945] 1966 and Popper 1957a respectively.
221. See also Latsis 1983.
222. Lakatos 1974a.
223. See Goodstein 1969.
224. Watkins 1974b.

Dear Paul,

Two nights ago I had terrible pains for three hours. Gillian got me a doctor who found me unconscious in the bathroom, where I fainted and knocked my forehead into two, and now at least I know that the two hemispheres, contrary to recent journalism, are equal. I was taken in an ambulance into hospital with a heart attack ('cardiac arrest').

In this modern hospital one is first seen by a sociologist & psychologist. They decided that I am completely Anglicised & my heart has therefore vanished. Then another chap took over who was an industrial psychologist. He discovered that for twenty years I have had no time to breathe, therefore no lung. (I forgot to say that the second hypotheses was pleurisy.)

Now I am here as a Great Lump of myalgic (= muscular) pain and I scream all the time. I asked for Beuler[225] & Freud's early works to read but I only got Trollope.

Didn't say that when I fainted Gillian fainted too and so she too was taken to the hospital with the ambulance. But she was let out as unworthy of attention. Walked home (no taxi).

I am now waiting for a Top Conference to decide whether a patient who, though clearly dying, terrorises the ward, should be discharged for Reason Unknown, or whether they should put me under glass and *observation* (!).

Little do they know that this is a free and antinductivist country and I shall leave tomorrow on foot. But tonight I wanted to see three more chaps die. Very funny how people come in and die.

All for now. I don't know whether Wendy has typed out my long letter to you. I instructed Greg that in case of my death your MS should be sent to Hannay with a POSTSCRIPT FROM THE THIRD WORLD by Imre Lakatos. I shall send it from there.

This is my first letter since I fainted away. You cannot complain, however do not spread the news because should I recover I do not want a Sick Image.

Tell John Searle that I'm asking for a salary raise since I now stop poisoning the students' Mind. All my lectures have been cancelled till 1975.

Yours fondly

Imre

31.1.74 5pm

PS I am still in the hospital but dishonourably discharged myself with Unknown Serious Illness. The doctors were very amused by my statement that beyond electrocardiogram and X-rays I prefer witchcraft. They will all now read *Against Method.* Be good, and don't go near hospitals.

225. This is likely a misspelling of Breuer.

* * *

Imre Lakatos died two days later. This is how Feyerabend later remembered him:

> Imre and I [. . .] differed in outlook, character, and ambition; yet we became really good friends. I was devastated and quite angry when I heard that Imre had died. "How can you do that to me?" I shouted at his shade. [. . .] I always felt that Imre's rationalism was not a matter of personal conviction but a political instrument that he would use or put aside as the situation demanded. He did have a sense of perspective. He genuinely admired Popper and wanted to form a movement around Popper's philosophy. Eventually he became disillusioned. [. . .]
>
> I still miss this outrageous, sensitive, ruthless, self-mocking but very humane individual. (Feyerabend, 1995, 130)

FOUR

Appendices

Appendix A: On Rearing Scholars

Imre Lakatos

Lakatos's enquiries into science, mathematics, history, and method were always related to pedagogical and political issues. (Even his *Proofs and Refutations* may be regarded as a pedagogical treatise which describes a better way of teaching mathematics through the recognition of its dialectical pattern.) This is especially true of some of his contributions to Hungarian literary and academic journals in the fifties.

In 1956 Lakatos was co-author of the Declaration of the National Committee of the Hungarian Academy of Science, which called for "the freedom of science from political and moral pressure" and in particular for "the freedom of the Hungarian scientific life from its Stalinist shackles" (Archive, 1.10). In the same period, during a discussion with friends, he is reported to have lost his temper when the question turned to defending Marxism: "You are talking about scientific method, why do you keep calling it Marxism?" Instead of preaching the dogmas of Communist orthodoxy, Lakatos claimed an active role for *dialectic* as an instrument of criticism, rather than as a rhetorical figure for empty scholasticism. This appeal to dialectic should be considered in connection with the prevalence of vulgar Marxism in organised working-class movements and of mechanistic materialism in an age of totalitarian systems which have threatened to destroy mankind. Hence, asserting the validity of dialectical rationality was, according to Lakatos, an attempt, on behalf of detractors who had not grasped the point of Hegel's logic, to condemn all forms of irrationality and decadence. In his defence of "dialectical rationality" as opposed to "irrationalistic mystification" Lakatos was probably influenced by György Lukács. Following this line of argument, *history* is not brought into the picture to 'explain' the necessary realisation of present-day society and, therefore, to vindicate the status quo, but rather in order to recognise that knowledge is fallible (for "no scientific theory, no theorem can conclude anything finally in the history of science"; see below).

Though he chose to leave his country for Cambridge following the uprising in late 1956, Lakatos would never completely turn his back on the teachings by which he had been brought up. Instead, he took with him the 'forbidden brew' of Hegelian-Marxist dialectic and made use of it in a creative way.

In "Tudományra Nevelésröl" ("On Rearing Scholars"; Archive, 1.9)—here published for the first time in the English translation by Ninon Leader—Lakatos argues in favour of a critical attitude, the abolition of censorship, and science as guide to the Party instead of the other way round. He delivered this impassioned speech at the Petöfi Circle pedagogy meeting in 1956. In it, he deals with the problems of educating a new generation of scholars by claiming a role for talent, curiosity, original thinking, autonomy, the right to doubt and dissent, demand for proofs, and respect for facts. Curiously enough, about twelve years later, in his letter to the director of the LSE (1978b, chap. 12), Lakatos was to argue against students' demands, supporting the principle of academic autonomy, along similar lines.

The problems concerning the education of a new generation of scholars have been discussed on many occasions recently. It seems that Eötvos College is about to be resuscitated, that research students will be better cared for. I am not going to comment on these facts directly; rather, I should like to raise some matters of principle, principles of vital importance for scientific education, principles which must be discussed and clarified, otherwise, I believe, we can expect only halfway measures which can be retracted any time.

The *first* question is that of *talent,* its sociological role and evaluation.

In *Golgotha,* a novel by Aleksey Tolstoy, a young Communist activist called Sarigin is doing his propaganda work among sailors. He is explaining that revolution is science. A sailor called Latiugin remarks that one has to have a talent for that. Whereupon the young propagandist rebuffs him: "You see, comrades, how Latiugin here is mistaken? Talent is a feature in us which is dangerous. It could easily lead us into bourgeois anarchism. Into individualism. . . . Let there be no more word about talent. It is a secondary thing. At the end of everything we can amuse ourselves with things like that, those of us, that is, who will live through all this."

Let us now pick up *Psychology* by Tieplov. This is a textbook in use at our universities. On page 158 we read: "In the conditions existing in the Soviet Union, the division of people into 'talented' and 'untalented' makes no sense. Nikolai Ostrovsky was right in saying: 'In our country only loafers are untalented. Because they do not wish to be talented.'"

There are two important points made in the above extracts: (1) Talent is dangerous for society because it can lead to bourgeois anarchism; (2) in the conditions presently existing in socialist countries, education and schooling are such overwhelmingly important factors compared with inborn or spontaneously developed talent that the latter can be *totally* neglected.

Let us remark first of all that the second view has some progressive historical roots. We find similar thoughts in the pedagogy of the Enlightenment, which directed its criticism against the feudal system's prerogatives by birth. For example, Helvetius, as we know, held that every human being is born with the same talent. According to him, the inequalities of minds are the inequalities of education.

Let us also bear in mind that there were several bourgeois ideological trends which used talent to justify class privileges and exploitation. In Fascist ideology, the Übermensch and Herrenvolk played an important, murderous role. Thus we can hardly deny our sympathy towards Aleksey Tolstoy's Sarigin and Nikolai Ostrovsky. Yet their romantic overstatements do not lack danger. In the history of revolutions it has often occurred that even the best ideas come to be primitively exaggerated and distorted. (I note here, by the way, that it would be interesting one day to examine and draw some general conclusions from the sectarian distortions of revolutions, right through from the Hussite revolution and the English and French bourgeois revolutions to the Russian revolution.)

In sectarian utopianism and demagogy, as we have all witnessed it in the past years, talent is related to the enemies of the people; on the one hand, the individual effect, sphere of influence or school of a talent are seen as clique systems, while on the other it is expected that no matter who is chosen at random to be cast into the machinery of state and party education, at the end of the assembly line out will pop the suitable leading personality, artist, scholar, scientist or politician—as ordered.

This is how those characters were produced, of whom Károly Csendes, former Deputy Head of the Party's Cadre Division, wrote in the July edition of *Party Life:* "We have to drag these comrades as some superfluous but expensive burden because, although they are in responsible positions, they are not competent to do anything."

A counter-selection has been going on for years at a national scale on this basis. Talented, courageous men of initiative were pushed more and more into the background of so-called "simple, colourless, decent, disciplined" men. When a post had to be filled or a prize given, it was always the latter type who moved up a rung, while the former moved down one. At the same time, when it came to sacking or even arresting someone, the same selective principles were at work, only this time operating in the opposite direction.

Could I be exaggerating? Hardly. We can continue reading from the article quoted above: "the requirement that the comrade who is appointed to a post should be talented became increasingly neglected. . . . We allowed untalented, incapable, ambitious lower-middle-class elements to occupy responsible posts."

But let me come back to the subject of bringing up scholars. Research programmes were reduced to substitutes for university training, universities to substitutes for secondary schools. Their main concern was to wrestle with the terror of the excessive number of drop-outs; to concentrate on the below-average student. No one cared about the development of the gifted student, no worthy task tested his prowess, no challenging demand was put to him to help him unfold. He could never learn what his capabilities were. Even his imagination was seldom fed by visits to foreign countries, by travelling, by contacts with different cultures. The romanticism which was later to characterise people such as Körösi Csomas, Lajos Csomas, Lajos Biros, Germanus, was restricted to 'party frames'. The feverish talent who did his research while starving no longer enflamed the 'cadre' of the evolving scientific phalanx: his lifetime profession had been decided for him by irrevocable orders, and if he happened to be dedicated to it this was only to be entered in his personal file as a tendency to bourgeois individualism.

This anti-giftedness view was apparent in secondary schools, too. No doubt it is important that the teacher give extra help to slow pupils, but it is, I believe, no less important that he give extra help also to gifted pupils. Yet the teacher who did this was branded by the comment that he surrounded himself with 'an élite clique', which was frequently qualified with the adjectives 'intellectual' or 'bourgeois'. When the 'clique' consisted unquestionably of working-class children, 'bourgeois' was replaced by 'bourgeois-like'.

So much for talent. Talent, which official pedagogy has not yet acknowledged as a pedagogical and psychological fact. So much for the education of the gifted, which has been contemptuously and antagonistically referred to the context of bourgeois class-pedagogy.

Here comes the *second* question. If education is to produce scholars—in whatever field—it must include, as one of its central elements, the training towards *original thinking;* it must help develop a reliance on individual judgement, sense of justice and truth, and conscience. In past years, however, there has been an ideological campaign against original thinking and against us believing our own sensory organs. It is enough to refer here to the unfortunately misunderstood or misinterpreted slogan: "The Party is our mind," or "eyes are not only the means, but also the restriction of seeing."

Another vital quality of future learned men has also been put in the dock, with "petty bourgeois" branded on its forehead: *curiosity*. Curiosity and interest were most brutally restricted within narrow, brain-stifling limits. The pile of strictly confidential stock in libraries was getting higher and higher. In the second-hand bookshop at Muzeum Körut—where they kept books on the index under lock and key in a back room, accessible only to a few privileged customers—there were eventually more books in the back room than in the outer, open room.

To train students and research students to respect facts, to demand exact thinking and proofs from them, is the basis of scientific education. Stalinism, on the other hand, branded these very demands as "bourgeois objectivism." Under the banner of party-minded science, a large (even, we could say, world-scale) attempt has been made to create fact-free and proof-free scholarship or science (for example, Lysenko's and Lepichinskaia's biology). The extermination of facts was often carried out under the pretext of a "Marxist" fight against empiricism—an invisible and frequently nonexistent "salient feature" was given first importance over the miserable and mostly unpalatable "phenomena." The victims of this fervid fight against formalism included logic, and many branches of applied mathematics (biometrics, econometrics) were anathematised. Dialectic was corrupted into scholastic sophism.

The history of science indicates that we ought to teach the future scholar to be modest, to be humble in his scientific claims, to be averse to all kinds of fanaticism. He ought to learn that what he does not understand, or what he disapproves of, still has a right to exist, and that no scientific theory, no theorem can conclude anything finally, in the history of science. Sectarian pedagogy, on the contrary, has been inciting to impatience and intolerance in science for the past ten years. The people who have been responsible for dethroning reason have declared the greatest scholars and scientists of the century to be lackeys of the bourgeoisie; they have dismissed prestigious scientific schools of thought, even certain branches of science altogether—e.g., genetics, cybernetics, mathematical logic, mathematical statistics—as bourgeois sciences. Young would-be Titans occupied themselves with "revealing the rotten bourgeois essence of Einstein's theory of relativity and of quantum-mechanics."

But while taking a stand against fanaticism in educating the young, we must also, of course, teach them to have a committed sense of truth; when we take a stand against the insolent self-assurance of ignorance, we must also teach them to act, and act with self-respect; so many splendid—and difficult—pedagogical tasks!

New, hitherto unfamiliar chapters ought to be included in pedagogical textbooks, such as "Methods for stimulating curiosity and developing it into interest," "How to teach people to think scientifically," "How to teach people the respect for facts" and—God forbid!—"How to teach people to doubt."

Will the time come when we shall find in our textbooks the fact and the explanation of why Marx's favourite motto was: "De omnibus dubitandum"? Or, why it is that Marx often used to say: "As far as I am concerned, I am not a Marxist"?

It is also relevant to ask at this point: when will logic be introduced as a compulsory subject in our secondary schools—as it is compulsory in the West and the Soviet Union? Let children learn what difference there is between a proved and an unproved statement. Of course, this might make the job of certain comrades more difficult when it comes to writing an article. We might even have fewer research students.

And finally: let us restore the *right to dissent.* It is vital for a scholar or an artist that, once he has found a truth, he be able to uphold it even when he is not understood, even if his views meet hostility. Maksim Gorky's actor is right in emphasising an important aspect of talent: "Talent is a man's faith in himself, in his strength . . ." Marx writes, in the last sentence of the Preface to *Das Kapital:* "I shall continue to guide myself by the maxim of the great Florentine: *'Segui il tuo corso e lascia dir le genti'* ['Follow your own bent, no matter what people say']."

The 'independence of science'—as long as we understand by this not an imaginary independence of reality and society, but that the scientist and scholar submit themselves only to facts and logic in their scientific judgements—remains a valid and crucial demand.

Are we denying the Party guidance of science? We cannot comment here in detail on the complex and important question of what is the right way in which the Party or the state ought to guide science policy. But there has to be a distinction between 'guiding' science and guiding science policy. One cannot deny the guiding role of the Party and state in matters of science policy, but *the Party cannot guide science.* On the contrary: *it is science that must guide the Party.* Party guidance of science—in the sense that the Party takes sides in battles between different scientific schools of thought and declares one right and another wrong—is reactionary and must be rejected. For example, in August 1948 the Central Committee of the Soviet Communist Party 'sanctioned' Lysenko's lecture directed against so-called 'Weismann-Morganism'. No Central Committee of any party can 'decide' or 'pass a resolution on' scientific discussions. Where such a procedure inevitably leads is well described, though unwittingly, by Trosin: "The discussion that took place at the Lenin Agricultural Academy of the Soviet Union at their August session of 1948 showed the strength and vitality of Michurin's biology and revealed the rotten bourgeois substance of the Weismann-Morgan theory. The complete obliteration of the Weismann-Morganism trend in the Soviet Union was made possible because the Central Committee of the Soviet Communist (Bolshevik) Party, and Comrade Stalin personally, supported the truly progressive, dialectical materialist trend in biology, represented by Michurin. In the Soviet Union the productive activity of the Michurinists enjoys the help and support of the Central Committee of the Party and of Comrade Stalin specifically. . . ." (D. M. Trosin, *Dialectics of development in Michurin's biology*).

The frightening atmosphere of Trosin's text above evokes one of Antonio Quarengo's letters, written in 1616: "Signore Galilei's discussions have dissolved into thin air, like the smoke from a witch's kitchen. The ecclesiastical court declared that to hold this view [the Copernican theory] is equivalent to deviating from the sacrosanct dogmas of the Church." And, reverting to our time, we might ask whether Quarengo's words do not apply to some Party decrees such as the ones on 'pedagogy' and on 'philosophy'. This "decree"—in Galilei's words—"had its origin not in judicious enquiry, but in passion none too well informed. . . . Advisors who were totally unskilled at astronomical observations ought not to clip the wings of reflective intellects by means of rash prohibitions" ("To the Discerning Reader," foreword to *Dialogue on the Great World Systems,* 1632, 5).

And how well Galileo Galilei knew these advisors:

> they draw hasty conclusions, they implant some statement in their minds, it makes no difference whether it is their own thought or borrowed from some person they regard as authority, and adhere to it so stubbornly that it is impossible to weed it out of their heads. Such arguments as seem to support their malformed idea, no matter how unfounded or nonsensical they may be, they greet with instant approval. On the other hand, no matter how rational and convincing the counter-arguments may be, they receive them not only reluctantly, but with uproar and hot rage. Some of them would go so far in their fury as to be ready to use any means to suppress and silence their adversaries. (*Dialogue,* 112)

> . . . what is more revolting in a public dispute, when someone is dealing with demonstrable conclusions, than to hear him interrupted by text (often written by someone for a quite different purpose) thrown into his teeth by an opponent? (*Dialogue,* 113)

It is not by chance that in the past years in Hungary Galileo Galilei's words were not granted, on the stage of science, the *nihil obstat.*

But let us come back to the right to dissent.

At the last Party Congress in China, Teng Xiao Ping talked about guaranteeing the right to dissent and remarked that if, perchance, truth happened to be on the side of a minority, this right would facilitate the recognition of that truth. This principle has enormous significance in science, where at no instance are new conceptions formulated by "the demand of the masses," but always by the single, solitary voice of a fragile scholar. It often takes many decades until his opinion becomes that of a majority.

That is to say, it would be good if our pedagogical textbooks devoted a chapter to "How to teach respect for the right to dissent." Bearing in mind that he who tramples upon a dissenting individual opinion is usually not interested in the opinion of the majority either.

Appendix B: Letters to the Director of the Department of Philosophy

Paul Feyerabend

The following two letters were written by Paul Feyerabend and addressed to Wallace I. Matson, director of the Department of Philosophy at the University of California at Berkeley, where Feyerabend was professor. Feyerabend sent copies to Lakatos, who kept them in his files along with his other letters. They are published here as illustrations of Feyerabend's attitude toward educational and social problems. The first letter focuses on the teaching of philosophy, its methods, its content, its standards, its modes of discussion, and its administration. The second letter tackles the political implications of the student revolt and questions the value of a university education.

---

26 January 1969

Dear Wally,

For a considerable time now I have been reading your communications, and I have always found myself in substantial agreement with them. [. . .] It is your intention to defend standards of evaluation which are more reasonable than those accepted by the administration. Excellent! Extremely laudable! Administrators must be kept in their place and must not be allowed to interfere with the activities of people who think and (hopefully) encourage others to think. But in trying to do this, you develop an impossible philosophy of knowledge, you uncritically swallow some results of the now quite popular correlation game and, worst of all, you appeal to standards which are even more inadequate than those you try to reject.

Philosophy, you say, is different from other disciplines. It is not a body of knowledge. There is no accumulation of facts. The search is more important than the results. And so on. Yet it has standards which are "every bit as rigorous" as the standards of other disciplines. They are determined, you say, "by [the] consensus of those in the discipline." Such a consensus is not difficult to achieve. "The pecking order is pretty determinate." Let us adopt these easily obtainable standards, you conclude, and our judgement of young (or ageing) philosophers will become more just and more realistic.

This argument assumes that it is good to conform, and that it is bad not to conform. It is good to conform not to the standards of physics, where conforming, according to you, means having diarrhoea of the pen, but to the very different standards of philosophy, which, because of the "small number" of philosophers in this country ("nearly all of them know all the others") can be both determined and enforced by a little friendly gossip.

But what philosopher who is worthy of this name will regard this as a criterion of excellence? Philosophy has very often turned against the status quo. I myself would think that this is one of the most eminent functions of philosophy: to critically examine existing standards, to expose their vicious character, to overthrow them by the suggestion of new standards. Bakunin's "I shall continue to be an impossible person as long as those who are now possible remain possible" has been the motto for many philosophers and should be for anyone aspiring to become a philosopher. This critical function is directed not only against the *content* of the doctrines discussed, but also against the prevalent *modes of discussion* (aphorisms instead of sustained argument; pamphlets instead of treatises; letters instead of articles; speeches instead of lectures; theatre instead of the classroom; confessions instead of investigations; seduction instead of persuasion, both individual and collective) as well as against the *way of life* prevalent among those in the profession—as in the inhabitants of the Thebaid against the professors in Alexandria—; today it would be nice if there were some people who absolutely refused to "work hard . . . read and read, teach and teach, talk and talk, argue and argue, and . . . write and write," thus showing that there is still some character left in this age of nervous professionalism. Why shouldn't we have a new Diogenes? And right here in our department? And why should we not defend him against some dean, even if this style of life does not at all agree with ours? I do not want to say that revolution in thought, style, action is the *essence* of philosophy. This would be quite incorrect. There *were* many philosophers, and there *are* still more philosophers who consciously or unconsciously define their tasks as the analysis and preservation of *what is.* Nor do I want to say that philosophy has any essence. All I want to point out is that your criterion of conformism, though at first sight quite liberal ("don't ask us to write a lot—there are other ways of being a philosopher") is, as a matter of fact, a conservative criterion ("and these other ways are agreed upon by the members of the profession") which at one stroke excludes a legitimate function of philosophy from the university. It is bound to eliminate what little life is left in academic philosophy.

So much for your criterion.

But, dear Wally, do you really intend to apply it as it stands? When you speak of the "consensus of those in the discipline," do you really mean to include, let us say, Marcuse, or Lichtman, or Heidegger, or Teilhard, or Brecht? For if you *do* intend to include these chaps, then all that talk about consensus is ever so much hot air: *there is no consensus* between these thinkers and, say, Strawson (there is no consensus, I would hope, between Grice and Lichtman). Hence, you must have defined your domain in a much more narrow fashion, and I cannot quite help suspecting (especially when reading your criteria as to what is and what is not philosophy) that a philosopher is, for you, a member of the Anglo-American school of logical empiricism and/or linguistic analysis, or else a man who, though dealing with other things also is at least capable of speaking the analytic lingo in a fashion sufficiently fluid to delight your (or Grice's, or Mates', etc., etc.) ears (at any rate, he must be able to speak English—otherwise he will hardly be "known" by "all the others"). There are, therefore, *two* reasons why your criterion of selection (consensus of the top dogs) must be rejected. Firstly, because it is a criterion of conformity. Philosophers who want to change the whole damn profession are at once excluded. Secondly, because the field is narrowed down even further; one is supposed to conform not to *all* of philosophy, but to a narrow, "professional," but actually quite provincial, subgroup of philosophers whose only merit consists in having found a few half-dead fleas in the rotting fur of a once great enterprise. (That it is the narrow criterion which is used, and not the wider one, is shown by the way in which invitations and advancements have been and are still being handled in this department; examples: Lichtman, Lakatos and, long ago, Popper.)

Now, quickly, a few words about the remainder of your draft.

To start with, I would strongly object to making philosophy a *special discipline* with special procedures and so on. This is a particular philosophical doctrine that is not held by everyone. Moreover, it is the (philosophical) attempt to *overcome* boundaries that has always led to progress, the last great examples being Bohr and Einstein who reformed physics by appealing to philosophical principles, and who enriched philosophy by the injection of physical results. The *correlation* you refer to (greatness vs length of publication list for 19th century *scientists*) is more than suspect. Don't you know that 19th century *philosophers* were very prolific too, and that even the minor philosophers were not happy unless they had volumes and volumes to their credit? Besides, scientists of the 19th century still had philosophical interests, so that a separator like you must look very carefully at the *content* of their papers and cannot rest content with their *number.* On the other hand, the 19th century witnessed a progress of philosophical *results* (Kant—Fichte—Hegel—Marx, for example) which looks almost scientific (in your sense of science). "The results of philosophy are certainly not cumulative as they are in mathematics or chemistry," you write—and you wisely omit physics for you have yourself written a paper which points to the close similarity between modern theoretical physics and the philosophy of the pre-Socratics. Now, physics, so an evaluator of disciplines would be inclined to say, is ahead of chemistry, hence is it not advisable to dissolve the 'results' of chemistry too, and in this way to make progress?

To sum up, the problem of an academic philosophy is twofold. It must fight the administration in order to be able to develop freely, according to whatever standards it sees fit to adopt at a given time. And it must fight its own prejudices, it must fight the very same standards which it uses as a weapon against the administration in order to be prevented from stagnating and in order to prevent other subjects from stagnating. (The withdrawal of philosophy into a "professional" shell of its own has had disastrous consequences. The younger generation of physicists, the Feynmans, the Schwingers, etc., may be very bright; they may be more intelligent than their predecessors, than Bohr, Einstein, Schrödinger, Boltzmann, Mach and so on. But they are uncivilised savages, they lack in philosophical depth—and this is the fault of the very same idea of professionalism which you are now defending.) This means, of course, that a philosophy department must occasionally be prepared to accept people who to all intents and purposes are worthless as "professional philosophers" and who cannot be expected to make a single contribution to "contemporary philosophy." It must occasionally be prepared to accept people who sneer at the professional standards of reading, lecturing, publishing, or even better, it must be prepared to accept people who are not at all aware of such standards. Should we now pick up every bum from the street who has something to say? Well, I for one would not object. I do not see much difference between the nonsense such a person is liable to produce and the professional nonsense that comes out of Oxford. Most likely, his nonsense would be more pleasant to listen to than the constipated academic product. But we need not go to such extremes. For we have very good help. We have excellent help. We have advisers who can help us where the situation seems impossible. We have our *students.* This brings me to the last point in this letter.

I know I come now to a subject on which I am in disagreement with almost all my colleagues. The attitude of most of you is that you are philosophers, that the students are not, that they do not know those fabulous tricks which you yourselves master and with such consummate skill, that they are lazy, stupid, unwilling to learn, that they must be trained to repeat the tricks after you, so that sometime in the future they may perhaps be able to become trainers themselves, modifying the tricks a little here and there (this is called "original research"), and being equally stern in the propagation of their knowledge (this is called "professional conscience"). I am sorry, but I see my task in an entirely different light. I see before me young people, who are capable of great new discoveries, who are capable of showing *us* where *we* have gone wrong, whose individuality has been almost obliterated by a crazy and competitive system of education, who have almost become grade-earning machines, whose initial curiosity has to a large extent been replaced by fear and the urge to please, but who perhaps can still discover what is left of their talents and make good use of them. Once, in a discussion in the Golden Bear, Benson Mates told me that the people of the younger generation don't know what they want, that they are undisciplined, that they are stupid, etc., etc. Why is this so? Because instead of being allowed to *learn,* they are beaten down so that all that is left is fear and aggressiveness.

I see myself as a *servant* of the students, and I think you should see yourselves as their servants, too. My way of serving them is to ask them what their interests are, or, if they don't know, to entertain them with a variety of ideas until they have discovered what their interests are, and then I ask them to tell me how they want me to further their interests. Where do you think the future development of mankind will come from? From senile writing machines like ourselves whose main achievement consists in having written a few silly articles which have found the praise of equally senile reading machines (I for one am ashamed of what I have written in the past), or from people who despite a disastrous education have still some spirit left? So what I would suggest is that student opinion be given a much larger, or, rather, a *decisive* voice in the hiring and the advancement of colleagues. They may choose someone whom everyone among us despises—but it is *their* life that is being influenced and they should have a right to influence it in any way they want. Will the "standard of philosophy" suffer from this? Well, Plato says that all things change for the worse; the worst thing alone will change for the better. The standards of philosophy, then, can only improve.

This note has loused up my whole Sunday. Well, I hope it is not superfluous. I also hope you will read it in the right spirit of patient amusement.

All the best, and write again.

Paul Feyerabend

6 February 1969

Dear Wally,

Having received your letter advising me that "it is against university policy to schedule regular class meetings off-campus" I tried, very much against my natural laziness, to obtain information about the relevant regulations and about university policy in general. I discovered that there is no written rule demanding that classes be held on campus and that even vice-chancellors have been in the habit of meeting seminars, or small classes, at home, or at some other congenial spot. There is, of course, the *expectation* that classes will be taught on campus and there is also a corresponding *custom;* however, this custom is neither clearly defined nor was it ever *firmly enforced.* As far as I could find out, the only argument that was ever used in its favour is the *convenience of students.*

Assuming that you are familiar with this situation and that you had it in mind when writing your letter, I interpret your request [. . .] as a suggestion addressed to me in the belief that it would be the best procedure under the given circumstances. Let me explain why I hold a different opinion and why I shall continue to hold classes off campus.

My first reason which, as I have just pointed out, coincides with the reason customarily put forth in this connection, is the convenience of the students.

To start with, there are many students in my class who support the strike, who do not wish to cross picket lines, and who would still like to attend classes. Moving the classes off campus solves their dilemma *provided* it does not create an opposite dilemma for those who are against the strike. I have asked some very vocal critics of the ideas and methods of TWLF.[1] They do not mind.

I would also like you to consider that many people and especially some very sweet girls have been frightened by the violence they have witnessed and by pressures exerted on them. Certainly they have a right to wish that they be instructed in less turbulent surroundings and to expect that one will comply with their wishes. It is not in my power to stop the violence. But it *is* in my power to move to a more peaceful spot. I conclude, then, that the criterion of convenience recommends staying off campus.

I must confess that it took me some time to arrive at this position. I am firmly against any restriction of personal freedom, even against that tiny infringement of it that occurs when one is forced to walk around an individual picket. I am also against any form of violence and coercion. I know this is a pretty unrealistic position to take especially in a complex society that can function only when people are either prepared, or forced, to restrict their freedom in many ways and that responds only to argument *plus* pressure, rarely to argument alone. But just because there are so many restrictions already, I am doubly mad at any *additional* restriction, even if those who impose it promise it is a restriction designed to *end* restrictions (things never turn out that way; see for example the history of the Russian revolution; cf. also Cohn-Bendit's analysis). Moreover, I look with great suspicion on people who believe in, make use of, defend, and try to enforce metaphysical symbolisms such as the sanctity of the cross, the inherent viciousness of an on-campus room, the sanctity of the American flag, or the sanctity of a picket line. Appealing to idols such as these seems to be a barbaric method for intelligent people to pursue. It was because of prejudices such as these that I continued to hold classes on campus long after the suggestion to move elsewhere had been made to me. I now think that I was wrong. I may be inclined to smile about the magic of the picket line—but I must also respect the attitude of those who believe in it. I may be critical of a strike as a means of achieving greater freedom—but I have no business interfering with the actions of people who have come to the conclusion that under the given circumstances a strike is the best method of bringing about progress. All this presupposes, of course, that violence does not become the *official* policy of the strikers (I have been assured that it is not the official policy of the TWLF and that people in responsible positions are trying everything to prevent it). For, despite some doubts even on this point, I am certainly not prepared yet to accommodate self-professed bullies and bomb-throwers ("Professor Feyerabend, could you please change the time of your class? I want to start a little fire in Dwinelle Hall at 2 P.M. and this is unfortunately the time when you are giving your fascinating lecture on revolutions in science"). But if violence should ever become an official policy, then there would be even more reason to move off campus in order to protect the audience as well as those strikers who continue to employ more peaceful procedures.

So much about the purely formal or administrative issues raised in your letter.

These issues are not the whole story, however. For I suspect that the motives behind your request, while admitting of a purely formal interpretation, are, to a large extent, *political.* I think they can be expressed in the following sequence of statements: (1) moving classes off campus means supporting the strike (a) with . . . (b) without the terror and vandalism that has occurred. (2) Supporting the strike is "bad" (a) because it is against campus regulations; (b) because it means working towards the dissolution of the very institutions to which we owe our intellectual and, of course, our financial existence (the latter more than the former). Hence, (3) putting classes off campus is "bad" because—and here we have the same (a)s and (b)s as above. Forgive me for presenting matters in such an overly schematic manner. My reasons are that I am addressing you as an *administrator* and that I want to show that even the political motivations for your request fail when looked at from a purely administrative point of view. I also want to make it clear to you that in your attitude towards the strike you confound reasons which are not connected and generally make a mess of things. But it is just in crises like the present one that we must be careful not to act, to condemn, to forbid, to request, for the wrong reasons. We must be careful not to generalise specific reasons and specific arguments, and we must watch ourselves lest we muddy up our thinking with assumptions which, though emotionally understandable, have no foundation, neither in fact nor in humanitarian attitude. With this in mind, let me now examine, one after the other, the statements at the beginning of the present paragraph.

To start with, it is obvious that (1) cannot possibly be correct (it is surprising how many people, students and faculty alike, believe it to be correct as a matter of course). Thus it is quite possible to move classes off campus for the protection of the students and *not* because one supports the strike. It is even possible to proceed in this way for the convenience of the strikers, i.e. because one respects their attitude, without adopting it, and because one accepts their aims (this is roughly the position in which I find myself, though with an important exception for which see below). These things are so clear and so obvious that I fail to comprehend how intelligent people or people who have an administrative obligation to be intelligent (deans, vice-chancellors, department chairmen) can possibly think otherwise. It is even less comprehensible to me how they can adopt the idea that moving classes off campus means supporting terror and vandalism. Yet this is precisely the position you took in our brief conversation about the matter. Even worse—you at once triumphantly held Wheeler Hall in front of my nose as if it had already been established that it was destroyed by arson, and on the explicit orders of the leaders of the TWLF. Is this the kind of thinking you are teaching your students? And, if it is, how can you be surprised at the results?

Next point: (2a) certainly cannot be correct especially in view of the fact that peaceful picketing, etc., has been asserted as a right by the academic senate when it passed the Tussman motion on February 3rd. On the whole, we are here in an administrative vacuum, for no definite regulations are available for the details of the case. (2b) I do not understand. I have not yet heard it stated as the official policy of the TWLF and of their supporters that they want to *destroy* the university. Surely, this would defeat their own aims, viz. to be able to participate to a larger extent than they have been participating so far. Karl Mack, when talking to my class, made it clear—what should have been obvious anyway—that the continued existence of the university is of great importance to all Third World people. After all, they want to *enter* it, they want to *learn* here and to prepare themselves for service in their community. There are of course some changes being suggested, but this is an altogether different affair. Look at it as you may—even the wider and more political context does not support your request to return classes to campus. It does, however, add a further argument in favour of off-campus instruction. I now turn to this final and, I think, most decisive argument.

We both participated in the Senate meeting of February 3rd, though most likely with very different emotions. For me this meeting showed an almost complete inability, on the part of the faculty, to understand the positive forces behind the present crisis and to express its sympathy with these forces in a way that could be understood by the strikers. I know that you will answer at once that you have no sympathy with thugs, bums, arsonists, with people whose intellectual and moral defects are comparable to those of the Nazis (your own words), etc. etc. Well, this is not the point I want to make. But as the "bums," "thugs," come up again and again, both in Sacramento and now apparently also in Moses Hall, let me briefly comment upon "them." There is no doubt that many participants of the strike, including people who on other occasions are perfectly charming and reasonable, have behaved like wild animals and have committed serious crimes, not in the *legal sense,* which does not interest me, but crimes *against humanity.* Now, after thousands of years of pain and suffering, after millions of small increments of consciousness, each of them paid for with human lives, after a continuous struggle for freedom and human dignity has shown some minor successes, we see—what? We see the most privileged beneficiaries of this struggle—*white university students*—dance around like madmen, spit at other and less fortunate people (the lot of a policeman is certainly not a fortunate one) and call them "pigs." Let no one tell me that the police are an instrument of oppression. Let us assume that they are such an instrument—is that not bad enough? Is it not bad enough that humans are turned into instruments for the oppression of others? Must their humanity be further reduced by calling them "pigs"? And is this the correct behaviour for those who demand self determination and human dignity for everyone? Or do they believe that human dignity can be restored only after it has been trampled underfoot *by everyone,* by the 'establishment' *as well as* by the foes of the 'establishment'? And if that is their philosophy, must they not be accused of a callousness that far surpasses the callousness of the supporters of the status quo? For these supporters are convinced that they are doing the right thing whereas they, the opponents, are aware—or at least *pretend to be aware*—that what is being done is wrong, but insist on doing it nevertheless. Or do they notice a wrong only when it occurs to *them?* These are some of my thoughts and they show you how much I agree with you *on the surface.*

But while you are satisfied with pointing out the moral defects of others whom you do not like, *I go on* and ask myself: *how is it possible* that young people who have attended the schools of the 'greatest country in the world', who are now studying at a top notch university, who are being instructed by the cream of the crop, by an intellectual élite, how is it possible that people who have been prepared in this truly marvellous fashion can lose their humanity so easily? Is it not because their education was something less than marvellous, is it not because we have failed to make the simple humanitarian virtues of charity and compassion attractive and strong enough to survive a little emotional strain? Is it not because humanitarian virtues *have almost no place* in this *phrontisterion* so that even we, the élite, the marvellous experts, frequently enough turn on our opponents in an equally vicious fashion (only, being a little bit older, a little bit weaker, a little bit scared, being tied down by a comfortable salary and a not so comfortable wife, we proceed less energetically, and certainly not with such abandon; this is the *only* reason why we can be called "more civilised"). Now, if this is the case, and Wally, *you* can hardly deny that it is, must we then not be grateful to those who, sensing that something is wrong, try to bring about a better world? Who are not content with the comfort provided by a beautiful *dream,* or by an abstract *idea* but who go on and try to *realise* this idea? And this is the point I really wanted to make. No doubt these people commit many errors and they are also impatient. But are *we* free from error? And is our so-called patience really positive virtue, or is it not rather a sign of mental and administrative indolence? No doubt they occasionally trample on the rights of others. But are we, in our civilised ways, less vicious? No doubt they lack kindness, understanding—but where could they possibly have learned those virtues? Is it not a miracle that there is still some kindness left in this younger generation considering that all their elders can do in a time of crisis is pass a self righteous motion about rights and, of course, law and order and spend their time discussing amendments and amendments and amendments? This, dear Wally, is how I see the situation. And therefore, I thought that in addition to considering the *convenience* of the students, I should make a gesture indicating that I have considered their *aims,* and their *intentions* also and that I am in sympathy with these aims: self-determination and increased dignity for an ever increasing number of people.

This, dear Wally, is a long reply to your short and rather abrupt note. But even the simplest action has numerous consequences. We, however, shall be able to advance only if we proceed not blindly, on the basis of an isolated rule, or of a sudden and unreflected burst of anger and official indignation, but considering as many of these consequences as possible.

Sincerely yours,

Paul Feyerabend

1. The *Third World Liberation Front* (TWLF) was a student organisation which included in its ranks Mexican, Filipino, Chinese, and Hispanic students. Along with the Afro-American student groups, it was responsible for one of the most belligerent university occupations of the time, which started at San Francisco State College in the autumn of 1968 and continued, with alternate ups and downs, until the spring of the following year.

Appendix C: The Intellectuals' Betrayal of Reason

Imre Lakatos

Like his reply to *Against Method,* Lakatos never completed "The Changing Logic of Scientific Discovery," his lifelong project of writing a history of the critical (and not so critical) interaction between methodological standards and scientific achievements. Lakatos had partially taken on this task in his paper "Newton's Effect on Scientific Standards" (1963–1964b), which reconstructs the battle between Cartesians and Newtonians over the standards of (scientific) "proofs" and (scientific) criticisms.

Lakatos's intention was to recapitulate the background of the problem and to assess the problem shift within the dispute between Cartesians and Newtonians. This would have helped to explain the success of Newtonian physics despite its violation of the standards commonly accepted by the scientific community. In particular, Lakatos pointed out the discrepancy between Newton's actual procedure (i.e., his method) and his *regulae philosophandi* (i.e., his theory of method), stated in explicit terms both to defend his discovery and defeat his rivals. Thanks to a peculiar historical paradox, the following generations were to accept not only the richness of Newton's results, but also the "poverty" of his theory of what made up his scientific achievement:

> In this sense one may say that Newton's theory of method created modern philosophy of science. But this move separated science and philosophy of science from 1686 to 1905 [when Einstein's theory superseded Newton's] or, rather, till 1934 [when Popper first considered the problem in this way]. Science, the schizophrenic genius, marched from victory to victory. Philosophy, unaware of the split

> between the wonderful Newtonian method as practised and the mad Newtonian method as professed, tried to clarify the professed method, and thus turned into *a mad study of madness.* ("The Rise of Defensive Positivism," Archive, 5.5; this passage is a different version of the concluding section of "Newton's Effect.")

Underlying this paradox is the fact that Newton took his methodological standards from past centuries dominated by the all-pervading idea that religious knowledge was certain and indubitable; by analogy, science was expected to respect similar standards. In fact, the split brought about by the Reformation (though against the original intentions of the founding fathers) raised the problem of *fallibilism.* How is the Church expected to be the authority on its own infallibility, given that the point in question is precisely whether the Church is the 'true' authority on religious matters? This question was to unleash a sceptical crisis not only in theology but also, shortly thereafter, in the sciences and in all other areas of human knowledge. (Consider, for instance, the shift in the criterion of truth from Erasmus to Descartes, as described by Popkin [1979].) Thus, sustaining Newton's views was the intellectual inconclusiveness and political destructiveness of the rival theological stands as well as the belief that science could achieve the certainty that theology at the time seemed painfully denied.

A century later, the stunning success of modern natural sciences, and primarily of Newton's research programme, was to breathe new life into the battle between the dogmatists and the sceptics. On the one hand, "some radical dogmatists, undeterred by the long series of defeats of reason in human affairs, have been trying to generalise Newtonian method to social, ethical and political problems." On the other, "some radical sceptics, undeterred by the long series of successes of Newtonian science, have been trying to show that all these successes were sham successes and even the best theories of the exact sciences were nothing more than irrational (if possibly 'great') beliefs." (See below.)

Lakatos refuses to side either with the radical dogmatists or with the radical sceptics ("the dogmatists tried to prove too much, the sceptics tried to explain too much"). Instead, he points to a new appraisal of this controversy by recognising "the basic unity of opposites (dogmatism-scepticism) and the possibility of their dialectical 'Aufheben' [superseding]" (Beck and Yourgrau 1970, 22). Yet Lakatos regards the "modern betrayal of reason", consisting in "the intellectual attack on the objective epistemological value of the exact sciences", as "criminal". If, on the one hand, Feyerabend has shown many different ways in which any principle of rationality can turn into a prison, and has consequently claimed along with the sceptics that the betrayal of 'reason' by man is preferable to the betrayal of man by reason, Lakatos, on the other hand, has challenged his rival's anarchism, insisting on the bias underlying the sceptico-dogmatist's point of view by showing how radical relativism ultimately leads to forms of intellectual surrender that maintain that *might is right.*

In order to propose his (fallibilist) stand as an alternative to dogmatism and scepticism, therefore, Lakatos himself—a worthy theologian of rationality—has to face the challenge of the Greek Pyrrhonians, whose arguments after lying forgotten for centuries had suddenly come to the forefront, especially in regard to the primary epistemological problem brought up by the Reformation. Luther's denial of the authority of the Church and his assertion of a new "rule of faith" (i.e., 'true' is what conscience is compelled to believe when reading Scripture) for determining religious truths constitutes a rather neat example of the "problem of the criterion" as put forward by Sextus Empiricus:

> In order to decide the dispute which has arisen about the criterion, we must possess an accepted criterion by which we shall be able to judge the dispute; and in order to possess an accepted criterion, the dispute about the criterion must be first decided. And when the argument thus reduces itself to the form of circular reasoning the discovery of the criterion becomes impracticable, since we do not allow the Dogmatic philosophers to adopt a criterion by assumption, while if they offer to judge the criterion by a criterion we force them to a regress ad infinitum. (Sextus Empiricus, *Outlines of Pyrrhonism,* quoted in Popkin 1979, 3)

Many of Feyerabend's criticisms of Lakatos's views in this volume are clearly a revival of this classical Pyrrhonian challenge. The problem of justifying the standards of theory appraisal does not arise as long as there is an unchallenged criterion, but once scientific revolutions are brought into the picture, epistemological scepticism may be back again (see, in particular, letter, IL to PF, 10 January 1974, and letter, PF to IL, 17 January 1974, this volume).

This hitherto unpublished paper, written sometime before 1967 (Archive, 8.2), partially reveals how Lakatos intended to meet this challenge.

Radical scepticism—the doctrine that knowledge is beyond the reach of human beings—was first formulated in antiquity. The attack of radical sceptics was directed against those who thought that knowledge *was* possible and that the search for truth distinguishes man from animal. Radical sceptics called them mockingly "dogmatists."*

The sceptics argued that what the dogmatists pretend is knowledge is in fact animal belief or deceitful ideology. They ridiculed the illusion of 'intellectual honesty'. They were the first intellectuals to betray reason. For the radical sceptic 'growth of knowledge' and 'intellectual standards' were phrases with no meaning. The history of ideas was the history of competing beliefs, devoid of objective rationality. "Verbal behaviour" was mere self-expression. Some sceptics excelled in the history of ideas: they wrote distinguished essays about human follies, under what influences they came about and disappeared. But they also did not think much of their own work which, in turn, was folly about folly. For a sceptic, of course, there can be no knowledge *about* beliefs, either: the history of beliefs cannot add up to more than a history of beliefs about beliefs. Intellectual history is then self-expression about self-expression.

There are many sorts of political philosophies compatible with radi-

* Few dogmatists ever held that the growth of knowledge is unlimited, that *all* problems are solvable. We may call these few "radical dogmatists."

cal scepticism. Some sceptics—the tolerant ones—believed that utopian dogmatism was responsible for the worst sufferings of mankind. They pointed out that people who claimed they knew moral, political and religious "truth," who also boasted of knowing which way "progress" lay, used inquisition and torture, bloody wars and genocide in order to verify their predictions in practice. Tolerant sceptics dreamt of a convention—a social contract—to decide how to restrain the human animal and minimise suffering. For them, happiness and welfare replaced truth: 'true' was what increased happiness. They argued that betrayal of reason (or rather, "reason") by man was better than betrayal of man by reason.

But these tolerant sceptics could not purge themselves of dogmatism. For in negotiating a good social contract they needed knowledge about the pragmatic consequences of the contract: when controversy developed they had to choose between turning dogmatist or resorting to force without argument. Thus, ultimately, there is only one type of political philosophy consistent with radical scepticism: the philosophy which equates right with might. This is how many sceptics became well-paid courtiers of the bloodiest tyrants in history. "Once the conception of objective truth is abandoned, it is clear that the question: 'what shall I believe?' has to be settled by the appeal to force and the arbitration of big battalions, not by the methods of either theology or science."*

When long quarrels between rival dogmatisms are intellectually inconclusive and altogether sterile, rational discussion and, indeed, communication itself break down; and especially when competing claims to knowledge lead to destruction, war and increased unhappiness, radical scepticism unfailingly raises its head. On its lips at first is the faint whisper of humanity protesting against false reason; but the mouth usually gets distorted in little time, and the whispering becomes animal howling. Or it becomes the unreserved apologetic of *any* law and order which is willing to allow the sceptic to enjoy his *ataraxia.***

* Russell, B., "The Ancestry of Fascism," *Let the People Think;* 1941, 77. Later Russell changed his mind and, after twenty odd years, proposed his famous slogan: "Remember your humanity and forget the rest. If you can do so the way lies open to a new paradise. If you cannot, nothing lies before you but universal death."

** Pyrrhonian scepticism advocates *ataraxia* (unperturbedness) as a way of life. This is achieved, given any problem, in three stages. First comes the stage that requires intellectual effort: the *antithesis.* This means the development of several alternative theories so that the domination of one alternative ceases. But this may be too difficult and the lazy sceptic may simply use standard techniques *(tropoi),* like infinite regress, to show the unprovenness of any solution offered. This brings about the *epoche,* the suspension of judgement which, after some mental concentration, brings about the *ataraxia.* Ancient tyrants loved sceptics.

Radical scepticism is then—as a dogmatist would put it—a false and immoral belief. It originates with some as a blind reaction against the outrages of dogmatism; with others as an ideology of "able and energetic men [who] have no outlet for their love of power" (ibidem). In its propaganda—a "valid sceptic argument" is, of course, impossible—it parasitises on dogmatism: it is the *ancient betrayal of reason.* I regard this ancient betrayal of reason as a venial sin.

It should be emphasised that most of what has been called "scepticism" in the history of thought is not the radical kind I have described, but a kind of "mitigated scepticism." "Mitigated sceptics" are in fact dogmatists who hold that *some kinds* of knowledge will be forever inaccessible. They are radical sceptics concerning *some kinds* of sources of knowledge, and dogmatists concerning others. Mystics trust revelation but reject the intellect and the senses as sources of knowledge. Intellectualists trust the intellect but not the senses, empiricists the senses but not the intellect. An influential brand of mitigated scepticism holds that religious and moral knowledge ('metaphysics') is impossible, but that knowledge—even if not 'ultimate knowledge'—about the external world and about logic is possible: this is scientific positivism. Scientific positivism is very new: it was motivated by two basic historic experiences; its sceptical aspect has its roots in the Thirty Years' War and the intellectual inconclusiveness and political destructiveness of competing camps of religious knowledge; whereas its dogmatic aspect was motivated by the stunning progress of modern natural sciences and primarily of the Newtonian research programme.

Some *radical dogmatists,* undeterred by the long series of defeats sustained by reason in human affairs, have been trying to generalise Newtonian method to social, ethical and political problems. Some *radical sceptics,* undeterred by the long series of successes of Newtonian science, have been trying to show that all these successes were sham successes and even the best theories of the exact sciences were nothing more than irrational (if possibly 'great') beliefs. The hallmark of the *modern betrayal of reason* is the intellectual attack on the objective epistemological value of the exact sciences. I regard this modern betrayal of reason as criminal.

Thus, since 1687 the whole history of the philosophy of science—and indeed of rationality—has revolved around a single central issue: what are the epistemological merits, if any, of Newtonian physics? The discussion brought about a radical shift of the whole problem. We may characterise this shift by an analogy from ethics.

Let us imagine that mankind has set up moral standards and maintained them for centuries without anybody being able to live up to them. Some claimed, hypocritically, that moral acts—and, indeed, moral men, did exist; others kept unmasking these reports and either claimed that the high moral standards would necessarily remain utopian forever, or set up programme after programme for making man moral and noble. Then, suddenly, a few men actually started behaving in a new way which soon struck most people as truly moral. First they hailed them as having finally realised the old dreams. But then it slowly dawned on them that although the new men's acts were undoubtedly virtuous, they did not in fact comply with the old standards. They tried to weaken the standards gradually, but frustrating scrutinies showed that, no matter how one weakens the standards, morality, as it was now being practised, could not satisfy them. Moreover, with the development of modern logic it became apparent that the old standards, even if watered down, could *never* be fulfilled for logical reasons. People had two ways out. The sceptics celebrated the end of all morality. But a new school of dogmatists accepted the *new moral facts* and devised *new moral standards* in the light of which their morality—a new morality—could be seen.

This was, then, *a revolution*—first in morality, then in moral theory. Not necessarily the last one: inconsistencies remained, new types of moral acts (expressed in "moral basic propositions") further upset the new standards. The problem of morality had shifted. Moreover, the revolutionaries set up a theory of moral progress which established standards for standard change; a theory in the light of which their revolution was not just a change in subjective fashion, but progress towards moral truth. (An aesthetic analogy would have served my purpose equally well. It is known that novel works of art may change aesthetic standards just as novel human acts may change moral standards.)

It seems to me that this analogy goes far in simulating the revolutionary problem-shift brought about in epistemology by the rise and fall of the Newtonian research programme.

Appendix D: Letter to His Editors

Imre Lakatos

In this poignant letter, written only two months before his death to the editors of the Boston Studies in the Philosophy of Science, Robert S. Cohen and Marx W. Wartofsky, Lakatos described his future editorial plans and his academic calendar (Archive, 12.2). About two years later, together with Paul Feyerabend, Cohen and Wartofsky were to be the editors of *Essays in Memory of Imre Lakatos* (1976).

---

10 December 1973

Dear Bob and Marx,

[. . .] I made a sort of balance sheet of what I want to achieve in the rest of my working life and this programme is part of it. To put my cards on the table, my plans for the next five years are as follows:

I should like to critically examine my vast backlog in unpublished manuscripts in the philosophy of science and the philosophy of mathematics and also in the history of science and mathematics and have them published in a form which at least partially satisfies my pedantry. I hope to get three books out in the next couple of years and also two volumes of substantial case-studies in scientific research programmes by my colleagues. (I have about eight essays which I think are on a par with Zahar's achievement.)

Institutionally I should try to expand Popperian influence in the London School of Economics. This was at its height in the late fifties and as Karl Popper became increasingly idiosyncratic and as people realised that naive falsificationism is an enemy and not an aid to the social sciences, this influence shrank to practically nothing. I hope that my methodology of scientific research programmes and its friendly reception by both the local Marxists and the local "bourgeois economists" will enable us to make our scientific method course available to practically all LSE students. I do sincerely hope that this is the best available method to instil some standards in the social and in the developing sciences. I hope I do not sound megalomaniac.

Further, I should like to establish contact between the LSE and one German and one American institution. The German institution will probably be the German Society for the Philosophy of Science which my friends will found in 1974. (There is no society like this yet.) The American institution with which I hope to establish some contact is of course Boston University. The reason for my trying to establish this contact is not simply proselytism but because I want two big doors to be opened to the Closed Society of the LSE Friends of the Open Society so that there should be an institutionalised encounter with different and no doubt ever new ideas from the United States and from the Continent. I think that the LSE, with its reputation and its standing, gives a good opportunity to do something towards achieving these aims.

In 1977 I intend to give up the editorship of the *BJPS* in order to devote two full years to completing my intellectual programme in the above-mentioned fields. Then, I want to take a year off, or perhaps even two, from my teaching and administrative work. That brings me to 1979 and after that I hope to be able to pursue an easier life.

I suppose that you are very well aware that it takes a lot of risky and perhaps unwise decisions to steer one's course between one's intellectual work, knowledge managership (this I suppose we all owe to our Socialist-Communist past) and our private life.

All this, rereading it, seems to be a bit pretentious. However, having recovered from my accident, and being in my early fifties, I think it is only proper to define to myself what I expect from life and I thought that I would put this to you both as close friends for comment. I am well aware that you have similar problems and similar planned solutions.

A note on the Greek conference. Because of recent events and the lack of proper information we postponed issuing official letters of invitation until early January. I do not see any reason why the conference should not proceed. I need not say that the new coup distressed me because I had some hopes about the last year of Papadopoulos's regime and I thought that things were improving. Even now I think that it would be a help to our more liberal-minded friends in Greece if a free conference were held there. The money is still clearly available but it cannot be moved outside Greece. I should perhaps add that Greek philosophy is now guided primarily by right-wing German Hegelian philosophers and to put in a presence of the Popperian scientific side and attach an Anglo-Saxon approach can now only be for the better. I hope that I am not wrong. Please do let me know what you think of all this. [. . .]

With warm greetings and very many thanks.

Yours sincerely,

Imre Lakatos

Imre Lakatos: Biography

Imre Lakatos was born in Debrecen, Hungary, in 1922 to Jewish parents. His real name was Imre Lipsitz. His father, a multilingual, highly educated wine merchant, separated from his wife when Imre was very young. Imre first studied at the Jewish *Réal Gymnasium* and later at Debrecen University, where he read physics and chemistry. At Debrecen University he met Éva Révész, whom he later married, and attended a seminar held by Árpád Szabó, a scholar of Greek and Latin philology and Greek mathematics. He graduated in mathematics, physics, and philosophy in 1944.

Anti-Jewish laws, the outbreak of the Second World War, and the German invasion of Hungary brought an end to both his studies and his active campaigning in the Marxist education groups he had organised at the university in 1944. In that same year his mother, grandmother, and uncle were forcibly moved to the Debrecen ghetto and were later killed in Auschwitz. Imre escaped with Eva and fled to Nagyvarad with false papers; here they were kept in hiding by non-Jewish families. He changed his name to the unquestionably Hungarian and less compromising Molnár. Among Jews in hiding and their non-Jewish host families and friends, he organised a semi-official Communist cell whose leaders effectively controlled the group members, giving permission for relationships, opening mail, and so forth. He is alleged to have led the group in deciding on the forced suicide of a young member by the name of Éva Iszák, a Jewish Rumanian anti-Fascist activist, fearing she would endanger the lives of the other members if captured.

Imre and Éva Révész returned to Debrecen in 1945 for a few months, to a ransacked house in which, having found several shirts monogrammed with the initials "IL," he once again changed his surname, to Lakatos (a common name among the Hungarian working class). After joining the legal Communist Party, he became very active in educating and converting non-Communists. (He even made a mysterious "conversion" to Calvinism, maybe as part of a deal that saw his godfather, Sándor Karácsony, become a Communist.)

Lakatos worked at the Ministry of Education, dealing with educational reform, and also played a key role in a small circle of élite Communist intellectuals, where his quick thinking and high mental calibre had a strong influence on the other members. He entered Eötvos Collegium (respected for its tradition of high-level classical education), and attended György Lukács's lectures on aesthetics at Budapest University, centred on Kant's *Critique of Judgement* and Hegel's *Phenomenolgy of Spirit.*

In 1947 Lakatos received his Ph.D. at Debrecen University with highest honours in philosophy, with physics and mathematics as specialties. His dissertation was entitled "On the Sociology of Concept Formation in Natural Science." In the following year he applied to study theoretical physics at Moscow University and, after an evaluation procedure including questioning on the Éva Iszák affair, was offered the place, possibly to remove him from Hungary, since he had been singled out as a sharp mind, and therefore not a comfortable fit with the dictatorial tendency of the new Hungarian government.

On his return to Budapest for a summer break, he found that he was banned from returning to Moscow, that his Party card would not be renewed, and that his flat had been let. He moved into a room offered him by the director of Eötvos Collegium. In 1950 he got news of a hearing on the Éva Iszák affair by the Central Control Committee of the Communist Party. Following the hearing, Lakatos was expelled from the Party, and Éva Révész and others taken for questioning. For Lakatos, interrogation, torture, and six weeks of solitary confinement in a basement cell at AVO (the State Security Police) headquarters were the prelude to a three-year internment in Recsk, a labour camp run by the AVO. Lakatos was released when the camp was closed in 1953 and returned to Budapest, possibly after having agreed to inform on certain intellectuals.

Lakatos obtained help from some friends and returned to academic work. Thanks to Tamás Lipták and to Alfred Rényi, director of the Mathematical Institute of the Hungarian Academy, he was offered a job at the library and later a research position. This gave him access to books censored for the general public.

He met Éva Pap, whom he married in 1955, and began to question his Communist faith—which had been left more or less intact by his experiences in the prison camp—thanks to his extensive reading of the classics of Western liberalism and possibly thanks to his talks with Miklós Gimes, foreign correspondent under Imre Nagy's pro-reform government.

Lakatos came into contact with the Hungarian school of mathematics—László Kálmár, George Pólya, Péter Rózsa, and Alfred Rényi—whose lectures in the form of Socratic dialogue were to influence the style of Lakatos's doctoral thesis at Cambridge.

In 1956 he returned to political activity in the Petöfi Circle discussion meetings on educational reform, which soon became much more outspoken than was originally intended. In one of these meetings Lakatos forcefully argued for encouragement of a critical attitude, absence of censorship, and science as a guide to the Party rather than the other way round (this volume, appendix A). Many of Lakatos's contributions to Hungarian literary and academic journals during this period reveal the sharpness, originality, forcefulness, clarity, and at the same time, ambiguity which was to distinguish all of his later works.

The October–November revolution saw Eva Pap's parents eager to leave the country, especially after the suicide of their son following the release from prison of several old school friends whom he had denounced in the old Stalinist days. The family escaped to Austria on foot and settled in Vienna with friends. Following an application letter in which he expressed his intention to study both probability theory and mathematical heuristics, Lakatos was awarded a Rockefeller Fellowship to study at Kings College, Cambridge, with R. B. Braithwaite and Timothy Smiley.

He completed his Ph.D. thesis, "Essays in the Logic of Mathematical Discovery," in the form of a philosophical dialogue which takes place in a secondary school classroom among several bright students and their unconventional professor. The subject of the dialogue, developed along a quasi-Platonic line, is a famous mathematical conjecture, Euler's theorem. The professor starts by illustrating the theorem along with instances of the applicability of the formula. The students raise several objections, mainly by describing certain 'exceptional' polyhedra which cast doubt on 'what the proof really proves', as well as on the theorem itself. In the questions and answers that follow, Lakatos gives a 'rational reconstruction'—via conjectures, proofs, and refutations—of the growth of mathematical knowledge from Euler's initial formulation to its topological transformation as put forward by Poincaré. The 'real' history is given in the footnotes. The essay should be read bearing in mind both Pólya's attempt to bring mathematical heuristics back to life and Popper's epistemological fallibilism. It is aimed at the traditional way of teaching and practising mathematics. The thesis upheld here is that developing mathematics is not equivalent to heaping up eternal and immutable truths, but resembles rather a more exciting and creative activity such as that of putting forward hypotheses, attempting to give them a strict 'proof', and subsequently criticising them stringently in order to find counterexamples to both the original conjecture (theorem) and the subsequent steps of the proof. This widening of the original Popperian fallibilism to include mathematical knowledge is far from uncritical: against the separation of "context of discovery" (of some interest for empirical psychology but totally apart from the logical analysis of knowledge) and "context of justification" (the only field subject to rational analysis), Lakatos proposes the intrinsic unity of the two and claims the right of mathematical heuristics to a space which is rational as opposed to psychologistic.

Meanwhile, in 1959, Lakatos met Karl Popper and attended his seminars at the London School of Economics, the same university where he would serve as lecturer in logic until the end of his career, excepting a brief stint at the University of California at San Diego in the early 1960s. The British government, without giving reasons, refused to grant Lakatos citizenship, and for the rest of his life he would remain without an official homeland.

At LSE Lakatos also first met Paul Feyerabend among Popper's students (and subsequent critics). Lakatos's interests focused more and more on the philosophy of the natural sciences, and in 1965 he organised a conference at Bedford College in London on logic, the philosophy of mathematics, and scientific method, the proceedings of which he edited and published in four volumes.

The student revolt of 1968 was for Lakatos reminiscent of the demands of Nazi students to suppress "Judaic liberal Marxist" doctrines, of the death sentences passed against scholars of genetics by the Soviet Communist Party, and of student canvassing against the teaching of Einstein's "bourgeois relativism" in Communist universities.

His first essay in philosophy of science, entitled "Changes in the Problem of Inductive Logic," also dates to 1968. In it he attacked Carnap from a point of view which he admitted was still Popperian. In 1970 the LSE created a chair especially for Lakatos, appointing him "Professor in Logic with Special Reference to the Philosophy of Mathematics." In that same year he published "Falsification and the Methodology of Scientific Research Programmes." "History of Science and Its Rational Reconstructions" followed in 1971, and shows Lakatos dealing with the interaction between the history and the philosophy of science by bringing together Hegelian historicism and Popperian fallibilism in an only apparently paradoxical way. In "Popper on Demarcation and Induction," Lakatos finally parted with Popper's views, which he had, at least according to Popper, already abandoned in his various "rereadings" of his teacher's ideas.

Lakatos died on 2 February 1974, following a severe heart attack. He was 51. Neither his reply to Feyerabend's *Against Method* nor the book he had planned to entitle "The Changing Logic of Scientific Discovery" were ever written. His ideas and projects, however, were to be gathered and upheld by a lively school of his students at LSE.

This biography is based on Jancis Long, "Lakatos in Hungary" (1998); John Watkins, "Lakatos Obituary," *The Times*, 6 February 1974; and J. R. Ravetz, "Imre Lakatos—Philosopher of Dialectic" (Archive, 11.2).

Paul Feyerabend: Biography

Paul Feyerabend was born in Vienna in 1924. His father was a civil servant; his mother, a seamstress, committed suicide in 1943. As a young man, Feyerabend was attracted by physics, mathematics, and astronomy (he was a passionate observer through the telescope he built with his father). His interests, however, extended much further than school subjects to drama, philosophy, cinema, singing, and opera. From an early age he seemed to accept any accidental event as a stimulus to widen his cultural horizons, approaching philosophy, for example, in consequence of having found a few philosophical texts among some second-hand books he had bought in bundles.

Four years after the *Anschluss,* he was drafted into the Nazi work service, and later entered the Wehrmacht and trained in Yugoslavia. Posted to the Russian front, he was awarded the Iron Cross for leading his men into a village under enemy fire and subsequently occupying it. Although Feyerabend later tended to dismiss the war as something of an interruption of his previous life, he was to bear its consequences for the rest of his life: he received a bullet wound in his spine that would leave him crippled. As he remembers in his autobiography: "During battle I often forgot to take cover. It was not out of bravery—I am a great coward and easily frightened—but out of excitement: flames on the horizon, shooting, indistinct voices, attacks from planes in the air and tanks on the ground—it was like a theatre and I acted accordingly. On one such occasion I got the Iron Cross, on another three bullets—one in my face, one in my right hand, the third in my spine."

In 1946 he was granted state funding to study singing and stage management at the Weimar Musikschule, and followed this by cultivating Italian, harmony, piano, and diction for a few months. The following year he decided to return to Vienna and applied to read history and sociology at the university there. He soon changed to lectures in theoretical physics with Hans Thirring, Karl Pzibram, and Felix Ehrenhaft Especially thanks to the nonorthodox view held by Ehrenhaft, which Feyerabend was to understand better in the course of the Alpbach seminar years later, he came into contact with the 'real' nature of scientific rationality and began to develop his epistemological anarchism.

At a meeting of the International Summer School of the Austrian College Society in Alpbach in 1948, Feyerabend first met Karl Popper. That same year he married for the first time. The Alpbach seminar eventually brought about his appointment as scientific secretary to the

Society, and also provided the opportunity for studying Marxism, especially through Walter Hollitscher.

In 1951 Feyerabend received his Ph.D. in philosophy; his dissertation on "basic statements" was supervised by Viktor Kraft, a former member of the Vienna Circle around whom a lively discussion forum on the reality of theoretical entities and the external world had formed, to which Wittgenstein himself was often invited as a visiting speaker. In the same period Feyerabend met Bertholt Brecht and turned down an offer to work as his production assistant (which he called "one of the greatest mistakes of my life," even though he was to say that, as with Marxism and the army, he would probably not have enjoyed the gregarious group mentality).

Meanwhile, he went on brief study trips to Copenhagen, Oslo, and Stockholm, where he met Niels Bohr. In 1952, Feyerabend left for Cambridge, hoping to study under Wittgenstein, but the latter died, and Feyerabend turned to the London School of Economics, where he crossed Popper's path once more. A year later, he declined the offer of a job as Popper's assistant and left for Vienna, where he worked with Arthur Pap and met Herbert Feigl.

The University of Bristol granted Feyerabend his first academic post as lecturer in philosophy of science for the term starting in 1955. Two subsequent years spent studying in Vienna reinforced his decision to cut all ties with what he later called the "Popperian Church."

From 1958 to 1990, the year he tendered his official resignation, Feyerabend was lecturer and then professor at the University of California at Berkeley, but he also spent time at other universities both in the United States (Yale, Minnesota) and abroad (London, Berlin, Auckland, Brighton, Kassel), wherever his restlessness and growing fame took him. ("America was the first country that gave me a vague idea of what a culture might be. By American culture I don't mean Thoreau, Dewey, Stevens, or Henry Miller, but Hollywood, vaudeville, musicals, wrestling, soap operas, stand-up comedy, Spillane, Chandler, Hammett—in short, show business and pulp.") Very early on he worked at the Minnesota Center for the Philosophy of Science in Minneapolis on a grant, with Ernest Nagel, Hilary Putnam, Adolf Grünbaum, Grover Maxwell, and Paul Meehl.

Between 1960 and 1970, the year the essay version of "Against Method" appeared, Feyerabend focused on the philosophy of physics, structured his methodological justification of the role of realism in science, and defined his views of the theoretical nature of observation and the incommensurability of scientific theories. Feyerabend's theoretical pluralism at this time gave way to a more radical form of "anarchism," possibly following more direct involvement in politics after the student revolt and an awakening of his interest in political philosophy and political questions about science.

During his visits to the LSE he met Imre Lakatos, who encouraged him to publish *Against Method* (1975) in order to systematically collect all the ideas expounded in his lectures. After having written what he himself referred to as the "stink bomb," Feyerabend took up a chair at the University of Auckland in New Zealand, and subsequently at Sussex. Lakatos died in 1974, and their joint project of publishing a debate volume, in which Lakatos was to have defended "Law and Order" against Feyerabend's anarchic attack, was never completed. *Against Method* was ultimately published as a patchwork collection of essays; the details of its production are set out fully in this volume.

In 1978 *Science in a Free Society* was published. It contains Feyerabend's views on the political consequences of his epistemological anarchism, and argues for the separation of science and state, and for the equal right to survival and access to power of traditions and "forms of life" other than scientific "rationality." Feyerabend thus attempted to dethrone science from its privileged position within Western culture.

He accepted a lecturing post at the Zürich Polytechnic in the summer semesters during the 1980s ("ten wonderful years of half-Berkeley, half-Switzerland"). His important papers of these years are collected in *Farewell to Reason* (1987), a sui generis apology for cultural relativism.

In January 1989 he married again and left California for Switzerland and Rome, where he lived with his wife, Grazia Borrini. He was stricken by a brain tumour, and died on 11 February 1994.

Bibliography

In the following bibliography, works are cited by their date of first publication, but the exact edition from which quotations are taken is also specified and its date given if different from the first edition. For example, where the source of a quotation is given as Lakatos 1970, 55, the bibliography entry shows that quotation is from Lakatos's "Falsification and the Methodology of Scientific Research Programmes," which was first published in 1970, but that this work was republished in Lakatos's *Philosophical Papers,* volume 1, published in 1978; it is the 1978 edition to which the page number refers. Some citations refer to the second or a subsequent edition if this differs significantly from the first and is more appropriate in the context. For example, in the case of Popper's *The Logic of Scientific Discovery,* the first German edition (*Logik der Forschung,* 1934) and the English revised version are both of interest, so this work is cited as Popper [1934] 1959, and the page references are to the later edition. In other cases, a range of dates is given. For example, Lakatos's "Proofs and Refutations" was published as separate papers in *The British Journal for the Philosophy of Science,* some of which appeared in 1963 and the rest in 1964. It is therefore cited as Lakatos 1963–1964.

Works by Imre Lakatos

1959–1961. What does a mathematical proof prove? Published as chapter 4 of Lakatos 1978b.

1960. Necessity, Kneale and Popper. Published as chapter 7 of Lakatos 1978b.

1961. Essays in the logic of mathematical discovery. Ph.D. thesis, University of Cambridge.

1961/1973. The method of analysis-synthesis. (This paper consists of two sections. The first is the final chapter of Lakatos 1961; the second is based on an address given at a conference in Jyväskylä, Finland, in 1973.) Published as chapter 5 of Lakatos 1978b.

1962. Infinite regress and foundations of mathematics. *Aristotelian Society Supplementary* 36:155–84. Republished as chapter 1 of Lakatos 1978b.

1963–1964a. Proofs and refutations. *British Journal for the Philosophy of Science* 14:1–25, 129–39, 221–43, 296, 342. Republished in revised form as part of Lakatos 1976a.

1963–1964b. Newton's effect on scientific standards. Published as chapter 5 of Lakatos 1978a.

1966. *Dokatatelstva I Oprovershenia.* Russian translation of "Proofs and refutations" by I. N. Veselovski. Moscow: Publishing House of the Soviet Academy of Science.

1967. A renaissance of empiricism in the recent philosophy of mathematics? In I. Lakatos (ed.), *The Problem in the Philosophy of Mathematics.* Amsterdam: North Holland. Republished in expanded form as chapter 2 of Lakatos 1978b.

1968a. Changes in the problem of inductive logic. In I. Lakatos (ed.), *The Problem of Inductive Logic.* Amsterdam: North Holland. Republished as chapter 8 of Lakatos 1978b.

1968b. Criticism and the methodology of scientific research programmes. *Proceedings of the Aristotelian Society* 69:149–66.

1970. Falsification and the methodology of scientific research programmes. In I. Lakatos and A. Musgrave (eds.), *Criticism and the Growth of Knowledge.* Cambridge: Cambridge University Press. Republished as chapter 1 of Lakatos 1978a.

1971a. History of science and its rational reconstructions. In R. Buck and R. Cohen (eds.), *P.S.A. 1970: In Memory of Rudolf Carnap.* Boston Studies in the Philosophy of Science, vol. 8. Dordrecht: Reidel. Republished as chapter 2 of Lakatos 1978a.

1971b. Replies to critics. In R. Buck and R. Cohen (eds.), *P.S.A. 1970: In Memory of Rudolf Carnap,* 174–82. Boston Studies in the Philosophy of Science, vol. 8. Dordrecht: Reidel.

1973. The problem of appraising scientific theories: Three approaches. Published as chapter 6 of Lakatos 1978b. (This paper was the basis for some lectures Lakatos gave at Alpbach in 1973.)

1974a. Popper on demarcation and induction. In P. A. Schilpp (ed.), *The Philosophy of Karl Popper.* La Salle, IL: Open Court. Republished as chapter 3 of Lakatos 1978a.

1974b. The role of crucial experiments in science. *Studies in the History and Philosophy of Science* 4:309–25. (Paper delivered at Pennsylvania State University in 1973, to which Professor Grünbaum replied.)

1974c. Anomalies versus "crucial experiments" (A rejoinder to Professor Grünbaum). Published as chapter 10 of Lakatos 1978b.

1976a. *Proofs and Refutations: The Logic of Mathematical Discovery.* Edited by J. Worrall and E. Zahar. Cambridge: Cambridge University Press.

1976b. Understanding Toulmin. *Minerva* 14:126–34. Republished as chapter 11 of Lakatos 1978b.

1978a. *Philosophical Papers.* Vol. 1, *The Methodology of Scientific Research Programmes.* Edited by J. Worrall and G. Currie. Cambridge: Cambridge University Press.

1978b. *Philosophical Papers.* Vol. 2, *Mathematics, Science and Epistemology.* Edited by J. Worrall and G. Currie. Cambridge: Cambridge University Press.

Lakatos, I. (ed.). 1967. *The Problem in the Philosophy of Mathematics.* Amsterdam: North Holland.

Lakatos, I. (ed.). 1968. *The Problem of Inductive Logic.* Amsterdam: North Holland.

Lakatos, I., and A. Musgrave (eds.). 1968. *The Problem in the Philosophy of Science.* Amsterdam: North Holland.

Lakatos, I., and A. Musgrave (eds.). 1970. *Criticism and the Growth of Knowledge.* Cambridge: Cambridge University Press.

Lakatos, I., and E. G. Zahar. 1976. Why did Copernicus' programme supersede Ptolemy's? In R. Westman (ed.), *The Copernican Revolution.* Los Angeles: University of California Press. Republished as chapter 4 of Lakatos 1978a.

Works by Paul Feyerabend

1955a. Carnaps Theorie der Interpretation theoretischer Systeme. *Theoria* 21:55–62.

1955b. Wittgenstein's philosophical investigations. *Philosophical Review* 64: 449–83. Republished in revised form as chapter 7 of Feyerabend 1981b.

1957. On the quantum theory of measurement. In S. Körner (ed.), *Observation and Interpretation,* 121–30. London: Butterworth. Republished in revised form as chapter 13 of Feyerabend 1981a.

1958a. An attempt at a realistic interpretation of experience. *Proceedings of the Aristotelian Society* 58:143–70. Republished in revised form as chapter 2 of Feyerabend 1981a.

1958b. Reichenbach's interpretation of quantum mechanics. *Philosophical Studies* 9:42. Republished as chapter 15 of Feyerabend 1981a.

1960a. Das Problem der Existenz theoretischer Entitäten. In E. Topitsch (ed.), *Probleme der Wissenschaftstheorie,* 35–72. Vienna: Springer. Republished in revised form as chapter 3 of Feyerabend 1978b.

1960b. Patterns of discovery. *Philosophical Review* 59:247–52.

1960c. Professor Bohm's philosophy of nature. *British Journal for the Philosophy of Science* 10:321–38. Republished in revised form as chapter 14 of Feyerabend 1981a.

1960d. On the interpretation of scientific theories. *Proceedings of the 12th International Congress in Philosophy, 1958 (Milan/Venice)* 5:151. Republished as chapter 3 of Feyerabend 1981a.

1961a. *Knowledge without Foundations.* Oberlin, OH: Oberlin College.

1961b. Niels Bohr's interpretation of quantum theory. In H. Feigl and G. Maxwell (eds.), *Current Issues in the Philosophy of Science.* New York: Holt, Rinehart and Winston.

1962a. Explanation, reduction, and empiricism. In H. Feigl and G. Maxwell (eds.), *Scientific Explanation, Space, and Time,* 28–97. Minnesota Studies in the Philosophy of Science, vol. 3. Minneapolis: University of Minnesota Press. Republished in revised form as chapter 4 of Feyerabend 1981a.

1962b. Problems of microphysics. In R. Colodny (ed.), *Frontiers of Science and Philosophy.* Englewood Cliffs, NJ: Prentice Hall. Partially republished as chapter 17, "Hidden variables and the argument of Einstein, Podolsky and Rosen," in Feyerabend 1981a.

1963a. How to be a good empiricist. In B. Baumrin (ed.), *Philosophy of Science.* Delaware Seminar in the Philosophy of Science, vol. 2. New York: Interscience.

1963b. Materialism and the mind-body problem. *Review of Metaphysics* 17:49–66. Republished in revised form as chapter 10 of Feyerabend 1981a.

1964a. A note on the problem of induction. *Journal of Philosophy* 61:349–53. Republished as chapter 12 of Feyerabend 1981a.

1964b. Review of A. C. Crombie (ed.), *Scientific Change. British Journal for the Philosophy of Science* 15:244–54.

1964c. Realism and instrumentalism: Comments on the logic of factual support. In M. Bunge (ed.), *The Critical Approach to Science and Philosophy.* New York: Free Press. Republished as chapter 11 of Feyerabend 1981a.

1964d. The structure of science. *British Journal for the Philosophy of Science* 16:237. Republished as chapter 3 of Feyerabend 1981b.

1965a. Problems of empiricism I. In R. G. Colodny (ed.), *Beyond the Edge of Certainty,* 145–260. University of Pittsburgh Series in the Philosophy of Science, vol. 2. Englewood Cliffs, NJ: Prentice Hall.

1965b. Reply to criticism: Comments on Smart, Sellars and Putnam. In R. S. Cohen and M. W. Wartofsky (eds.), *Boston Studies in the Philosophy of Science,* vol. 2: *Proceedings of the Boston Colloquium for the Philosophy of Science, 1962–1964.* New York: Humanities Press. Republished in revised and expanded form as chapter 6 of Feyerabend 1981a.

1965c. On the "meaning" of scientific terms. *Journal of Philosophy* 12:266. Republished as chapter 5 of Feyerabend 1981a.

1967a. On the improvement of the sciences and the arts, and the possible identity of the two. In R. S. Cohen and M. W. Wartofsky (eds.), *Proceedings of the Boston Colloquium for the Philosophy of Science 1964/1966,* 387–415. *Boston Studies in the Philosophy of Science.* Dordrecht: Reidel.

1967b. Bemerkungen zur Geschichte und Systematik des Empirismus. In P. Weingartner (ed.), *Grundlagen der Wissenschaften und ihre Wurzeln in der Metaphysik.* Salzburg: Georg Pustet. Republished in revised form as chapter 11 of Feyerabend 1978b.

1968. Science, freedom and the good life. *Philosophical Forum* 1:127–35.

1968–1969. On a recent critique of complementarity. *Philosophy of Science,* 35:309, 36:82. Republished as chapter 16, "Bohr's World View," of Feyerabend 1981a.

1969a. Linguistic arguments and scientific method. *Telos* 2:43. Republished as chapter 9 of Feyerabend 1981a, and in an expanded German version as chapter 7 of Feyerabend 1978b.

1969b. Science without experience. *Journal of Philosophy* 66:791. Republished as chapter 7 of Feyerabend 1981a.

1970a. Against method: Outline of an anarchistic theory of knowledge. In M. Radner and S. Winokur (eds.), *Analyses of Theories and Methods of Physics and Psychology,* 17–130. *Minnesota Studies in the Philosophy of Science,* vol. 4. Minneapolis: University of Minnesota Press.

1970b. Classical empiricism. In R. E. Butts and J. W. Davies (eds.), *The Methodological Heritage of Newton.* Toronto: Toronto University Press. Republished as chapter 2 of Feyerabend 1981b.

1970c. Consolations for the specialist. In I. Lakatos and A. Musgrave (eds.), *Criticism and the Growth of Knowledge.* Cambridge: Cambridge University Press.

1970d. Experts in a free society. *The Critic* 29(2) November–December: 4–8.

1970e. In defence of classical physics. *Studies in the History and Philosophy of Science* 1(1):59–85.

1970f. Problems of empiricism II. In R. G. Colodny (ed.), *The Nature and Function of Scientific Theories.* University of Pittsburgh Series in the Philosophy of Science, vol. 4. Pittsburgh, PA: University of Pittsburgh Press.

1972. Von der beschränkten Gültigkeit methodologischer Regeln. *Neue Hefte für Philosophie,* 2–3, Göttingen. Republished as chapter 10 of Feyerabend 1978b.

1973. Die Wissenschaftstheorie—eine bisher unbekannte Form des Irrsinns? *Proceedings of the German Conference of Philosophy,* Kiel, 1972. Hamburg: Felix Meiner. Republished as chapter 12 of Feyerabend 1978b.

1974a. Popper's *Objective Knowledge. Inquiry* 17:475–507. Republished as chapter 9 of Feyerabend 1981b.

1974b. Die Wissenschaft in einer freien Gesellschaft. In W. C. Zimmerli (ed.), *Wissenschaftskrise und Wissenschaftskritik,* 109–19. Basel: Schwabe. Republished as chapter 14 of Feyerabend 1978b.

1975a. *Against Method: Outline of an Anarchistic Theory of Knowledge.* London: New Left Books.

1975b. How to defend society against science. *Radical Philosophy* 2:4–8. Republished in Hacking 1981.

1975c. Imre Lakatos. *British Journal for the Philosophy of Science* 26:1–8.

1975d. "Science": The myth and its role in society. Afterword: Theses on anarchism. *Inquiry* 18:167–81.

1975e. Let's make more movies. In C. J. Bontempo and S. J. Odell (eds.), *The Owl of Minerva: Philosophers on Philosophy,* 201–10. New York: McGraw Hill.

1976a. On the critique of scientific reason. In C. Howson (ed.), *Method and Appraisal in the Physical Sciences.* Cambridge: Cambridge University Press. Republished as chapter 10 of Feyerabend 1981b.

1976b. Logic, literacy and Prof. Gellner. *British Journal for the Philosophy of Science* 27:167–81. Republished in a revised and expanded version as chapter 2 of the third part of Feyerabend 1978a.

1977a. Marxist fairytales from Australia. *Inquiry* 20:372–97. Republished as chapter 3 of the third part of Feyerabend 1978a.

1977b. Changing patterns of reconstruction. *British Journal for the Philosophy of Science* 28:351–69.

1978a. *Science in a Free Society.* London: New Left Books.

1978b. *Die Wissenschaftstheorie und die Autorität der Wissenschaf.* Braunschweig: Friedr. Vieweg & Sohn.

1978c. The Gong Show—Popperian style. In G. Radnitzky and A. Andersson (eds.), *Progress and Rationality in Science,* 387–92. Dordrecht: Reidel.

1979. Dialogue on method. In G. Radnitzky and A. Andersson (eds.), *Structure and Development of Science,* 63–131. Dordrecht: Reidel.

1980a. *Erkenntnis für freie Menschen.* Frankfurt am Main: Suhrkamp.

1980b. Democracy, elitism and scientific method. *Inquiry* 23:3–18.

1981a. *Philosophical Papers.* Vol. 1, *Realism, Rationalism and Scientific Method.* Cambridge: Cambridge University Press.

1981b. *Philosophical Papers.* Vol. 2, *Problems of Empiricism.* Cambridge: Cambridge University Press.

1984a. Philosophy of Science 2001. In R. S. Cohen and M. W. Wartofsky (eds.), *Methodology, Metaphysics and the History of Science,* 137–47. Boston Studies in the Philosophy of Science, vol. 84. Dordrecht: Reidel.

1984b. Xenophanes: A forerunner of critical rationalism? In G. Andersson (ed.), *Rationality in Science and Politics,* 95–109. Dordrecht: Reidel.

1987. *Farewell to Reason.* London, New York: Verso.

1988. *Against Method: Outline of an Anarchistic Theory of Knowledge.* 2d ed. London: Verso.

1991. *Three Dialogues on Knowledge.* Oxford: Blackwell.

1993. *Against Method.* 3d rev. ed. London: Verso.

1995. *Killing Time: The Autobiography of Paul Feyerabend.* Chicago: University of Chicago Press.

1996. *These zum Anarchismus: Artikel aus der Reihe "Unter dem Pflaster liegt der Strand."* Berlin: Kramer.

Feyerabend, P. K., and H. Albert. 1997. *Briefwechsel.* Edited by Wilhelm Baum. Frankfurt am Main: Fischer Taschenbuch Verlag.

Other Works Cited

Achinstein, P. 1971. *Law and Explanation: An Essay in the Philosophy of Science.* London: Oxford University Press.

Agassi, J. 1974. Modified conventionalism is more comprehensive than modified essentialism. In P. A. Schilpp (ed.), *Philosophy of Karl Popper,* 693–96. La Salle, IL: Open Court.

———. 1975. *Science in Flux.* Dordrecht: Reidel.

Ahonen, G. 1989. On the empirical content of Keynes' "General Theory." *Ricerche Economiche* 43(1–2):256–69.

———. 1990. Commentary on D. Wade Hands' "Second thoughts on 'Second thoughts: Reconsidering the Lakatosian progress of *The General Theory*'." *Review of Political Economy* 2:94–101.

Albert, H. [1969] 1984. *Treatise on Critical Reason.* Princeton, NJ: Princeton University Press.

Avenarius, R. 1888–1890. *Kritik der Reinen Erfahurung.* 2 vols. Leipzig.

Ayer, A. J. [1936] 1946. *Language, Truth and Logic.* 2d rev. ed. London: V. Gollancz. Reprint, Harmondsworth, England: Penguin Books, 1990.

Backhouse, R. E. 1993. Lakatosian perspectives on general equilibrium analysis. *Economics and Philosophy* 9(2):271–87.

———. 1994. The Lakatosian legacy in economic methodology. In R. E. Backhouse (ed.), *New Directions in Economic Methodology.* London, New York: Routledge.

———. 1997. *Truth and Progress in Economic Knowledge.* Aldershot: Edward Elgar.

Backhouse, R. E., D. Hausman, U. Maki, and A. Salanti (eds.). 1997. *Economics and Methodology: Crossing Boundaries.* London: Macmillan.

Bartley, W. W. III. 1968. Theories of demarcation between science and metaphysics. In I. Lakatos and A. Musgrave (eds.), *The Problem in the Philosophy of Science.* Amsterdam: North Holland.

Beck, A. D., and W. Yourgrau (eds.). 1970. *Physics, Logic and History.* New York, London: Plenum.

Berlin, I. 1939. Verifiability in principle. *Proceedings of the Aristotelian Society* 39:225–48.

Bernal, J. D. 1954. *Science in History.* 4 vols. London: Watts.

Birner, J. 1990. *Strategies and Programmes in Capital Theory: A Contribution to the Methodology of Theory Development.* Ph.D. thesis, University of Amsterdam.

Blaug, M. 1976. Kuhn versus Lakatos, or paradigms versus research programmes in the history of economics. In S. Latsis (ed.), *Method and Appraisal in Economics,* 149–80. Cambridge: Cambridge University Press.

———. 1980. *A Methodological Appraisal of Marxian Economics.* Amsterdam: North Holland.

———. [1980] 1992. *The Methodology of Economics.* 2d ed. Cambridge Surveys of Economic Literature. Cambridge: Cambridge University Press.

———. 1990. Reply to D. Wade Hands' "Second thoughts on 'Second thoughts: Reconsidering the Lakatosian progress of *The General Theory*'." *Review of Political Economy* 2:102–4.

———. 1991. Second thoughts on the Keynesian revolution. *History of Political Economy* 23:171–92.

———. 1994. Why I am not a constructivist: Confessions of an unrepentant Popperian. In R. E. Backhouse (ed.), *New Directions in Economic Methodology.* London, New York: Routledge.

Bloor, D. 1976. *Knowledge and Social Imagery.* London: Routledge and Kegan Paul.

Bohr, N. 1913a. On the constitution of atoms and molecules. *Philosophical Magazine* 26:1–25, 476–502, 857–75.

———. 1913b. The spectra of helium and hydrogen. *Nature* 92:231–32.

Born, M. 1949. *Natural Philosophy of Cause and Chance.* Oxford: Oxford University Press.

———. 1962. *Einstein's Theory of Relativity.* New York: Dover.

Buck, R., and R. Cohen (eds.). 1971. *P.S.A. 1970: In Memory of Rudolf Carnap.* Boston Studies in the Philosophy of Science, vol. 8. Dordrecht: Reidel.

Bunge, M. (ed.) 1964. *The Critical Approach to Science and Philosophy: Essays in Honor of K. R. Popper.* London: Free Press.

Campbell, R., and T. Vinci. 1983. Novel confirmation. *British Journal for the Philosophy of Science* 34:315–41.

Carnap, R. 1936–1937. Testability and meaning. *Philosophy of Science* 3:419–71, 4:1–40.

Cartwright, N., J. Cat, L. Fleck, and T. E. Uebel. 1996. *Otto Neurath: Philosophy between Science and Politics.* Cambridge: Cambridge University Press.

Church, A. 1949. Review of Ayer, *Language, Truth and Logic,* 2d revised and re-set edition. *Journal of Symbolic Logic* 14(1):52–53.

Clark, P. 1976. Atomism versus thermodynamics. In C. Howson (ed.), *Method and Appraisal in the Physical Sciences,* 41–104. Cambridge: Cambridge University Press.

Cohen, I. B. 1971. *Introduction to Newton's "Principia."* Cambridge, MA: Harvard University Press

———. 1974. Newton's theory vs. Kepler's theory and Galileo's theory: An example of a difference between a philosophical and historical analysis of science. In Y. Elkana (ed.), *The Interaction between Science and Philosophy,* 299–338. New York: Humanities Press.

———. 1980. *The Newtonian Revolution.* Cambridge: Cambridge University Press.

Cohen, R. S., P. K. Feyerabend, and M. W. Wartofsky (eds.). 1976. *Essays in Memory of Imre Lakatos.* Boston Studies in the Philosophy of Science, vol. 39. Dordrecht, Boston: Reidel.

Cohen, R. S., and M. W. Wartofsky (eds.). 1983. *Epistemology, Methodology and the Social Sciences.* Dordrecht: Reidel.

Cohn-Bendit, D. and G. Cohn-Bendit. 1968. *Obsolete Communism: The Left-Wing Alternative.* Translated from the French by Arnold Pomerans. London: Deutsch.

Colodny, R. G. (ed.). 1972. *Paradigms and Paradoxes.* University of Pittsburgh Series in the Philosophy of Science, vol. 5. Pittsburgh, PA: University of Pittsburgh Press, 1972.

Couvalis, G. 1989. *Feyerabend's Critique of Foundationalism.* Aldershot: Avebury.

Crombie, A. C. [1952] 1969. *The History of Science from Augustine to Galileo.* London: Heinemann. Rev. ed., Harmondsworth: Penguin Books.

Dahrendorf, R. 1995. *A History of the London School of Economics and Political Science 1895–1995.* Oxford: Oxford University Press.

Dalla Chiara, M. L. (ed.). 1981. *Italian Studies in the Philosophy of Science.* Boston Studies in the Philosophy of Science, vol. 47. Dordrecht: Reidel.

de Grazia, A. (ed.). 1966. *The Velikovsky Affair.* London: Sidwick and Jackson.

de Marchi, N., and M. Blaug (eds.). 1991. *Appraising Economic Theories.* Aldershot: Edward Elgar.

de Santillana, G. 1961. *The Crime of Galileo.* London: Mercury Books.

Descartes, R. 1637. *Discourse on Method and Other Writings.* Harmondsworth: Penguin Books, 1960.

Dorling, J. 1968. Length contraction and clock synchronization: The empirical equivalence of the Einstein and Lorentzian theories. *British Journal for the Philosophy of Science* 19:67–69.

———. 1971. Einstein's introduction of photons: Argument by analogy or duction from the phenomena? *British Journal for the Philosophy of Science* 22:1–8.

———. 1979. Bayesian personalism, the methodology of research programmes, and Duhem's problem. *Studies in History and Philosophy of Science* 10:177–87.

———. 1982. Further illustrations of the Bayesian solution of Duhem's problem. Mimeo.

Drake, S. 1978. *Galileo at Work: His Scientific Biography.* Chicago, London: University of Chicago Press.

Duhem, P. 1892. Quelques réflexions au sujet des théories physiques. In P. Duhem, *Prémices philosophiques,* 1–39. Leiden: Brill, 1987.

———. 1906. *La théorie physique: Son objet et sa structure.* Chevalier et Rivière, Paris. English translation of 2d ed. (1914), *The Aim and Structure of Physical Theory* (Princeton, NJ: Princeton University Press, 1954).

———. 1908. *To Save Phenomena.* Chicago: University of Chicago Press, 1969.

Dusek, V. 1998. Lukacs and Brecht as Teachers of Feyerabend and Lakatos. In *History of the Human Sciences,* vol. 11, no. 2: 25–44.

Einstein, A. 1905. Über einen die Erzeugung und Verwandlung des Lichtes betreffenden heuristischen Gesichtspunkt. *Annalen der Physik,* serie 4, vol. 17: 132–48.

———. 1951. *Albert Einstein: Philosopher-Scientist.* Edited by P. A. Schilpp. New York: Tudor.

Einstein, A., H. Born, and M. Born. 1969. *Briefwechsel 1916–1955.* München: Nymphenburger Verlagshandlung.

Elkana, Y. (ed.). 1974. *The Interaction between Science and Philosophy.* New York: Humanities Press.

Evans, E. E. 1913. The spectra of helium and hydrogen. *Nature* 92:5.

Evans-Pritchard, E. E. 1937. *Witchcraft, Oracles and Magic among the Azande.* Oxford: Clarendon Press.

———. 1965. *Theories in Primitive Religion.* Oxford: Clarendon Press.

Eysenck, H. J. 1971. *Race, Intelligence and Education.* London: Temple Smith.

Feigl, H. 1950. The mind-body problem in the development of logical empiricism. *Revue internationale de philosophie* 4:64–83.

———. 1970–1971. Some crucial issues of mind-body monism. *Synthèse* 22: 295–312.

Feigl, H., and P. E. Meehl. 1974. The determinism-freedom and body-mind problems. In P. A. Schilpp (ed.), *Philosophy of Karl Popper.* La Salle, IL: Open Court.

Festinger, L., H. W. Riecken, and S. Schacher. 1956. *When Prophecy Fails.* Minneapolis: University of Minnesota Press.

Fisher, R. M. 1986. *The Logic of Economic Discovery: Neoclassical Economics and the Marginal Revolution.* Brighton: Wheatsheaf.

Fock, W. A. (ed.) 1962. *Philosophische Probleme der Modernen Naturwissenschaft.* Berlin.

Forman, P. 1969. The discovery of diffraction of x-rays by crystals: A critique of the myths. *Archive for History of Exact Sciences* 6:72–81.

Foucault, M. 1961. *Folie et déraison: Histoire de la folie à l'âge classique.* Paris: Plon. Reprinted with a new preface and appendix as *Histoire de la folie à l'âge classique* (Paris: Gallimard, 1972).

Fowler, A. 1913. The spectra of helium and hydrogen. *Nature* 92:95.

Frankel, H. 1979. The career of continental drift theory: An application of Imre Lakatos' analysis of scientific growth to the rise of drift theory. *Studies in History and Philosophy of Science* 10(1):21–66.

Galilei, G. 1632. *Dialogue on the Great World Systems.* Chicago: University of Chicago Press, 1963.

Gardner, M. R. 1982. Predicting novel facts. *British Journal for the Philosophy of Science* 33:1–15.

Gavroglu, K., Y. Goudaroulis, and P. Nicolacopoulos (eds.). 1989. *Imre Lakatos and Theories of Scientific Change.* Dordrecht: Kluwer.

Gellner, E. 1974. Lakatos obituary. *The Times,* 8 February.

———. 1975. Beyond truth as falsehood: Review of Feyerabend *Against Method. British Journal for the Philosophy of Science* 26:331–42.

Geymonat, L. 1957. *Galileo Galilei.* Torino: Einaudi. English translation, London, New York: McGraw Hill, 1965.

Giedymin, J. 1970. The paradox of meaning variance. *British Journal for the Philosophy of Science* 21:257–68.

———. 1971. Consolations for the irrationalist? *British Journal for the Philosophy of Science* 22:39–48.

———. 1973. Logical comparability and conceptual disparity between Newtonian and relativistic mechanics. *British Journal for the Philosophy of Science* 24: 270–76.

———. 1978. Introduction: Radical conventionalism, its background and its evolution: Poincaré, Le Roy and Ajdukiewicz. In J. Giedymin (ed.), *Kazimierz Ajdukiewicz: The Scientific World-Perspective and Other Essays 1931–1963.* Dordrecht: Reidel.

———. 1982. *Science and Convention: Essays on Henri Poincaré's Philosophy of Science and the Conventionalist Tradition.* Oxford: Pergamon.

Gillies, D. 1988. Non-Bayesian confirmation theory and the principle of explanatory surplus. In A. Fine and J. Leplin (eds.), *P.S.A. 1988,* vol. 2, 373–80. East Lansing, MI: Philosophy of Science Association.

———. 1990. Bayesianism versus falsificationism. *Ratio* 3:82–98.

———. 1993. *Philosophy of Science in the Twentieth Century.* Oxford: Blackwell.

——— (ed.). 1992. *Revolutions in Mathematics.* Oxford: Oxford University Press.

Giorello, G. 1981. Intuition and rigor: Some problems of a "logic of discovery" in mathematics. In M. L. Dalla Chiara, *Italian Studies in the Philosophy of Science.* Boston Studies in the Philosophy of Science, vol. 47. Dordrecht: Reidel.

———. 1992. The "fine structure" of mathematical revolutions: Metaphysics, legitimacy and rigour. The case of the calculus from Newton to Berkeley and Maclaurin. In D. Gillies (ed.), *Revolutions in Mathematics,* 134–68. Oxford: Oxford University Press.

Glymour, C. 1980. *Theory and Evidence.* Princeton, NJ: Princeton University Press.

Goodman, N. 1955. *Fact, Fiction and Forecast.* London: Athlone Press.

Goodstein, R. L. 1969. Empiricism in mathematics. *Dialectica* 23:50–57.

Grosser, M. 1962. *The Discovery of Neptune.* Cambridge, Mass.: Harvard University Press. Reprint, New York: Dover, 1979.

Grünbaum, A. 1960. The Duhemian argument. *Philosophy of Science* 32:75–87.

———. 1966. The falsifiability of a component of a theoretical system. In P. K. Feyerabend and G. Maxwell (eds.), *Mind, Matter and Method: Essays in Honor of Herbert Feigl,* 273–305. Minneapolis: University of Minnesota Press.

———. 1971. Can we ascertain the falsity of a scientific hypothesis? *Studium Generale* 22:1061–93.

Hacking, I. 1972. Review of *L'Archéologie du savoir. Cambridge Review* 2: 166–70.

———. 1979. Imre Lakatos' philosophy of science. *British Journal for the Philosophy of Science* 30:381–410.

———. 1981. *Scientific Revolutions.* Oxford: Oxford University Press.

Hall, R. 1971. Can we use the history of science to decide between competing methodologies? In R. Buck and R. Cohen (eds.), *P.S.A. 1970: In Memory of Rudolf Carnap. Boston Studies in the Philosophy of Science,* vol. 8. Dordrecht: Reidel.

Haller, R. 1991. The Neurath principle: Its grounds and consequences. In T. Uebel (ed.), *Rediscovering the Vienna Circle: Austrian Studies on Otto Neurath and the Vienna Circle.* Dordrecht: Kluwer Academic Publishers.

Hallet, M. 1979. Towards a theory of mathematical research programmes. *British Journal for the Philosophy of Science* 30:1–25, 135–59.

Hands, D. W. 1985. Second thoughts on Lakatos. *History of Political Economy* 17:(1):1–16.

———. 1990. Second thoughts on "Second thoughts": Reconsidering the Lakatosian progress of *The General Theory. Review of Political Economy* 21:69–81.

———. 1993. Popper and Lakatos in economics methodology. In U. Mäki, B. Gustaffson, and C. Knudsen (eds.), *Rationality, Institutions and Economic Order,* 61–75. London: Routledge and Kegan Paul.

Hausman, D. M. 1992. *The Inexact and Separate Science of Economics.* Cambridge: Cambridge University Press.

Havas, P. 1964. Four-dimensional formulation of Newtonian mechanics and their relation to special and general relativity. *Review of Modern Physics* 39:490.

Hegel, G. W. F. 1807. *Phenomenology of Spirit.* Oxford: Clarendon Press, 1977.

Heilbron, J. L., and T. S. Kuhn. 1969. The genesis of Bohr's atom. *Journal of Historical Studies in the Physical Sciences* 1:211–90.

Hempel, C. G. 1965. *Aspects of Scientific Explanation.* New York: Free Press.

Hempel, C. G., and P. Oppenheim. 1945. A definition of "degree of confirmation." *Philosophy of Science* 12:98–115.

Hesse, M. 1963. A new look at scientific explanation. *Review of Metaphysics* 17: 98–108.

Holton, G. 1973. *Thematic Origins of Scientific Thought: Kepler to Einstein.* Cambridge, MA: Harvard University Press.

Howson, C. 1973. Must the logical probability of laws be zero. *British Journal for the Philosophy of Science* 24:153–63.

———. 1988. Accommodation, prediction and Bayesian confirmation theory. In A. Fine and J. Leplin (eds.), *P.S.A. 1988,* vol. 2, 381–92. East Lansing, MI: Philosophy of Science Association.

———. 1997. *Logic with Trees: An Introduction to Symbolic Logic.* London, New York: Routledge.

——— (ed.). 1976. *Method and Appraisal in the Physical Sciences.* Cambridge: Cambridge University Press.

Howson, C., and P. Urbach. 1989. *Scientific Reasoning: The Bayesian Approach.* La Salle, IL: Open Court.

Hübner, K. 1978. *Kritik der wissenschaftlichen Vernunft.* Freiburg-München: Verlag Karl Alber.

Hume, D. 1748. *An Enquiry Concerning Human Understanding and Concerning the Principles of Morals.* Edited by A. Selby-Bigge. Oxford: Clarendon Press, 1927.

Jensen, A. R. 1969. How much can we boost IQ and scholastic achievement? *Harvard Educational Review* 39:1–123.

———. 1972. *Genetics and Education.* London: Methuen.

———. 1973. *Education Differences.* London: Methuen.

Kadvany, J. 1995. The mathematical present as history. *Philosophical Forum* 26(4):263–87.

Keynes, J. M. 1936. *The General Theory of Employment, Interest and Money.* London: Macmillan, 1973.

Kidd, H. 1969. *The Trouble at L.S.E 1967–1968.* London: Oxford University Press.

Kline, M. 1972. *Mathematical Thought from Ancient to Modern Times.* Oxford, New York: Oxford University Press.

Koertge, N. 1971. Inter-theoretic criticism. In R. Buck and R. Cohen (eds.), *P.S.A. 1970: In Memory of Rudolf Carnap.* Boston Studies in the Philosophy of Science, vol. 8. Dordrecht: Reidel.

Koestler, A. 1959. *The Sleepwalkers.* London: Hutchinson.

Koetsier, T. 1991. *Lakatos' Philosophy of Mathematics: A Historical Approach.* Amsterdam, London, New York: North-Holland.

Kraft, V. 1953. *The Vienna Circle.* New York: Philosophical Library.

———. 1974. Popper and the Vienna Circle. In P. A. Schilpp (ed.), *The Philosophy of Karl Popper,* 185–204. La Salle, IL: Open Court.

Kuhn, T. 1957. *The Copernican Revolution: Planetary Astronomy in the Development of Western Thought.* Cambridge, MA: Harvard University Press.

———. [1962] 1970. *The Structure of Scientific Revolutions.* 2d ed. Chicago: University of Chicago Press.

———. 1970a. Logic of discovery or psychology of research? In I. Lakatos and A. Musgrave (eds.), *Criticism and the Growth of Knowledge,* 1–23. Cambridge: Cambridge University Press.

———. 1970b. Reflection on my critics. In I. Lakatos and A. Musgrave (eds.), *Criticism and the Growth of Knowledge,* 231–78. Cambridge: Cambridge University Press.

———. 1978. *Black-Body Theory and the Quantum Discontinuity 1894–1912.* Oxford: Clarendon Press.

Kvasz, L., G. Kampis, and M. Stoeltzner (eds.). In press. *The Philosophy of Imre Lakatos—Its Roots, Content and Limitations.* Publication of the Vienna Circle Institute. Dordrecht, Boston, London: Kluwer.

Larvor, B. 1998. *Lakatos: An Introduction.* London: Routledge.

Latsis, S. 1972. Situational determinism in economics. *British Journal for the Philosophy of Science* 23:207–45.

———. 1983. The role and status of the rationality principle in the social sciences. In R. S. Cohen and M. W. Wartofsky (eds.), *Epistemology, Methodology and the Social Sciences.* Dordrecht: Reidel.

———. 1976. *Method and Appraisal in Economics.* Cambridge: Cambridge University Press.

Laudan, L. 1977. *Progress and Its Problems.* Berkeley: University of California Press.

———. 1989. If it ain't broke, don't fix it. *British Journal for the Philosophy of Science* 40:369–75.

Lecky, W. E. H. 1910. *History of the Rise and Influence of the Spirit of Rationalism in Europe.* 2 vols. London: Longmans Green.

Lenin, V. I. 1908. *Collected Works.* Vol. 13, *Materialism and Empirio-Criticism.* London: Martin Lawrence, 1927.

Lichtmein, G. 1970. *Lukács.* London: Collins.

Long, J. 1998. Lakatos in Hungary. *Philosophy of Social Sciences* 28(2):244–311.

Lukács, G. 1962. *The Destruction of Reason.* London: Merlin Press. Reprint 1980.

Mach, E. 1886. *Analysis of Sensations.* New York: Dover, 1959.

———. 1905. *Erkenntnis und Irrtum: Skizzen zur Psychologie der Forschung.* Leipzig: J. A. Barth. English translation for the Vienna Circle Collection, *Knowledge and Error: Sketches on the Psychology of Enquiry* (Dordrecht, Boston: Reidel, 1976).

Machamer, P., and S. Lunsford. 1975. Review of W. Shea, *Galileo's Intellectual Revolution* (London: Macmillan, 1972). *British Journal for the Philosophy of Science* 26:81–82.

Magee, B. (ed.). 1971. *Modern British Philosophy.* London: Secker and Warburg.

Mäki, U., B. Gustaffson, and C. Knudsen (eds.). 1993. *Rationality, Institutions and Economic Order.* London: Routledge and Kegan Paul.

Marcuse, H. 1941. *Reason and Revolution.* Oxford: Oxford University Press. Reprint with supplementary chapter, London: Routledge and Kegan Paul, 1967.

Masterman, M. 1970. The nature of a paradigm. In I. Lakatos and A. Musgrave (eds.), *Criticism and the Growth of Knowledge,* 59–89. Cambridge: Cambridge University Press.

McMullin, E. 1971. The history and philosophy of science: A taxonomy. In R. Stewuer (ed.), *Historical and Philosophical Perspectives of Science,* 12–67. Minnesota Studies in the Philosophy of Science, vol. 5. Minneapolis: University of Minnesota Press.

Merton, R. K. 1973. *The Sociology of Knowledge, Theoretical and Empirical Investigation.* Chicago, London: University of Chicago Press.

Mészáros, I. 1972. *Lukács' Concept of Dialectic.* London: Merlin Press.

Mill, J. S. 1843. *A System of Logic Ratiocinative and Inductive.* London: J. W. Parker.

———. 1859. *On Liberty.* London: Parker.

Miller, A. 1974. On Lorentz's methodology. *British Journal for the Philosophy of Science* 25:29–45.

Miller, A. I. 1981. *Albert Einstein's Special Theory of Relativity: Emergence (1905) and Early Interpretation (1905–1911).* Reading, MA: Addison-Wesley.

Mondadori, M. 1989. Preface to *La logica dell'incerto,* by B. de Finetti, vii–xxiv. Milano: Il Saggiatore.

Mongin, P. 1988. Problèmes de Duhem en théorie de utilité expèrée. *Fundamenta Scientiae* 9:299–327.

Moseley, F. (ed.). 1995. *Heterodox Economic Theories: True or False?* Aldershot: Edward Elgar.

Motterlini, M. 1995. Has Lakatos really gone a long way towards epistemological anarchism? *Epistemologia* 18:215–32.

———. 1999. Professor Imre Lakatos between the Hegelian Devil and the Popperian deep blue sea. In L. Kvasz, G. Kampis, and M. Stoeltzner (eds.), *The Philosophy of Imre Lakatos: Its Roots, Content and Limitations.* Publication of the Vienna Circle Institute. Dordrecht, Boston, London: Kluwer. In press.

Murphy, N. 1989. Another look at novel facts. *Studies in History and Philosophy of Science* 20(3):385–88.

Musgrave, A. 1974. Logical versus historical theory of confirmation. *British Journal for the Philosophy of Science* 25:1–23.

———. 1976. Why did oxygen supplant phlogiston? In C. Howson (ed.), *Method and Appraisal in the Physical Sciences,* 181–210. Cambridge: Cambridge University Press.

———. 1978. Evidential support, falsification, heuristics and anarchism. In G. Radnitzky and A. Andersson (eds.), *Progress and Rationality in Science,* 180–201. Dordrecht: Reidel.

———. 1993. *Common Sense, Science and Scepticism.* Cambridge: Cambridge University Press.

Nagel, E. 1961. *The Structure of Science: Problems in the Logic of Scientific Explanation.* London: Routledge & Kegan Paul, 1979.

Neurath, O. 1935. Pseudorationalismus der Falsifikation. *Erkenntnis* 5:353–65. English translation, "Pseudorationalism of Falsification," in Neurath 1983.

———. 1983. *Philosophical Writings (1913–1945).* Edited by R. S. Cohen and M. Neurath. Vienna Circle Collection, vol. 16. Dordrecht: Reidel.

Newton-Smith, W. 1981. *The Rationality of Science.* London: Routledge and Kegan Paul.

Nunan, R. 1984. Novel facts, Bayesian rationality and the history of continental drift. *Studies in History and Philosophy of Science* 15(4):267–307.

O'Hear, A. 1980. *Karl Popper.* London: Routledge and Kegan Paul.

——— (ed.). 1995. *Karl Popper: Philosophy and Problems.* Cambridge: Cambridge University Press.

Oakeshott, M. 1993. *Morality and Politics in Modern Europe: The Harvard Lectures.* S. R. Letwin (ed.). London-New Haven: Yale University Press.

Oldroyd, D. 1986. *The Arch of Knowledge: An Introductory Study of the History of the Philosophy of Science.* New York, London: Methuen.

Pais, A. 1982. *Subtle is the Lord: The Life and Science of Albert Einstein.* Oxford: Oxford University Press.

———. 1991. *Niels Bohr's Times: In Physics, Philosophy and Polity.* Oxford: Oxford University Press.

Pera, M. 1989. Methodological sophisticationism: A degenerating project. In K. Gavroglu, Y. Goudaroulis, and P. Nicolacopoulos (eds.), *Imre Lakatos and Theories of Scientific Change,* 169–87. Dordrecht: Kluwer.

Pitt, J. (ed.). 1985. *Change and Progress in Modern Science.* Dordrecht: Reidel.

Planck, M. 1900a. Über eine Verbesserung der Wienschen Spektralgleichung. *Verhandlungen der Deutschen Physikalischen Gesellschaft* 2:202–4.

———. 1900b. Zur Theorie des Gesetzes der Energieverteilung im Normalspektrum. *Verhandlungen der Deutschen Physikalischen Gesellschaft* 2:237–45.

Poincaré, H. 1890. Sur le problème des trois corps et les équations de dynamique. *Acta mathematica* 13:1–270. Republished in Poincaré 1916–1956, vol. 7, 262–479.

———. 1891. Le problème des trois corps. *Revue générale des sciences pures et appliquées* 2:1–5. Republished in Poincaré 1916–1956, vol. 7, 529–37.

———. 1898. Sur la stabilité du système solaire. *Annuaire du Bureau des Longitudes,* B1–B16. Republished in Poincaré 1916–1956, vol. 7, 538–47.

———. 1902. *La science et l'hypothèse.* Paris: Flammarion. Authorized English translation by G. B. Halsted, *Science and Hypothesis,* in Poincaré 1913.

———. 1905. *La valeur de la science.* Paris: Flammarion. Authorized English translation by G. B. Halsted, *The Value of Science,* in Poincaré 1913.

———. 1908. *Science et méthode.* Paris: Flammarion. Authorized English translation by G. B. Halsted, *Science and Method,* in Poincaré 1913.

———. 1913. *The Foundations of Science.* Lancaster, PA: The Science Press. 2d ed. 1946.

———. 1916–1956. *Oeuvres.* 11 vols. Paris: Gauthier-Villars.

Polanyi, M. 1958. *Personal Knowledge: Towards a Post-Critical Philosophy.* London: Routledge.

———. 1967. *The Tacit Dimension.* London: Routledge.

Pólya, G. 1954. *Mathematics and Plausible Reasoning.* 2 vols. Princeton, NJ: Princeton University Press.

———. 1962. *Mathematical Discovery.* New York: Wiley.

Popkin, R. 1968. Scepticism, theology and the scientific revolution in the seventeenth century. In I. Lakatos and A. Musgrave (eds.), *The Problem in the Philosophy of Science.* Amsterdam: North Holland.

———. 1970. Scepticism in the study of history. In A. D. Beck and W. Yourgrau (eds.), *Physics, Logic and History.* New York, London: Plenum.

———. 1979. *The History of Scepticism from Erasmus to Spinoza.* Berkeley: University of California Press.

———. 1993. *The High Road to Pyrrhonism.* Edited by R. A. Watson and J. E. Force. Indianapolis, Cambridge: Hackett.

Popper, K. R. [1932–1933] 1979. *Die Beiden Grundprobleme der Erkenntnistheorie.* Tübingen: J. C. B. Mohr.

———. [1934] 1959. *Logik der Forschung.* Vienna: Springer. Expanded English edition, *The Logic of Scientific Discovery* (London: Hutchinson).

———. 1940. What is dialectic? *Mind* 49:403–26. Republished as chapter 15 of Popper 1963.

———. [1945] 1966. *The Open Society and Its Enemies.* 2 vols. 2d rev. ed. London: Routledge.

———. 1949. Naturgesetze und theoretische Systeme. In S. Moser (ed.), *Gesetz und Wirklichkeit,* 43–60. Vienna: Tyrolia Verlag. Republished as appendix I of Popper 1972.

———. [1956] 1982a. *Postscript to the Logic of Scientific Discovery.* Vol. 2, *The Open Universe: An Argument for Indeterminism,* edited by W. Bartley III. London: Hutchinson. (*Postscript to the Logic of Scientific Discovery* was written in 1956.)

———. [1956] 1982b. *Postscript to the Logic of Scientific Discovery.* Vol. 3, *Quantum Theory and the Schism in Physics,* edited by W. Bartley III. London: Hutchinson.

———. [1956] 1983. *Postscript to the Logic of Scientific Discovery.* Vol. 1, *Realism and the Aim of Science,* edited by W. Bartley III. London: Hutchinson.

———. 1957a. *The Poverty of Historicism.* London: Lowe and Brydone.

———. 1957b. The Aim of Science. *Ratio* 1, no. 1:24–35. Republished as chapter 5 of Popper 1972.

———. 1963. *Conjectures and Refutations.* London: Routledge and Kegan Paul.

———. 1967. Quantum mechanics without "the observer." In M. Bunge (ed.), *Quantum Theory of Reality.* Berlin, Heidelberg, New York: Springer Verlag.

———. 1972. *Objective Knowledge.* Oxford: Clarendon Press.

———. 1974a. Intellectual autobiography. In P. A. Schilpp (ed.), *The Philosophy of Karl Popper,* 3–183. La Salle, IL: Open Court.

———. 1974b. Replies to my critics. In P. A. Schilpp (ed.), *The Philosophy of Karl Popper,* 961–1197. La Salle, IL: Open Court.

Preston, J. 1997. *Feyerabend: Philosophy, Science, and Society.* Cambridge: Polity Press,

Putnam, H. 1974. The "corroboration" of theories. In P. A. Schilpp (ed.), *The Philosophy of Karl Popper,* 221–40. La Salle, IL: Open Court.

———. 1981. *Reason, Truth and History.* Cambridge: Cambridge University Press.

Quine, W. V. O. 1960. *Word and Object.* Cambridge, MA: Harvard University Press.

———. 1969. *Ontology, Relativity and Other Essays.* New York: Columbia University Press.

———. 1970. On the reasons for the indeterminacy of translation. *Journal of Philosophy* 67:178–83.

Radnitzky, G., and A. Andersson (eds.). 1978. *Progress and Rationality in Science.* Boston Studies in the Philosophy of Science, vol. 58. Dordrecht: Reidel.

Rand, A. 1970. Kant versus Sullivan. In *The Ayn Rand Library,* vol. 1, *Philosophy: Who Needs It?* New York: Signet Books, 1982.

Ravetz, J. 1971. *Scientific Knowledge and Its Social Problems.* Oxford: Clarendon Press.

Redhead, M. L. G. 1978. Ad hocness and the appraisals of theories. *British Journal for the Philosophy of Science* 24:355–61.

———. 1980. A Bayesian reconstruction of the methodology of scientific research programmes. *Studies in History and Philosophy of Science* 11:341–47.

———. 1986. Novelty and confirmation. *British Journal for the Philosophy of Science* 37:115–18.

Reichenbach, H. 1959. *Modern Philosophy of Science.* London, New York: Routledge and Kegan Paul.

Ritchie, A. D. 1926. Induction and probability. *Mind* 35:301–18.

Rosenberg, A. 1976. *Microeconomics Laws: A Philosophical Analysis.* Pittsburgh, PA: University of Pittsburgh Press.

———. 1986. Lakatosian consolations for economics. *Economics and Philosophy* 2:217–39.

Roseveare, N. T. 1982. *Mercury's Perihelion from Le Verrier to Einstein.* Oxford: Clarendon Press.

Rutherford, E. 1911. The scattering of α and β particles by matter and the structure of the atom. *Philosophical Magazine* 21:669–75, 680, 683–84, 688.

Sabra, A. I. 1967. *Theories of Light: From Descartes to Newton.* Reprint, Cambridge: Cambridge University Press, 1981.

Salanti, A. 1991. Roy Weintraub's studies in appraisal: Lakatosian consolations or something else? *Economics and Philosophy* 7:221–34.

———. 1994. On the Lakatosian apple of discord in the history and methodology of economics. *Finnish Economic Papers,* vol. 7, no. 1.

Savage, W. (ed.). 1990. *The Justification, Discovery and Evolution of Scientific Theories.* Minnesota Studies in the Philosophy of Science, vol. 14. Minneapolis: University of Minnesota Press.

Schaffner, K. 1974. Einstein versus Lorentz: Research programmes and logic of comparative theory evaluation. *British Journal for the Philosophy of Science* 25: 45–78

Schick, F. 1969. Review of Lakatos (ed.) *Problems of Inductive Logic. British Journal for the Philosophy of Science* 20:269–72.

Schilpp, P. A. (ed.). 1974. *The Philosophy of Karl Popper.* La Salle, IL: Open Court.

Schwinger, J. 1986. *Einstein's Legacy.* New York: Scientific American Books.

Shea, W. R. 1972. *Galileo's Intellectual Revolution.* London: Macmillan.

Shockley, W. 1971a. Negro IQ deficit: Failure of a "malicious coincidence" model warrants new research proposals. *Review of Educational Research* 41: 227–48.

———. 1971b. Models, mathematics, and the moral obligation to diagnose the origin of Negro IQ deficit. *Review of Educational Research* 41:369–77.

———. 1972a. Dysgenics, geneticity, raceology: A challenge to the intellectual responsibility of educators. *Phi Delta Kappa* 53:297–307.

———. 1972b. A debate challenge: Geneticity is 80% for white identical twins' IQ's. *Phi Delta Kappa* 53:415–19.

Sneed, J. D. 1971. *The Logical Structure of Mathematical Physics.* Dordrecht: Reidel. 2d ed. rev., 1979.

Stegmüller, W. [1973] 1976. *Theorienstrukturen und Theoriendynamik.* Berlin, Heidelberg, New York: Springer. English translation, *The Structure and Dynamics of Theories* (Berlin: Springer).

Swinburne, R. 1973. *An Introduction to Confirmation Theory.* London: Methuen.

Szabó, A. 1958. "Deiknymi" als mathematischer Terminus für "Beweisen." *Maia* 10:1–26.

———. 1960. Anfänge des Euklidischen Axiomensystems. *Archive for the History of Exact Sciences* 1:27–106.

———. 1978. *The Beginning of Greek Mathematics.* Boston, London: Reidel.

Thomas, K. 1971. *Religion and the Decline of Magic: Studies in Popular Beliefs in Sixteenth and Seventeenth Century England.* London: Weidenfeld and Nicolson.

Thomason, N. 1992. Could Lakatos, even with Zahar's criterion for novel fact, evaluate the Copernican research programme? *British Journal for the Philosophy of Science* 43:161–200.

Toeplitz, O. [1949] 1963. *Die Entwicklung der Infinitesimal-rechnung,* edited by Gottfried Köthe. English translation, *The Calculus: A Genetic Approach,* translated by Louise Lange. Chicago: University of Chicago Press.

Toraldo di Francia, G. 1976. *L'indagine del mondo fisico.* Torino: Einaudi.

Trotsky, L. 1937. *The Revolution Betrayed.* Garden City, NY: Doubleday.

Tugendhat, E. 1992. *Ethik und Politik.* Frankfurt am Main: Suhrkamp.

Uebel, T. (ed.) 1991. *Rediscovering the Vienna Circle: Austrian Studies on Otto Neurath and the Vienna Circle.* Dordrecht: Kluwer Academic Publishers.

Urbach, P. 1974. Progress and degeneration in the "IQ debate." *British Journal for the Philosophy of Science* 25:99–135, 235–59.

———. 1978. The objective promise of a research programme. In G. Radnitzky and A. Andersson (eds.), *Progress and Rationality in Science,* 99–113. Dordrecht: Reidel.

———. 1989. The Bayesian alternative to the methodology of scientific research programmes. In K. Gavroglu, Y. Goudaroulis, and P. Nicolacopoulos (eds.), *Imre Lakatos and Theories of Scientific Change,* 399–411. Dordrecht: Kluwer.

Velikovsky, I. 1950. *Worlds in Collision.* 18th ed., London: Book Club Associates, 1973.

Vint, J. 1993. *Capital and Wages: A Lakatosian History of Wages Fund Doctrine.* Aldershot: Edward Elgar.

Watkins, J. 1952. Political tradition and political theory: An examination of Professor Oakeshott's political theory. *Philosophical Quarterly* 2:232–37.

———. 1958. Confirmable and influential metaphysics. *Mind,* n.s. 67:344–65.

———. 1974a. Lakatos Obituary. *The Times,* 6 February.

———. 1974b. *The Unity of Popper's Thought.* In P. A. Schilpp (ed.), *The Philosophy of Karl Popper,* 371–412. La Salle, IL: Open Court.

———. 1979. Imre Lakatos. In *International Encyclopedia of the Social Sciences Biographical Supplement.* New York: Macmillan/Free Press.

———. 1984. *Science and Scepticism.* Princeton, NJ: Princeton University Press.

———. 1997. Karl Raimund Popper. *Proceedings of the British Academy* 94: 645–84.

Weintraub, E. 1985. *General Equilibrium Analysis: Studies in Appraisal.* Cambridge: Cambridge University Press.

Westfall, R. S. 1980. *Never at Rest: A Biography of Isaac Newton.* Cambridge: Cambridge University Press.

Westman, R. (ed.) 1976. *The Copernican Revolution.* Los Angeles: University of California Press.

Whewell, W. 1831. Review of John F. W. Herschel's *Preliminary Discourse* (1830). *London Quarterly Review,* vol. XLV, no. 90, July, 374–407.

———. 1840. *Philosophy of the Inductive Sciences, Founded upon their History.* 2 vols. London: Frank Cass, 1967.

Whitt, L. A. 1992. Indices of theory promise. *Philosophy of Science* 59:612–34.

Will, C. 1986. *Was Einstein Right?* New York: Basic Books.

Wisdom, J. O. 1952. *Foundations of Inference in Natural Science.* London: Methuen.

Wittgenstein, L. 1958. *Philosophical Investigations.* Oxford: Oxford University Press.

Worrall, J. 1976. Thomas Young and the "refutation" of Newtonian optics: A case-study in the interaction of philosophy of science and history of science. In C. Howson (ed.), *Method and Appraisal in the Physical Sciences,* 107–79. Cambridge: Cambridge University Press.

———. 1978a. The ways in which the methodology of scientific research programmes improves on Popper's methodology. In G. Radnitzky and A. Andersson (eds.), *Progress and Rationality in Science,* 45–70. Dordrecht: Reidel.

———. 1978b. Research programmes, empirical support, and Duhem's problem: Replies to criticism. In G. Radnitzky and A. Andersson (eds.), *Progress and Rationality in Science,* 321–38. Dordrecht: Reidel.

———. 1985. Scientific discovery and theory confirmation. In J. Pitt (ed.), *Change and Progress in Modern Science,* 301–31. Dordrecht: Reidel.

———. 1988. The value of a fixed methodology. *British Journal for the Philosophy of Science* 39:263–75.

———. 1989a. Why both Popper and Watkins fail to solve the problem of induction. In F. D'Agostino and I. C. Jarvie (eds.), *Freedom and Rationality: Essays in Honor of John Watkins.* Dordrecht: Kluwer.

———. 1989b. Fix it and be damned: A reply to Laudan. *British Journal for the Philosophy of Science* 40:376–88.

———. 1990. Scientific revolutions and scientific rationality: The case of the "elderly holdout." In W. Savage (ed.), *The Justification, Discovery and Evolution of Scientific Theories,* 319–54. Minneapolis: University of Minnesota Press.

———. 1991. How to deduce theories from phenomena: Newton on scientific method. Mimeo.

———. 1995. Revolution in permanence. In A. O'Hear (ed.), *Karl Popper: Philosophy and Problems.* Cambridge: Cambridge University Press.

Yourgrau, W., and W. Chandler. 1968. A new formalized version of the verifiability principle. *Ratio* 10(1):54–63.

Zahar, E. 1973. Why did Einstein's programme supersede Lorentz's? *British Journal for the Philosophy of Science* 24:95–123, 223–62. Republished in Howson 1976, 211–75.

———. 1983. The Popper-Lakatos controversy in the light of *Die Beiden Grundprobleme der Erkenntnistheorie. British Journal for the Philosophy of Science* 34:149–71.

———. 1989. *Einstein's Revolution: A Study in Heuristics.* La Salle, IL: Open Court.

Zolo, D. 1989. *Reflexive Epistemology: The Philosophical Legacy of Otto Neurath.* Boston Studies in the Philosophy of Science, vol. 118. Dordrecht: Kluwer.

Index